AI and Precision Medicine in Infectious Disease Management

This book presents technical advancements of artificial intelligence (AI)–based technologies and computational biology approaches, highlighting their applications in developing smart healthcare and precision medicine for managing infectious diseases. It discusses cutting-edge AI tools, particularly machine and deep learning models, and generative AI to assist in disease prediction and prevention, disease surveillance and diagnosis, drug discovery, drug design and development, healthcare, personalized medicine, pharmacology, and vaccine development. The contents fit the goal of establishing "Good Health and Well-Being," which is one of the "Sustainable Development Goals (SDGs) 2030."

Key Features:

- Includes AI/omics approaches in precision medicine and smart healthcare
- Summarizes AI-driven drug and vaccine developments and AI tools for infectious disease prevention for clinicians, researchers, and policymakers
- Proposes AI-based strategies for disease prediction, surveillance, and diagnosis

AI in Clinical Practice

The Impact of Artificial Intelligence in Radiology
Adam Eltorai, Ian Pan and Henry H Guo

AI and Precision Medicine in Infectious Disease Management
Jen-Tsung Chen

For more information about the series, please visit https://www.routledge.com/AI-in-Clinical-Practice/book-series/AICLINICAL

AI and Precision Medicine in Infectious Disease Management

Edited by
Jen-Tsung Chen

CRC Press
Taylor & Francis Group
Boca Raton London New York

CRC Press is an imprint of the
Taylor & Francis Group, an **informa** business

Designed cover image: Shutterstock Id: 2312941681

First edition published 2026
by CRC Press
2385 NW Executive Center Drive, Suite 320, Boca Raton FL 33431

and by CRC Press
4 Park Square, Milton Park, Abingdon, Oxon, OX14 4RN

CRC Press is an imprint of Taylor & Francis Group, LLC

ISBN: 978-1-041-01621-2 (hbk)
ISBN: 978-1-041-01620-5 (pbk)
ISBN: 978-1-003-61569-9 (ebk)

DOI: 10.1201/9781003615699

Typeset in Palatino
by SPi Technologies India Pvt Ltd (Straive)

Contents

Biography of the Editor

Dr. Jen-Tsung Chen is a professor of cell biology at the National University of Kaohsiung in Taiwan, where he teaches cell biology, genomics, proteomics, plant physiology, and plant biotechnology. Dr. Chen's research interests include bioactive compounds, chromatography techniques, plant molecular biology, plant biotechnology, bioinformatics, and systems pharmacology. He is an active editor of academic books and journals to advance the exploration of multidisciplinary knowledge involving plant physiology, plant biotechnology, nanotechnology, ethnopharmacology, and systems biology. He serves as an editorial board member in reputed journals, including *Plant Methods*, *GM Crops & Food*, *Plant Nano Biology*, *Biomolecules*, *International Journal of Molecular Sciences*, and as a guest editor in *Frontiers in Plant Science*, *Frontiers in Pharmacology*, *Journal of Fungi*, and *Current Pharmaceutical Design*. Dr. Chen published books in collaboration with Springer Nature, CRC Press/Taylor & Francis Group, and CABI, and he is handling book projects for some more international publishers on topics of bioinformatics, drug discovery, epigenetics, nanotechnology, bioengineering, plant functional genomics, plant speed breeding, and CRISPR-based plant genome editing. In 2023 and 2024, Elsevier and Stanford University recognized Dr. Chen as one of the "World's Top 2% Scientists." Dr. Chen received the Springer Nature Editor of Distinction Award in 2025.

Contributors

Jothi Dheivasikamani Abidharini
Department of Human Genetics and Molecular
   Biology
Bharathiar University
Coimbatore, India

Kazi Asraf Ali
Department of Pharmaceutical Technology
Maulana Abul Kalam Azad University
   of Technology, West Bengal,
Haringhata, Nadia, India

Krishnamoorthy Amirthamba
PG and Research Department of Biotechnology
Kongunadu Arts and Science College
Coimbatore, India

Arumugam Vijaya Anand
Department of Human Genetics and Molecular
   Biology
Bharathiar University
Coimbatore, India

Arif Nur Muhammad Ansori
Postgraduate School
Universitas Airlangga
Surabaya, Indonesia
Virtual Research Center for Bioinformatics and
   Biotechnology
Surabaya, Indonesia

Gunasekaran Arthi
Department of Human Genetics and Molecular
   Biology
Bharathiar University, Coimbatore, India

Mohammed Jaffer Shakeera Banu
Division of Molecular Biology
Microbiology Laboratory, Coimbatore, India

Amlan Bishal
Department of Pharmaceutics
Bharat Technology
Howrah, India

Arthi Boro
Department of Human Genetics and Molecular
   Biology
Bharathiar University
Coimbatore, India

Rideb Chakraborty
Department of Pharmaceutics
SRM College of Pharmacy, SRMIST
Kattankulathur, India

Pallavi Chand
Department of Pharmaceutics
Uttaranchal Institute of Pharmaceutical
   Sciences, Uttaranchal University
Dehradun, India

Ashish Singh Chauhan
Department of Pharmaceutics
Uttaranchal Institute of Pharmaceutical
   Sciences, Uttaranchal University
Dehradun, India

Sabyasachi Choudhuri
IKS Health Pvt. Ltd.
Hyderabad, India

Tanushree Das
Eminent College of Pharmaceutical Technology
Barasat, India

Udayakumar Devika
Department of Biotechnology
University of Calicut, Malappuram, India

Stuti Ghosh
Department of Biological Sciences
Bose Institute
Kolkata, India

Chowdhury Mobaswar Hossain
Department of Pharmaceutical Technology
Maulana Abul Kalam Azad University
   of Technology, West Bengal,
Haringhata, Nadia, India

Jatin Jangra
Department of Pharmaceutical Engineering and
   Technology
Indian Institute of Technology (Banaras Hindu
   University)
Varanasi, India

Amit Kotal
Department of Pharmaceutical
   Technology
Maulana Abul Kalam Azad University of
   Technology, West Bengal,
Haringhata, Nadia, India

Rajnish Kumar
Department of Pharmaceutical Engineering and
   Technology
Indian Institute of Technology (Banaras Hindu
   University)
Varanasi, India

Arul Sampath Kumar
PG and Research Department of Biotechnology
Kongunadu Arts and Science College
Coimbatore, India

Mani Manoj
Department of Human Genetics and Molecular
    Biology
Bharathiar University
Coimbatore, India

Loganathan Murugesan
Department of Human Genetics and Molecular
    Biology
Bharathiar University
Coimbatore, India

Subhrajyoti Nandy
Department of Pharmaceutical Technology
Maulana Abul Kalam Azad University of
    Technology, West Bengal,
Haringhata, Nadia, India

Irina Negut
National Institute for Lasers
Plasma and Radiation Physics
Măgurele, Romania

Cláudia S. Oliveira
Universidade Católica Portuguesa, CBQF –
    Centro de Biotecnologia e Química Fina –
    Laboratório Associado, Escola Superior de
    Biotecnologia
Rua Diogo Botelho, Porto, Portugal

Jeyabal Philomenathan Antony Prabhu
Department of Human Genetics and Molecular
    Biology
Bharathiar University
Coimbatore, India

Anindya Pradhan
Dr. B. C. Roy Post Graduate Institute of
    Pediatric Sciences
Kolkata, India

Arvind Raghav
Department of Pharmaceutics
Teerthanker Mahaveer College of Pharmacy,
    Teerthanker Mahaveer University
Moradabad, India

Rajamanikkam Ramya
PG and Research Department of Biotechnology,
    Kongunadu Arts and Science College
Coimbatore, India

Asirvatham Alwin Robert
Department of Endocrinology and Diabetes
Prince Sultan Military Medical City,
    Ministry of Defense,
Riyadh, Kingdom of Saudi Arabia

Sudipto Saha
Department of Biological Sciences
Bose Institute
Kolkata, India

Md Adil Shaharyar
Bioequivalence Study Centre, Department of
    Pharmaceutical Technology
Jadavpur University
Kolkata, India

Sivaramakrishnan Sharmili
Department of Biomedical Science
Alagappa University
Karaikudi, India

Anuradha Singh
Department of Chemistry
SS Khanna Girls Degree College (A Constituent
    College of the University of Allahabad)
Prayagraj, India

Srikamali Sundar
PG and Research Department of Biotechnology
Kongunadu Arts and Science College
Coimbatore, India

Preface

In recent years, advanced bioinformatics and computational and systems biology have made major technical advancements, chiefly in artificial intelligence (AI) for advancing tools or platforms of identification, prediction, simulation, and modeling. AI algorithms, including machine learning, deep learning, and generative AI, that are enhanced by accelerated computing and are integrated with multiple omics tools, have started to level up approaches, leading to smart healthcare and precision medicine.

This book provides knowledge of AI technologies by summarizing their methods and applications in combination with data science for combating global infectious diseases and is highly expected to enhance global health. This book covers promising ways to advance disease prevention, prediction, surveillance and diagnosis, drug discovery, design, and development, healthcare, personalized medicine, pharmacology, and vaccine development, assisted by AI-based approaches. This book shares AI-enabled strategies for managing infectious diseases, such as HIV infections and tuberculosis.

This book presents knowledge based on advanced tools of computational biology and AI models and therefore can be an ideal reference for students, teachers, professors, and researchers in biomedical science and infectious diseases particularly, the subtopics related to disease prediction and surveillance, diagnosis, drug discovery, drug design, drug development, pharmacology, disease prevention and management, and healthcare.

1 Controlling Emerging Infectious Diseases

Latest Overview of Epidemiology and Therapeutics

*Arthi Boro, Mani Manoj, Gunasekaran Arthi, Udayakumar Devika,
Krishnamoorthy Amirthamba, Mohammed Jaffer Shakeera Banu,
Jeyabal Philomenathan Antony Prabhu, and Arumugam Vijaya Anand*

1.1 INTRODUCTION

Emerging infectious diseases (EIDs) are diseases that have newly emerged in a population or are undergoing a rapid rise in incidence or spread over space (Ramon-Torrell, 2023). This category encompasses not only newly undescribed pathogens but also well-studied infectious agents that have adapted or enhanced their pathogenicity, leading to outbreaks in new populations or in new geographic regions. One of the most frequently cited examples of an EID is the human immunodeficiency virus (HIV), which emerged as a major global health issue in the early 1980s, and the severe acute respiratory syndrome coronavirus (SARS-CoV), which was initially described in 2003. These pathogens were not common or reported before their emergence, indicating their impact as notable additions to the infectious disease arena (Pushparaj et al., 2022; Arumugam et al., 2020; Meyyazhagan et al., 2020; Shanmugam et al., 2020; Kuchi Bhotla et al., 2020; Kuchi Bhotla et al., 2021).

On the other hand, re-emerging infectious diseases are those which had previously been brought under control or declining but re-emerged, most often because of factors like antimicrobial resistance (AMR), changes in vector behavior, weaknesses in public health infrastructure, or changes in human behavior. The reoccurring pandemics with Influenza A viruses – depicted by events in 1918, 1957, and 1968 – provide a good example of the category, depicting the cyclical pattern of certain pathogens (Georgiev, 2009).

The majority of drivers of emergence and re-emergence are environmental, such as demographic change, greater global travel and trade, urbanization, and spin-off from technological innovation and medical research (Abdallah, 2018; Velayuthaprabhu and Archunan, 2005; Velayuthaprabhu et al., 2007; Alagendran et al., 2010; Velayuthaprabhu et al., 2011; Velayuthaprabhu et al., 2013; Velayuthaprabhu et al., 2016; Varghese et al., 2020; Lee et al., 2021; Mohd, Balasubramanian, et al., 2021; Mohd, Kumar, et al., 2021; Gundappa et al., 2022; Meyyazhagan et al., 2022; Sangeetha, Anand, & Begum, 2022; Sangeetha, Nargis Begum et al., 2022; Ramya et al., 2023; Paranitharan et al., 2025). Significantly, a large percentage of EIDs are zoonotic, transmitted from animals to humans, most often through arthropod vectors or direct animal contact. Flaviviruses and arenaviruses are excellent examples, demonstrating the close interdependence of animal and human health (Louten, 2016). The shared notion of new infectious diseases is in direct association with EIDs, such as newly emerged pathogens or those which have gained new virulence factors, resistance factors, or ecological niches (Chen, 2022).

An examination of previous epidemics such as SARS (2003), Ebola (2013–2014), COVID-19 (2019 and onwards), and Monkeypox (2022) discloses an enduring disconnect between scientific knowledge, public health response, and social response. The SARS epidemic demonstrated global systems' vulnerability to zoonotic transmission and reasserted the evergreen value of proven public health practices, such as quarantine and contact tracing, which were later a central part of the response to the COVID-19 pandemic (Bartos, 2020). The Ebola virus epidemic in West Africa underscored major shortcomings in risk communication, community engagement, and health system preparedness. Cultural issues, historic distrust, and misinformation drove resistance to public health interventions, as in earlier epidemics (Colgrove, 2014). Similarly, the COVID-19 pandemic is a recent counterpart of the 1918 influenza pandemic, especially in the deployment of non-pharmaceutical interventions, such as social distancing, mask mandates, and lockdowns.

Scholars like Snowden have argued that pandemics, historically, have been great drivers of social and cultural transformation, and have impacted such domains as religion, public policy, and the arts (Snowden, 2019). Historical trends like these validate the importance of open and adaptive government in the management of public health crises, where the fine balance between building public trust and avoiding panic is key (Colgrove, 2014). To this point, the cyclical nature of outbreak control is emphasized, and with each event, lessons are learned for future events.

DOI: 10.1201/9781003615699-1

1.1.1 Historical Perspectives and Landmark Outbreaks

The history of control strategies for infectious diseases illustrates a remarkable shift from traditional containment methods to sophisticated, interdisciplinary, and predictive methods. Figure 1.1 illustrates a timeline of major emerging infectious outbreaks, contextualizing how these global health threats have evolved over time and driven innovation in disease control strategies. Historically, spatial methods like quarantine, isolation, and mass vaccination campaigns (e.g., smallpox and poliomyelitis) were the primary methods for disease eradication and control (Verweij & Dawson, 2011). The increasing complexity of pathogen evolution, especially the evolution of drug-resistant strains, has, however, necessitated more advanced strategies. The evolutionary epidemiology science has emerged as an essential science that examines the co-evolution of pathogens and hosts under varying selective pressures, and hence informs us about virulence, transmission dynamics, and host immunity (Feng et al., 2006). In parallel, advances in machine learning (ML) and artificial intelligence (AI) have revolutionized the horizon of outbreak prediction. Predictive models driven by deep learning and ensemble methods can handle vast multidimensional datasets with ease and detect patterns and predict outbreaks with greater accuracy (Rahman et al., 2023).

In addition, the incorporation of control theory, and even more so the paradigm of reachability, has led to new paradigms for individualized control of chronic infectious diseases like HIV. Such models complement intervention strategies based on dynamic system theory, thus demonstrating the utility of mathematical modeling in clinical decision-making (Anelone & Spurgeon, 2017).

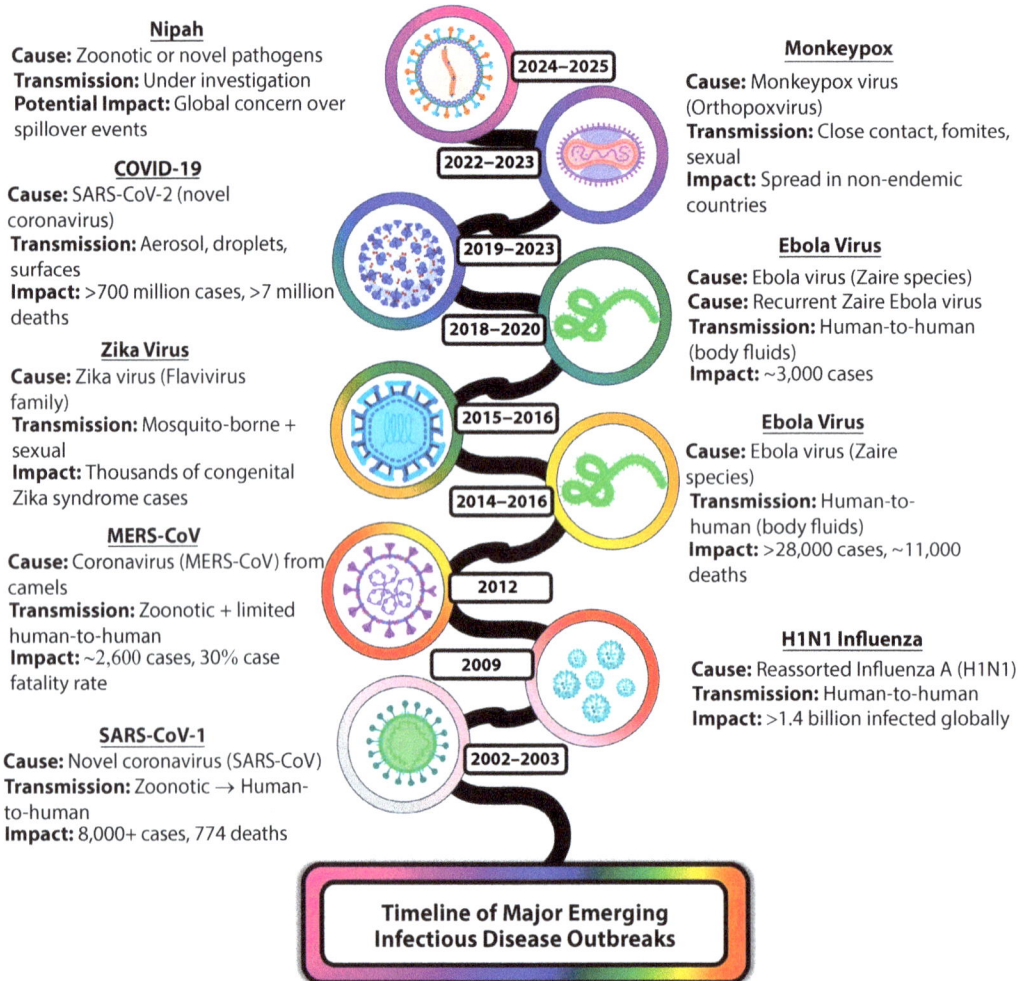

Figure 1.1 Timeline of emerging infectious disease outbreaks.

Such interdisciplinarity-appealing to clinicians, ecologists, epidemiologists, and data scientists-emphasizes the need to combine traditional public health strategies and cutting-edge technological devices to respond to the threat of emerging and re-emerging infectious diseases (Restif & Graham, 2015).

1.1.2 Eco-social and One Health Perspectives

One Health and Eco-social models are useful models for explaining the complex and interdependent nature of infectious disease emergence. One Health promotes a holistic approach to health that is appreciative of interdependencies across animal, human, and environmental systems. It is particularly relevant to zoonotic disease control, emerging pathogens, and AMR due to increased permeability across ecological, agricultural, and clinical spaces (Bashetti et al., 2024).

Eco-social drivers such as climate change, urbanization, loss of biodiversity, and globalization influence disease patterns by altering host-pathogen interactions, vector distributions, and transmission routes (Jones et al., 2017). For example, deforestation and fragmentation have increased human-wildlife contact and thus increased the risk of zoonotic spillover. Hence, public health interventions need to adopt a systems-based agenda combining insights from ecology, veterinary sciences, and environmental policy. The One Health approach also allows cross-sectoral intervention, including vector control, environmental sanitation, vaccination, and wildlife surveillance (Ellwanger et al., 2021). A greater vision-conjoining eco health with planetary health is also needed during the Anthropocene, an era marked by hyperanthropogenic footprints on planetary systems. Transnational health initiatives, collaborative governance, and interdisciplinarity are key to enhancing resilience and sustainability (Shaikh, 2018). Despite the global trend toward the management of non-communicable diseases (NCDs), infectious diseases continue to bear a significant burden, particularly in low- and middle-income countries (LMICs). Zoonotic diseases, responsible for a significant proportion of EIDs, are often neglected in global health discussions. Non-typhoidal *salmonellosis*, *brucellosis*, and *anthrax* cause at least ten million disability-adjusted life years (DALYs) every year; nonetheless, the surveillance and burden estimation process are still intermittent (di Bari et al., 2022). Neglected tropical diseases (NTDs) like schistosomiasis, lymphatic filariasis, leishmaniasis, and intestinal nematode infections are of major public health concern. NTDs disproportionately affect poor and vulnerable communities, perpetuating cycles of health inequities and poverty (Pisarski, 2019). Infectious diseases alone caused 704 million DALYs worldwide in 2019, and the leading causes were tuberculosis, malaria, and HIV/AIDS (Naghavi et al., 2024). The sub-Saharan region of Africa bears an out-of-proportion burden, reflecting systemic issues regarding health infrastructure, access to healthcare, and prevention efforts. The World Health Organization's International Health Regulations (IHR) aim to enhance global surveillance and response capacity, highlighting the need for international coordination and resource allocation to counter emerging and ongoing health threats (Holmes et al., 2017).

1.2 DRIVERS OF EMERGENCE

The evolution of biological systems, including pathogen evolution and zoonotic transmission of diseases, is governed by a complex interplay of ecological, environmental, and anthropogenic drivers. Deforestation, land-use change, urbanization, and intensified agriculture all disrupt natural ecosystems, increase human-wildlife interaction, and increase the spread of infections from animals to people (Swei et al., 2020). By changing the geographic range and vector ecology of disease vectors, climate change intensifies these disruptions and affects the dynamics of vector-borne disease transmission, including Lyme disease and dengue fever (Gibb et al., 2024). Globalization processes like rapid global mobility and international trade enhance cross-border transmission of pathogens and raise the risk of EIDs (Ghatak et al., 2020).

Pathogen evolution, specifically through genetic mutation to enhanced virulence or AMR, lies at the core of infectious disease emergence and re-emergence. Such evolution enables pathogens to infect new hosts and inhabit new ecological niches, rendering it ever more difficult to perform disease control and surveillance (Engering et al., 2013). This dynamic is illustrated in Figure 1.2. These processes of evolution are not occurring in isolation; rather, they are occurring within broader socioecological contexts. Environmental degradation, for example, results in landscape fragmentation and mosaic ecosystem development, wherein human populations increasingly interact with wildlife, enabling the transmission of zoonotic and vector-borne diseases (Machalaba & Karesh, 2017). Exotic pet trade, wildlife trade, and release of invasive species into new environments by humans also lower natural ecological barriers that otherwise impede disease transmission significantly.

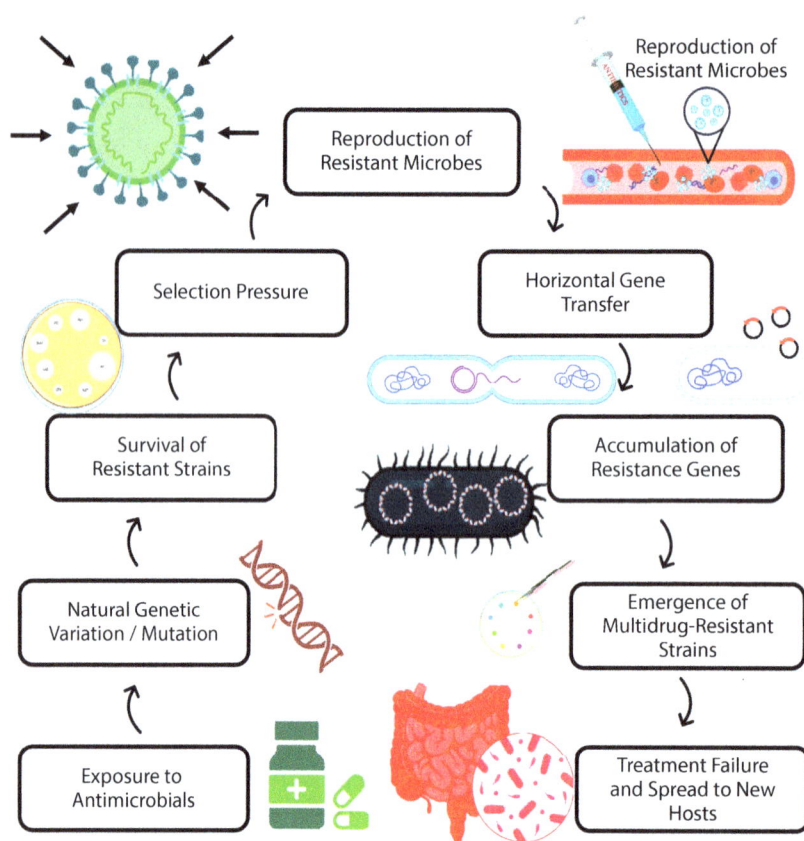

Figure 1.2 Evolution of AMR: A stepwise mechanism.

Social determinants, particularly poverty, poor hygiene, poor nutrition, and under-resourced health systems, drive the infectious disease burden in vulnerable populations. Aside from elevating susceptibility to infection from diseases like leprosy, malaria, and tuberculosis, these determinants also complicate the delivery of proper prevention and treatment interventions (Garcia & da Silva, 2016). Climate change also aggravates these determinants by altering pathogen ecology and lowering the resilience of communities already facing systemic inequities (Flahault, 2007; Branda et al., 2024). At the same time, international migration and air travel facilitate pathogen dispersal and the displacement of vulnerable populations, as in the case of H1N1 influenza and HIV/AIDS transmission (Dawson-Hahn et al., 2021; Coltart & Behrens, 2012).

Health inequities amplify the disease burden by providing fertile ground for pathogen transmission and increasing morbidity and mortality rates within marginalized groups. Healthcare, education, and socioeconomic disparities contribute to unequal infection rates, such as the incidence of respiratory syncytial virus (RSV) among Indigenous children and the disproportionate impact of COVID-19 among socioeconomically vulnerable populations (Ulanova, 2025; Husain et al., 2023). These inequities underscore the need to include structural determinants in disease prevention models, along with biological and environmental determinants (Quinn & Kumar, 2014).

Anthropogenic environmental changes such as intensive agriculture, infrastructure development, road construction, and dam construction interfere with ecosystems and allow for human incursion into previously undisturbed habitats. Human incursion enhances contact among wildlife, livestock, and human beings, thus encouraging the risk of cross-species transmission of disease agents. The majority of remote and resource-constrained areas have poor access to care service and surveillance systems, which leads to underreporting and delayed detection of outbreaks, which significantly hinders public health responses (McFarlane et al., 2013; Gibb et al., 2024). Prevention of emergence and transmission of infectious diseases calls for an integrated, multi disciplinary response based on the One Health paradigm. It is essential to improve health surveillance systems, implement ecosystem-based interventions that are tailored to specific geographic contexts, and coordinate policy among the health, agriculture, and environment sectors as key steps toward addressing the complex problems of disease emergence. Furthermore, to ensure global health security, it is imperative to integrate climate adaptation, sustainable land-use practices, and governance changes with an emphasis on mitigation and resilience. It is advisable to identify and act on the interconnected social, ecological, and biological drivers of disease emergence to avoid future epidemics and foster both human health and environmental integrity (Shanks et al., 2022; Martinus et al., 2023).

1.3 MOLECULAR EPIDEMIOLOGY AND PATHOGEN DYNAMICS

Next-generation sequencing (NGS) has emerged as an essential tool for the molecular diagnosis in EIDs, with real-time data greatly enhancing public health action. The ability of NGS to rapidly sequence pathogen genomes has proved invaluable, not least because of the COVID-19 pandemic, during which millions of SARS-CoV-2 genomes were sequenced in near-real time to enable continuous monitoring of viral evolution and transmission (Aulicino & Kimata, 2024). It is not only required for epidemics with symptomatic infection but also for infections with high asymptomatic rates of transmission, including HIV and HCV, where traditional diagnostic procedures are inadequate (Horsburgh et al., 2024). Coupled with epidemiological information, NGS allows a One Health approach that recognizes the interconnectivity of human, animal, and environmental health and thereby increases the scope and impact of disease surveillance.

Despite challenges like high cost, the requirement of highly skilled technical expertise, and the need for the development of a strong bioinformatics infrastructure, the role of NGS in enabling outbreak management and driving precision medicine efforts cannot be overestimated. The genomic information acquired through NGS enables the detection of emerging variants, drug resistance emergence, and variation in virulence – each of which is critical for intervention in a timely manner. A critical component of such surveillance is the construction of phylogenetic trees, which map the evolutionary history of the pathogens. Not only do the trees track the history of infecting organisms' relationships, but they also convey critical information about their temporal and spatial distribution, particularly by phylodynamic modeling, which extracts key epidemiological parameters from genetic information (Edwards & Rhodes, 2021).

Current advances, including scalable platforms such as Phylowave, facilitate lineage identification by shared fitness traits on an automated basis, rendering real-time surveillance more accurate in terms of pathogen transmission (Lefrancq et al., 2025). This line of inquiry also informs pathogen evolution, adaptation, and host interaction, with predictive signs of future outbreaks. In addition, phylogenetic analysis informs pathogen diversity and co-evolution in human hosts, facilitating more sophisticated public health measures. Platforms like GISAID and Nextstrain have made it much easier to implement genomic epidemiology, combining pathogen genomic information with spatiotemporal data to provide high-precision transmission dynamics.

This genomics-driven epidemiological approach is now standard for outbreak investigations, allowing for the characterization of transmission routes, source identification of infection, and the dissemination of AMR. For example, worldwide surveillance of resistant *Staphylococcus aureus* and *Escherichia coli* strains has been greatly facilitated by the integration of genomic information with conventional epidemiological techniques. Visualization software packages, including the suggested "phylepic" chart, make it easier to understand by correlating phylogenomic information with epidemiological timelines, thereby providing an integrated image of outbreaks which conventional methods cannot (Suster et al., 2024). An example is the COVID-19 pandemic, where data-sharing platforms such as GISAID guided the global public health policy by providing near-real-time access to information on variants (Wang et al., 2022).

Besides tracking outbreaks, genomic epidemiology has great potential to further precision medicine by individualizing treatment according to the genetic constitutions of the pathogen and host. There are, however, challenges of enormous scale, particularly the standardization of the data, issues of privacy, and incorporation of the technology into established public health networks, particularly in low- and middle-income nations. Still, ongoing development and availability of genomic tools affirm their great value to the strengthening of global health systems in the context of new infectious threats.

1.4 SURVEILLANCE, EARLY DETECTION, AND PREDICTIVE MODELING

The advances in scientific research have led to impressive advances in surveillance, early detection, and predictive modeling, especially in the case of infectious diseases. Understanding the processes by which diseases spread over networks is central to epidemiological forecasting since the topology of transmission networks reveals possible infection routes and facilitates the control of disease spread (Danon et al., 2011). The study of worldwide transportation networks has played a key role in unveiling hidden patterns in noisy outbreaks, enabling the identification of channels through which cross-border disease transmission is facilitated (Brockmann, 2017). Digital surveillance networks such as ProMED-mail, founded in 1994, have become crucial for the real-time monitoring of emerging and re-emerging infectious diseases. Figure 1.3 illustrates an AI-driven digital surveillance ecosystem for emerging infectious disease monitoring. These networks operate round the clock, identifying unusual health events affecting humans, animals, and plants, and transmit critical information through online platforms and social media (Carrion & Madoff, 2017).

Syndromic surveillance systems have also been implemented to enable early outbreak detection from the analysis of real-time health information, even without a confirmed diagnosis. The systems detect symptomatic and behavioral patterns in populations, which can signal concealed bioterrorist attacks or naturally occurring epidemics. This approach has become increasingly popular in North America and Europe, supported by automated data systems that enhance early detection (Henning, 2004; Mandl et al., 2004; May et al., 2009). Through the advancement of collaboration among healthcare providers, public health agencies, and academic researchers, syndromic surveillance has become a valuable tool for public health. Moreover, sentinel networks are tasked with tracking outbreaks using structured medical data and electronic records gathered from hospitals and insurers, thereby enabling timely detection of disease patterns (Platt et al., 2009; Bai et al., 2017).

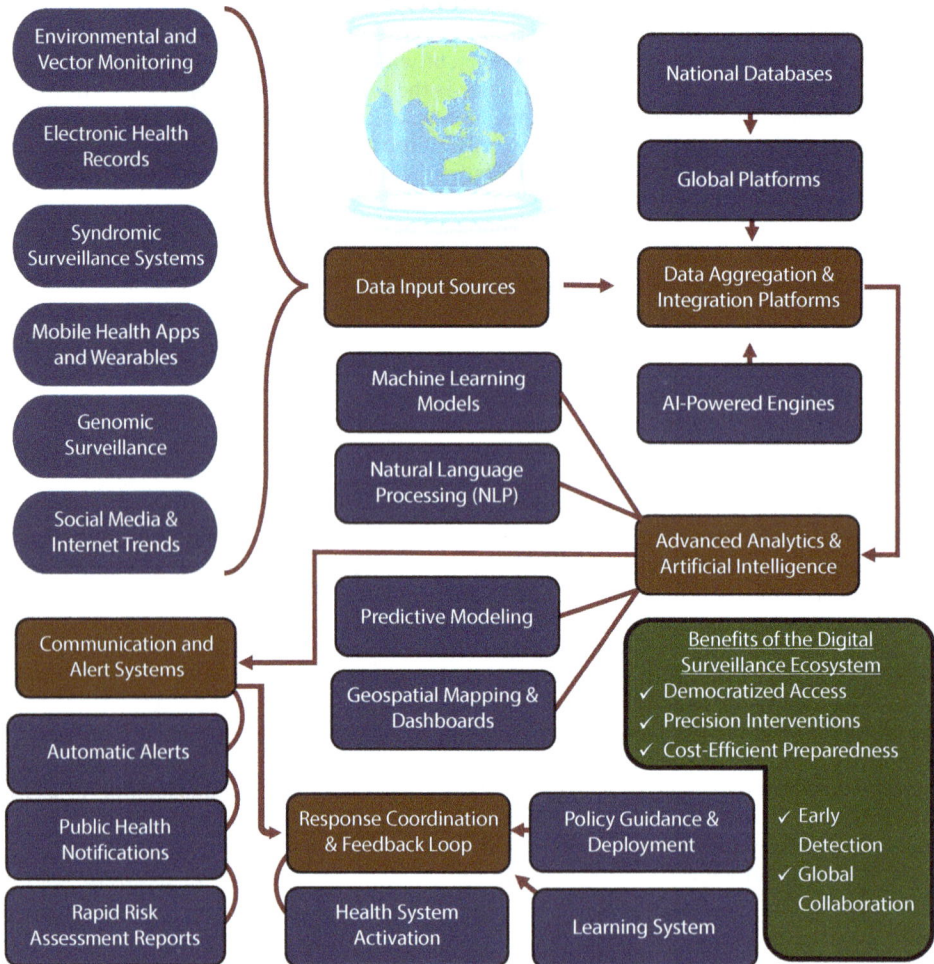

Figure 1.3 Digital surveillance ecosystem for emerging infections.

The advent of AI and big data analytics revolutionized predictive modeling in disease surveillance. Through the integration of disparate data sources such as social media, electronic health records, mobility data, and environmental sensors, AI makes it possible to provide insight into real-time modeling to detect early warning signs of impending outbreaks. Underpinned by deep learning algorithms, spatiotemporal analysis, and natural language processing, such models identify patterns and predict the onset of diseases (Mustapha et al., 2023; Jiao et al., 2023). A classic example of such convergence is BlueDot, a Canadian company using ML with deep transport network data to identify outbreak hotspots and predict the global spread of disease (MacIntyre et al., 2023; Beshwari et al., 2020). Similarly, HealthMap, an AI-based platform launched in 2006, geographically detects and identifies new infectious diseases using real-time data from digital media. It draws on a continuously updated lexicon and natural language tools to reconstruct epidemiological patterns from electronic medical records and online alerts (Adeoye et al., 2025; Rana et al., 2014).

Wastewater-based epidemiology has become an upcoming approach for early detection and community-level surveillance. Wastewater biomarkers have been examined by researchers to detect pathogens and quantify public exposure to various substances, including pharmaceuticals and environmental pollutants. The non-invasive method allows for almost real-time public health trend monitoring and early outbreak notification (Lorenzo & Picó, 2019; Sims & Kasprzyk-Hordern, 2020; Mao et al., 2020). When coupled with environmental biosensors that use biological components like enzymes or antibodies to detect specific environmental agents, the approach is a fast, cost-effective, and portable device for detecting air, soil, and water disease markers (Van Dorst et al., 2010; Justino et al., 2017).

The integration of electronic health records (EHRs) with bio signal monitoring greatly improves the detection and surveillance systems. EHRs aggregate large amounts of patient health data and enable longitudinal monitoring of clinical presentation, which, when combined with analytical methods, enhance the composition of personalized and timely interventions for infectious and chronic diseases (Menachemi & Collum, 2011; Kohli & Tan, 2016). Bio signal monitoring, especially cardiovascular parameters like heart rate variability, offers ongoing, non-invasive assessment of reactions of the autonomic nervous system to changes in health and thus acts as an added tool for surveillance and early detection of disease (Rajendra Acharya et al., 2006).

1.5 HOST-PATHOGEN INTERACTION AND IMMUNOBIOLOGY

The development of infectious diseases is a significant global disease threat, in good part due to the extensive range of immune evasion mechanisms utilized by pathogens. Through such processes, pathogens can efficiently repress and blunt host immune responses, thus optimizing their survival and transmission potential. For instance, the pathogenic fungus *Candida auris* utilizes interleukin-1 receptor antagonist (IL-1Ra) to repress neutrophil function, thus enabling it to survive on cutaneous surfaces by evading IL-1 receptor-mediated immune responses (Balakumar et al., 2024). On the other hand, bacterial pathogens utilize mechanisms that disrupt the immunological synapse, leading to the inhibition of T-cell activation and memory formation – both of which are crucial elements of effective adaptive immunity (Capitani & Baldari, 2022). Similarly, positive-sense, single-stranded RNA viruses, including coronaviruses, possess advanced mechanisms to evade detection by the innate immune system, including camouflaging the viral RNA and disrupting interferon production; both are fundamental for the induction of antiviral immunity (Nelemans & Kikkert, 2019).

Differences observed in host susceptibility to infectious pathogens are largely determined by genetic determinants, especially the Human Leukocyte Antigen (HLA) system. HLA molecules are critical for antigen presentation and initiation of immune response and therefore play a significant role in determining the courses of most infectious diseases, such as COVID-19, tuberculosis, and HIV (Medhasi & Chantratita, 2022; Alves et al., 2006). Some HLA alleles have been identified as risk or protective factors, influencing the severity and course of the disease. Some class I and II HLA alleles, for example, are prone to severe COVID-19, which suggests that genetic variation is the prime driver of variation in disease response across populations. Advances in the science of immunogenomics, especially through the integration of multi-omics technologies, have increased our knowledge of host-pathogen interactions and can potentially be harnessed to develop personalized therapeutic approaches (Kwok et al., 2021; Alves et al., 2006).

Apart from genetic considerations, the gut microbiome has been confirmed as a host-pathogen interaction determinant. Dysbiosis, defined by the disruption of microbial networks, compromises the protective role of the microbiota, thereby facilitating the establishment of opportunistic pathogens. The imbalance of microbes has been linked to a range of health states, such as gastrointestinal infection, inflammatory bowel disease, and systemic diseases of obesity and cancer (Khan et al., 2024).

The gut-lung axis is a typical example of the extensive influence of dysbiosis, where changes in gut microbiota affect respiratory health, thereby complicating disease courses. In addition, competitive interactions between pathogens and resident microbiota may modulate the evolution of pathogenic virulence, underscoring the importance of microbial ecology in disease processes (Stevens et al., 2024). Microbiome-directed interventions, such as probiotics and fecal microbiota transplantation, have demonstrated potential in restoring microbial balance and suppressing infection (Kamel et al., 2024).

In this context, the mechanism of trained immunity has been accorded considerable attention as a novel strategy for boosting host defense mechanisms. In contrast to adaptive immunity, which is antigen-specific, trained immunity is defined by epigenetic and metabolic reprogramming of the innate immune cells and thus enables an amplified, nonspecific response to subsequent infection (Hajishengallis & Chavakis, 2024). This mechanism provides considerable advantage in situations involving EIDs, especially in situations where the potential for accelerated vaccine development could be impeded. Compounds like β-glucan have been found to enhance the induction of trained immunity, thus enhancing immune reactivity and potentially enhancing vaccine efficacy. Additionally, through the induction of trained immunity, there is potential to reverse the impact of AMR and environmental modification in infectious disease epidemiology by inducing a rapid, broad-spectrum immune response. Be that as it may, the benefits provided through trained immunity must be weighed against the risk of chronic inflammation or autoimmune disease depending on overactivation. Recent innovation, such as the use of nanoparticle-based delivery platforms, has permitted the induction of controlled and sustained trained immunity, thus opening new landscapes of preventive measures against a broad range of pathogens (Esser-Kahn, 2023; Hajishengallis et al., 2019). Cumulatively, the complex interplay between pathogen evasion mechanisms, host genetic predisposition, microbiome integrity, and new immunological concepts like trained immunity underpin the complexity in the fight against EIDs.

1.6 THERAPEUTIC STRATEGIES: CURRENT AND FUTURE

1.6.1 Antiviral and Antimicrobial Therapies

Host-directed medicinal drugs and broad-spectrum antivirals are a potential therapeutic strategy for the treatment of viral infections, especially for recently emerged and re-emerged viruses. The majority of traditional antiviral drugs target viral proteins, a method that tends to result in rapid resistance generation; however, host-directed antivirals (HTAs) might overcome this limitation by targeting common host factors vital for viral replication as well. As an example, inhibitors like FLS-359 targeting sirtuin 2 have broad-spectrum antiviral activity against a range of RNA and DNA viruses, including coronaviruses and herpesviruses (Roche et al., 2023). These host-directed drugs are disrupting cellular functions common to viruses across the board, thus reducing the possibility of resistance generation through viral mutation. Additionally, dihydroorotate dehydrogenase (DHODH) inhibitors have antiviral activity against SARS-CoV-2, which is employed to illustrate the therapeutic potential of HTAs in countering a range of viral threats (Zheng et al., 2022). These drugs inhibit nucleotide biosynthesis, thus inhibiting viral replication at its source. Additionally, drug induction of innate immunity like Gilteritinib can induce a robust antiviral effect through the induction of interferon regulatory factors, thus augmenting host defense against a broad range of viral pathogens (Maarifi et al., 2022). These observations highlight the strategic benefits involved in the targeting of host biology as a therapeutic target, which is a more integrated and potentially superior strategy for antiviral therapy. Overall, a combination of HTAs with traditional antiviral therapy has great potential to significantly broaden the repertoire of therapeutic strategies available for the treatment of viral infections (He et al., 2024).

The drug repurposing has been the go-to strategy in fighting the COVID-19 pandemic, with a characteristic of quick action on the basis of drugs with well-proven safety profiles. The strategy has made it possible for one to attain fast-tracked clinical deployment when most critically needed in emergencies. Remdesivir, originally for Ebola, was one of the first drugs repurposed for COVID-19 and approved for emergency use for its antiviral activity, although controversial ability to lower mortality (Saini et al., 2024). Ivermectin, also a repurposed drug, has provided encouraging in vitro findings with successful SARS-CoV-2 replication inhibition, although the achievement of therapeutic concentrations in human beings has proven difficult (Eweas et al., 2021). Despite initial doubts about its efficacy, ivermectin has multiple mechanisms of action, including antiviral, anti-inflammatory, and anti-cytokine actions, supported by numerous studies confirming to lower mortality, hospitalization rate, and viral clearance. Interestingly, numerous meta-analyses and landmark clinical trials uncovered positive impacts, although experiences are different (Khadka et al., 2020). The drug repurposing strategy has the benefits of being cost-effective and swift, both of which are needed during emergencies, but with some drawbacks, for example, low success rates and potential side effects (Verma et al., 2023). Despite the fact that drugs such as remdesivir and ivermectin were promising, their deployment indicates the importance of undertaking extensive clinical trials to prove their efficacy and safety in the pandemic setting (Babalola & Ajayi, 2023).

The process of drug resistance in infectious disease is a complex challenge, driven by a wide range of biological and ecological drivers. A perfect example of this is antibiotic resistance, which is a clinical and ecological process that affects the microbiosphere and changes bacterial populations through processes like horizontal gene transfer and mutations (González-Candelas et al., 2011). Resistance evolution is also driven by the excessive use of antibiotics and the spread of resistance genes in the environment. *Mycobacterium tuberculosis* resistance is maintained by mechanisms involving compensatory evolution, efflux pumps, and target mimicry, thus illustrating the intricacies of the resistance evolution process and the need for novel drug targets (Al-Saeedi & Al-Hajoj, 2017). The existence of resistant pathogens is a global evolutionary response to the selective pressure of drugs, and population biological models have been used to demonstrate their utility in explaining the processes in a range of pathogens (Zur Wiesch et al., 2011). The models allow for the prediction of resistance patterns and facilitate the rational design of therapeutic interventions. Bacteria utilize a diverse array of mechanisms, like target mutations and efflux pump function, to gain resistance to antibiotics, leading to the emergence of multi-drug-resistant strains that pose a formidable challenge to existing treatment regimens (Zhang & Cheng, 2022). Combating drug resistance requires a two-pronged strategy: speeding up drug discovery and adopting evolutionary management strategies to slow down the evolution of resistance. While the discovery of novel pharmacological agents is crucial, evolutionary management with interventions like regulations to limit the spread of resistance could be a more realistic long-term option (McClure & Day, 2014). New treatment approaches like phage therapy and CRISPR-Cas precision therapy are being explored for their ability to target specifically multi-drug-resistant bacterial strains by interfering with the resistance mechanisms at the level of genes (Zhang & Cheng, 2022).

The multi-targeted systems pharmacology approach is a new strategy to the treatment of infectious diseases based on the intricacy of biological networks for maximization of therapeutic efficacy as well as safety. The strategy is most applicable in the fight against the threat of multi-drug-resistant pathogens, as it targets multiple sites simultaneously and hence decreases the likelihood of resistance development, as well as increases the efficacy of the treatment (Alberca & Talevi, 2020). The combination of multi-omics data with pharmacokinetic/pharmacodynamic models and patient-specific dosing schemes is aimed at increasing the optimization of antimicrobial treatment, taking into consideration patient heterogeneity, pathogen, and drug properties through personalized dosing regimens (Yow et al., 2022). Systems pharmacology allows for the overall comprehension of drug mechanisms, making it possible to identify and optimize therapeutics based on receptor biology, network theory, and state-of-the-art data analysis. A clear example of this is the study of *Momordica charantia*, where network pharmacology was applied to profile its antiviral activity through cooperative interactions with multi-components, targets, and pathways, hence showing the potential of natural products in multi-target drug discovery (Das et al., 2023). The shift from one-target one-drug to multi-target multi-drug is the key importance of computational and experimental approaches to the identification of drug-target interactions, which is central to the development of new therapeutics against complex diseases (Nishamol & Gopakumar, 2015). The new strategy is an integrated and dynamic approach to the treatment of infectious diseases, particularly in the age of increasing drug resistance and the emergence of novel pathogens.

1.6.2 Immunotherapeutics and Biologics

Monoclonal antibodies, nanobodies, and immune adhesins have increasingly been identified as effective therapeutic agents in the treatment of infectious diseases, each with their own advantages. Monoclonal antibodies have traditionally been used as a result of their high specificity and effectiveness in targeting pathogens but are limited by their large size, which can inhibit tissue penetration and delivery efficiency (Kozani et al., 2021). On the other hand, nanobodies, derived from heavy-chain-only antibodies, provide a strong alternative because of their small size, higher stability, and simple production process. These advantages provide for efficient tissue penetration, making nanobodies promising drug candidates for virus neutralization and the treatment of antibiotic-resistant bacteria. These have been successfully used in vaccine design, rapid diagnostic tests, and targeted therapies, as well as determination of microbial structure and virulence mechanisms (Rizk et al., 2024). Approval of caplacizumab, the first therapeutic derived nanobody, indicates their promise in therapy, and ongoing research is evaluating their use in detection and treatment of infections like SARS-CoV-2 (Alexander & Leong, 2024). Furthermore, nanobodies can be engineered into multivalent or fusion proteins, further enhancing their therapeutic effectiveness against infectious pathogens (Mei et al., 2022). Their high solubility, low immunogenicity, and cost-effectiveness make nanobodies a multifaceted tool for the overcoming of infectious disease challenges (Sanaei et al., 2020).

Cytokine modulators and immune checkpoint inhibitors are components of complex interplay that regulates immune responses in infectious diseases. Immune checkpoint inhibitors, PD-1 and CTLA-4, have vast potential in revitalizing immune responses to chronic infections like HIV and hepatitis B by restoring function in exhausted T cells and potentially reversing viral latency (Wykes & Lewin, 2018). In sepsis, the inhibitors improve both innate and adaptive immune responses by blocking suppressive mechanisms, leading to increased clearance of pathogens and survival (Patil et al., 2017). In severe COVID-19, characterized by the dominance of cytokine storms and immune exhaustion, checkpoint inhibitors can potentially enable the restoration of immune function by suppressing the overexpression of inhibitory checkpoints (Abdolmohammadi-Vahid et al., 2024). The use of checkpoint inhibitors in combination with cytokine modulators, however, is also effective if the inflammatory pathways are well-regulated. Targeting molecules like IDO, TGFβ, and adenosine has the potential to suppress the suppressive activity of regulatory T cells and myeloid-derived suppressor cells, thereby enabling increased efficacy of checkpoint inhibitors and the promotion of an enhanced immune response (DeVito et al., 2018).

Peptide vaccines enhance immune responses to infectious diseases by using well-characterized peptide antigens that are specifically engineered to induce defined immune responses. Peptide vaccines provide a safer option compared to conventional whole-organism vaccines, which possess the ability to induce autoimmune or allergic reactions (Nevagi et al., 2018). Peptide vaccines are especially efficient in inducing cytotoxic T-cell responses, crucial for the elimination of virus-infected cells and are enhanced by the use of adjuvants and sophisticated delivery systems (Rosendahl Huber et al., 2014). For instance, peptide conjugation with Toll-like receptor 7/8 agonists and poly(β-amino ester) nanoparticles enhances delivery efficiency and stability, enhancing DNA vaccine performance (Yang et al., 2024). Moreover, lipid conjugation of peptides has the potential to induce innate-like T cells, enhancing the immune response (Hayman et al., 2018). Despite these advantages, adjuvants and structural alterations are used to improve stability and efficacy since peptide vaccines have limited immunogenicity and rapid breakdown. New formulations and chemical modifications are being investigated to eradicate these limitations and enhance their therapeutic use (Bojarska et al., 2024). Overall, peptide vaccines are promising agents in infectious disease prevention and treatment, but optimization is necessary (Rosendahl Huber et al., 2014).

CRISPR-Cas-based antimicrobials offer a state-of-the-art strategy for the treatment of infectious disease, especially that caused by antibiotic-resistant bacteria. The CRISPR-Cas system, which was originally a bacterial adaptive immune system, may be modified to specifically target and eliminate bacterial genomes with high specificity and flexibility (Mayorga-Ramos et al., 2023). The system can inactivate chromosome genes or cut out plasmids involved in resistance, biofilm, and virulence, directly interfering with the pathogenicity processes. The technology, however, is still in its infancy, with problems such as delivery strategies and bacterial resistance to CRISPR-Cas systems still to be fully resolved (Duan et al., 2021). Potential delivery strategies include bacteriophages, nanoparticles, and conjugative plasmids, each with their own merits and demerits. Firms such as Locus Biosciences and Eligo Bioscience are developing CRISPR-based therapeutics, although most of the work is preclinical (Palacios Araya et al., 2021). Beyond antibacterial therapy, CRISPR-Cas systems also hold potential for antiviral therapy, enhancing precision in detection and treatment of infection. More work is needed to overcome the current challenges and realize the full clinical potential of CRISPR-Cas-based antimicrobials (Zhang, 2024).

1.6.3 Advanced Drug Delivery Platforms

Nanocarriers, spearheaded by liposomes and dendrimers, have been successful drug delivery tools in infectious diseases. Delivery systems improve drug bioavailability, reduce side effects, and sustain therapeutic concentration at infection sites. Liposomes made of lipid bilayers have been widely employed in the treatment of HIV, tuberculosis, and malaria, among others. Their physicochemical characteristics like lipid composition, particle size, zeta potential, and lamellarity may be engineered for controlled and targeted drug release, hence enhancing selectivity as well as reducing toxicity (Seele, 2025). In malaria, for instance, liposome-mediated delivery has been reported to surpass drug resistance and enhance therapeutic efficacy through the selective targeting of infected cells (Nassimbwa Kabanda, 2024).

Dendrimers are a highly versatile class of nanocarriers with intricately branched, tree-like structures that enable precise functionalization, thus improving targeted drug delivery while reducing cytotoxicity (Theochari et al., 2020). The nanocarriers efficiently address key challenges in the treatment of infectious diseases, such as the need for sustained release of the drug and the reduction of frequent dosing, by providing long-term drug availability at the infection site (Shah et al., 2025). The application of nanotechnology in drug delivery systems specifically in the use of liposomes and dendrimers, marks a considerable advancement in the field of infectious disease therapeutics (Dunuweera et al., 2019).

1.6.3.1 Smart Drug Release Systems

Stimuli-responsive drug delivery, particularly temperature- and pH-responsive, represents a major advance in the targeted treatment of infectious diseases. In order to facilitate targeted medication delivery at the infection site, new technologies take use of the pathogenic microenvironment, which is characterized by low pH and high temperatures. For instance, dual pH- and thermo-responsive hydrogels have been prepared from polymers that dynamically respond to the environmental conditions, thus enabling modulated and targeted drug delivery (Bucătariu et al., 2024). Similarly, mesoporous silica nanoparticles with the Poly(Dimethylaminoethyl Methacrylate) (PDMAEMA) modification release drugs in response to acidic microenvironments and elevated temperatures, characteristic features of infected tissue, and minimize off-target effects and systemic toxicity (Ramezanian et al., 2023).

These systems usually utilize processes such as protonation and deprotonation of ionizable moieties and temperature-induced conformational changes to control drug release (Shah et al., 2018). The pH-sensitive nanomaterials are particularly useful for the treatment of bacterial infections and organ-specific diseases due to their ability to alter their surface properties in response to local acidity to produce maximum therapeutic effects (Meher & Poluri, 2020). The use of stimulus-responsive biomaterials therefore improves the specificity and safety of therapy by allowing drug activation under disease-specific conditions only (Wanakule & Roy, 2012).

1.6.3.2 Innovative Delivery Platforms

Transdermal microneedles and inhalable antivirals present groundbreaking strategies for infectious disease prevention and treatment, circumventing constraints related to conventional delivery systems. Microneedle-delivery-mediated transdermal drug delivery systems (TDDS) are painless and minimally invasive, enhancing patient compliance and ensuring effective vaccine and biologic delivery. Through bypassing stratum corneum and delivery of drugs to immunologically active dermal layers, microneedles increase bioavailability and immune activation (Wang et al., 2021). Diverse microneedle geometries, like solid, hollow, and dissolvable, support the delivery of multiple therapeutic agents such as peptides, proteins, and vaccines (Goswami et al., 2024). The systems also support sustained release of drugs and minimize systemic toxicity, making them suitable for the treatment of systematic and localized infections (Jamaledin et al., 2020).

In comparison, inhalable antiviral platforms allow for direct drug delivery into the respiratory system with the benefit of fast onset and reduced systemic exposure. This is especially useful in the therapy of respiratory infections like influenza and COVID-19 where timely and localized therapy is essential. Both transdermal and inhalable platforms are directly tied to personalized medicine, with greater accuracy, reduced side effects, and better therapeutic impact. These technologies are the future of infectious disease therapy, especially in enhancing global access and adherence to advanced therapeutics (Ghosh et al., 2024).

1.7 VACCINE SCIENCE AND IMMUNIZATION STRATEGIES

The arena of vaccine development against infectious diseases has seen stunning advancement with the advent of state-of-the-art technologies, such as mRNA, DNA, viral vectors, inactivated vaccines, and virus-like particles (VLPs). Among these, mRNA vaccines have come to be a game-changer, with quick development timelines, high immunogenicity, and scalability. Their success in COVID-19 offers the promise of their potential for use against other pathogens such as influenza, rabies, and Zika virus. The malleability of the mRNA platform, coupled with its cost-effectiveness for production, makes it a high-value commodity in the global vaccine development market. In contrast, VLPs mimic the structural features of native viruses without any loss of genetic material, thereby conferring a robust and secure immunogenic profile. VLP-based vaccines have shown high efficacy and safety, as seen in licensed vaccines against human papillomavirus (HPV) and hepatitis B. While mRNA and VLP technologies are the flavor of the times, DNA and vector-based vaccines remain high-value alternatives in their ability to stimulate long-term immune responses and production viability. Together, these new platforms represent the dynamism inherent in modern vaccinology, as it addresses both endemic and emerging infectious challenges (Le et al., 2022; Ahamad et al., 2024; Zhang et al., 2023).

The integration of AI and ML with reverse vaccinology (RV) has considerably increased the efficiency and accuracy of vaccine discovery. AI-based epitope prediction has played a pivotal role in delineating immunologically relevant regions across entire proteomes, thus enabling systematic design of multiepitope vaccines. These computational methods complement the discovery of conserved antigenic targets based on the heterogeneity of HLA profiles and offering cross-protection against multiple serotypes. Deep learning algorithms have considerably improved the predictive accuracy of B-cell epitope discovery with significant implications in the development of vaccines against SARS-CoV and SARS-CoV-2. Such algorithms, which attain diagnostic accuracy of up to 82%, outperform conventional algorithms by incorporating structural and evolutionary information to improve immune target prediction (Shi et al., 2024). Methods such as Vaxign-ML demonstrate the efficacy of supervised ML models in predicting protective bacterial antigens with improved accuracy (Ong et al., 2020). Thus, AI-based approaches reduce the time-consuming processes of conventional vaccine development and accelerate the translation of in silico predictions to viable candidates, thus enabling a swift response to the dynamic situations of infectious diseases (Bhakta et al., 2021; Ponne et al., 2024).

Vaccine stability along the supply chain is a compelling challenge, especially in regions characterized by limited infrastructure. Traditional cold chain requirements 2°C to 8°C are logistically burdensome and costly. Thermostable vaccines are an option to consider in low-resource settings, where stable refrigeration is not an option. Vaccine development with the ability to withstand high temperatures is a distribution logistics paradigm shift. For example, the Gardasil vaccine remains stable for longer than a decade at 25°C, and co-lyophilization development has enabled the RTS,S/AS01 malaria vaccine to maintain stability at 37°C for six months and brief durations of 45°C without loss of activity (Fortpied et al., 2020). Thermostable vaccine technologies reduce dependency on energy-intensive refrigeration technology and enable more equitable access through simplification of storage and transportation procedures. Logistical intricacies faced in the global COVID-19 vaccine distribution, especially with formulations requiring ultra-low temperatures, have highlighted the strategic importance of thermostable vaccine technologies in planning for the next pandemic crisis (Chen & Kristensen, 2009; Schnetzinger et al., 2024; Seprényi et al., 2022).

Mucosal immunization is an underexploited but very promising strategy, of special relevance to infections entering via mucosal surfaces. Vaccines given by the intranasal, oral, and other mucosal routes share the advantage of eliciting both topical and systemic immunity, thus effectively entrapping pathogens at the point of entry. Targeting mucosal-associated lymphoid tissue, the strategy maximizes immunological coverage of the respiratory and gastrointestinal tracts and thereby provides broader protection than parenteral immunization alone could deliver. Mucosal vaccines, however, are faced with daunting formulation and delivery issues. Protein antigens present usually exhibit poor immunogenicity and stability in the chemical environment of the mucosal surface. These limitations are currently being overcome by advancements in biomaterials, nanotechnology, and mucoadhesive delivery methods, which are improving antigen stability, targeting, and absorption. Current mucosal vaccines (e.g., those being prepared for cholera, typhoid, and influenza) demonstrate feasibility; however, further optimization in vaccine design is still required for wider application. The social impact of non-invasive delivery methods, further coupled with the convenience of rapid mass immunization, also underscores the importance of the strategies in global health programs (Azegami et al., 2014; Van der Ley & Schijns, 2023; Wang et al., 2015; Rhee, 2020).

The COVID-19 pandemic has accentuated the problem of vaccine equity, exposing profound inequalities in global access to life-saving health interventions. The COVAX facility-collectively spearheaded by WHO, GAVI, and UNICEF-was intended to enable an equitable supply of vaccines, especially for LMICs. While it has mitigated some inequalities, structural barriers in global health governance, procuring systems, and logistic systems have constrained its overall performance. By early 2022, only 61% of doses allocated to 148 countries were delivered, even though low-income regions were prioritized through subsidies (Pushkaran et al., 2024). In response to this reality, GAVI's strategic plan (Gavi 6.0) advocates for a transition to regional autonomy, with a focus on local manufacturing capacity and sustainable finance models. Such initiatives as the BioNTech mRNA factory in Rwanda are examples, intended to decentralize production and empower LMICs. Yet, vaccine nationalism and the concentration of manufacturing capacity among high-income countries continue to constrain equitable access. Future policies need to prioritize capacity building, transparent governance, and public-private sector collaborations to create a resilient and equitable global vaccine infrastructure (Alamgir & Sajjad, 2024; Pushkaran et al., 2023).

1.8 INNOVATIVE DIAGNOSTICS AND POINT-OF-CARE TECHNOLOGIES

Recent advances in new diagnostic technologies and point-of-care systems have significantly transformed the horizon of modern healthcare by greatly improving diagnostic accuracy, timeliness, access, and economic efficiency. These advances are primarily driven by the interaction between new scientific knowledge and the development of sophisticated laboratory instrumentation (Cancrini & Iori, 2004). One of the advances in the area is the loop-mediated isothermal amplification (LAMP) technique, which allows the amplification of nucleic acids under isothermal conditions with high specificity using six primers. Its ease of use, quick execution, and energy savings make it a suitable candidate for low-cost, field-portable diagnostics (Notomi et al., 2015; Becherer et al., 2020).

CRISPR-based diagnostic platforms have also further transformed diagnostics. SHERLOCK (Specific High-sensitivity Enzymatic Reporter unLOCKing) utilizes the collateral cleavage activity of CRISPR-Cas systems with isothermal amplification for ultra-sensitive and portable detection of RNA and DNA targets. The system enables multiplexed detection of viral and bacterial infections in clinically important samples (Kellner et al., 2019; Joung et al., 2020; Zahra et al., 2023). DETECTR (DNA Endonuclease Targeted CRISPR Trans Reporter), a CRISPR-Cas12a-based diagnostic platform, enables quick, low-cost, and reliable detection of viral infections. It utilizes Cas12a activation upon target RNA recognition with subsequent pre-amplification to increase specificity and sensitivity-more efficient than with conventional PCR-based assays in certain instances (Petri & Pattanayak, 2018).

In addition, multiplex immunoassays and biosensors have become important platforms for parallel multiple biomarker detection from a single biological sample. These platforms utilize electrochemical sensing schemes, such as methylene blue-conjugated secondary antibodies, to measure proteins at the picogram level, thus allowing cost-effective high-throughput diagnostics (Jones et al., 2019). Their use in the biopharmaceutical firm is manifested in the accelerated screening of drug-associated biomarkers. While their use in the clinic is in its infancy, these sensors represent a viable method for non-invasive and accurate disease monitoring (Ahsan, 2022; Rosa et al., 2022).

AI has made a great difference in the practice of radiology and computer diagnosis. Under the pretext of increased diagnostic precision and better patient care, AI-driven image analysis software advances customized medicine by addressing the drawbacks of conventional diagnostic methods. The integration of AI with radiologic imaging improves not only the accuracy of interpretations but also treatment planning and decision-making (Oyeniyi & Oluwaseyi, 2024; Bhure & Shete, 2024).

Wearable biosensors and telemedicine platforms constitute the core of impelling diagnostic innovation. Characterized by their cost-effectiveness, non-invasive nature, and simplicity, these wearable devices allow continuous real-time monitoring of physiological parameters. These technologies are representative of the growing concern regarding health issues and have been utilized in monitoring heart rate, oxygen saturation, and other biomarkers (Chakrabarti et al., 2022). Telemedicine has become a central strategy in closing gaps in healthcare access, especially in rural and underserved communities. Its use during the COVID-19 pandemic demonstrated its capacity to lower infection risk without sacrificing the quality of healthcare services. Successful use of telemedicine relies on a range of critical factors, including EMRs, audiovisual communication platforms, provider-patient communication, IT support, training, and suitable coding infrastructure (Franc et al., 2011; Smith et al., 2020).

Finally, systems that yield results directly from samples are becoming more established in outbreak environments with the potential for on-site rapid diagnostic testing. These closed, integrated systems integrate nucleic acid extraction, amplification, and detection into a single device, thus reducing the potential for contamination and the need for highly specialized laboratory infrastructure. They provide the potential for the replacement of traditional real-time PCR, especially in field and resource-limited environments (Koo et al., 2023; Hauner et al., 2024).

1.9 ONE HEALTH AND PLANETARY HEALTH APPROACHES

Surveillance of zoonotic disease in wildlife and livestock is a fundamental component of infectious disease management because animal reservoirs and transmitters are central to the spread of infectious diseases. The interface between wildlife and livestock is increasingly seen as a hotspot for the emergence of zoonotic disease due to drivers such as globalization, habitat fragmentation, and climate change. The drivers facilitate the transmission of viral families such as *Bunyavirales*,

Coronaviridae, and *Flaviviridae*, and bacterial pathogens such as *Mycobacterium* and *Brucella* (Meurens et al., 2021). Surveillance systems are vital for early detection and response in case of threats, recalling that around 75% of EIDs in humans originate from animals, mostly wildlife (Merianos, 2007). In Europe, disease epidemiology related to wildlife requires strong continent-wide surveillance systems to provide protection for public, livestock, and wildlife health (Yon et al., 2019). The One Health approach, which covers human, animal, and environmental health, is increasingly advocated as a model for successful anticipation and prevention of the emergence of disease through the detection of pathogens that can overcome the species barrier (Morse, 2014). Even though wildlife is ecologically heterogeneous, effective surveillance programs have the potential to improve early detection and control of diseases that threaten humans and domestic animals (Artois et al., 2009). Successful prevention and control require national and international cooperation to improve surveillance systems, supported by cross-sectoral coordination and specialist research (Merianos, 2007).

Management of zoonotic diseases relies heavily on coordination among the veterinary, environmental, and human health sectors. One Health has been very popular worldwide as a strategy for improving health security by strengthening early detection, prevention, and control of zoonotic and emerging diseases (Belay et al., 2017). Integrative One Health is most urgently needed in highly populated and rapidly changing areas like East Asia and the Pacific, where new diseases are most prevalent. There are, however, issues with the application of One Health, particularly in low- and middle-income economies like India, where barriers include the lack of policy prioritization, departmental overlapping jurisdictions, and weak institutional infrastructure (Asaaga et al., 2021). In Indonesia, limited knowledge and understanding of the One Health concept by health and veterinary practitioners further hinder its application, necessitating education and training programs to fill knowledge gaps (Thursina et al., 2024). Furthermore, international collaboration and concerted action to fight global disease determinants like climate change and resource extraction are critical in reducing the global burden of infectious diseases (Richards, 2013). To operationalize One Health fully, there is a need to enhance intersectoral communication, enhance institutional frameworks, and invest in workforce capacity building.

Zoonotic disease spillover is anticipated by the knowledge of the intricate interactions among pathogens, wildlife, and humans and their environment. Cross-species disease transmission is modeled using sophisticated modeling tools, such as continuous-time stochastic models, which link reservoir host dynamics with human epidemiology. Most zoonotic spillover events are not self-sustaining in the human population, yet their expansion highlights the necessity of anticipatory strategies to prevent pandemics (Saldaña et al., 2024). Percolation models characterize spillover as a multistep event, using directed graphs to model the influence of drivers such as pathogen load and host susceptibility (Washburne et al., 2019). Ecosystem interfaces, such as rural-forest edges, are high-risk zones for spillover as a result of ecological fragmentation and increased human-wildlife contact. The idea of "permeability" helps to construct a generic spillover theory by quantifying the probability of host-pathogen interaction in such interfaces (Borremans et al., 2019). SARS-like coronavirus spillover risk assessments reveal landscape use change, wildlife host density increase, and human exposure pattern changes to form regional risk clusters, usually where access to healthcare is poor. This highlights the necessity of merging public health planning with environmental and land-use policy to minimize outbreak risk (Muylaert et al., 2023). All these models and concepts offer useful tools to identify and manage spillover hotspots prior to the development of outbreaks.

The emergence of infectious diseases is closely linked with ecological processes and the crossing of planetary boundaries. Global environmental changes ranging from climate change, deforestation, and urbanization, are restructuring ecosystems in ways that increase the emergence and spread of infectious diseases by disrupting prevalent pathogen-host interactions (Carlson et al., 2025). One Health provides a platform that recognizes the need to respond to these changes through cross-sectoral management of climate-sensitive diseases. Importantly, climate change destabilizes ecosystem services that modulate disease transmission, thus making control efforts challenging toward new threats (Dovie et al., 2022). Anthropogenic ecosystem degradation, particularly in transitional zones where natural environments intersect human settlements, increases the risk of zoonotic diseases. Studies indicate that areas with moderate ecological integrity are highly vulnerable to newly observed EIDs (Marcolin et al., 2024). Furthermore, loss of biodiversity and the general impacts of the Anthropocene increase the spread of zoonotic and vector-borne diseases, showing the need for integrated strategies that incorporate biodiversity conservation, disease surveillance, and resilient health systems (Carlson et al., 2025). Interdisciplinary strategy that addresses the environmental, social, and biological drivers of disease remains critical to reduce vulnerability and ensure the long-term health of human communities and ecosystems.

1.10 ANTIMICROBIAL RESISTANCE (AMR): THE SILENT PANDEMIC

Infectious disease-related AMR is increasingly linked to complex bacterial defense mechanisms such as biofilm development, efflux pump activity, and horizontal gene transfer (HGT). HGT enables the transmission of antibiotic resistance genes (ARGs) among bacterial populations and is especially evoked in biofilm communities, where transformation, transduction, and conjugation are facilitated by the intimate cell-to-cell contact and trapping of DNA within the extracellular polymeric substance (EPS) matrix (Michaelis & Grohmann, 2023). Biofilms provide structural support and enhance metabolic heterogeneity to create favorable conditions for gene exchange and survival of resistant populations. Efflux pumps, which are embedded in bacterial membranes, also enhance AMR by actively pumping the antibiotic from the cell, thus reducing intracellular drug concentrations and compromising therapeutic efficacy. Efflux pumps enhance the stability of biofilms in pathogens like *Staphylococcus aureus*, thus enhancing bacterial resistance to therapy (Sinha et al., 2024). The synergism of these processes is of serious concern; biofilms impede not only the penetration of antimicrobials but also facilitate a microenvironment for survival of persister cells and upregulation of resistance genes, including those coding for efflux systems (Sedgley & Dunny, 2015). The coexistence of these mechanisms together presents tremendous challenges to clinical management, thus necessitating the need for targeted interventions to disrupt biofilms, inhibit efflux function, and suppress gene transfer pathways to adequately curtail AMR (Burmeister, 2015).

In the context of NTDs, AMR presents unique challenges informed by both proximate and systemic factors. Proximally, the sheer and largely unregulated application of antimicrobials in mass drug administration (MDA) programs for NTD control has significantly accelerated the accumulation of resistance (Akinsolu et al., 2019). This is exacerbated by the lack of precision in diagnostic assays and repeated application of a narrow repertoire of therapeutic agents, which constitutes selective pressure on resistant microbes. Distally, socio-environmental determinants of economic disadvantage, limited access to healthcare, poor sanitation, and poor health systems exacerbate the problem by triggering widespread proliferation of resistant microbes (Nguyen-Thanh et al., 2024). Environmental reservoirs of resistance, such as hospital effluent and agricultural runoff, further exacerbate the problem by dispersing resistant strains and genetic material across extended ecosystems (Irfan et al., 2022). For this reason, multi-pronged action must be employed to address AMR in the context of NTDs. Intensification of surveillance systems and stewardship programs can regulate misuse of antimicrobials, while public awareness initiatives can promote education and behavioral change. No less important is also the necessity to strengthen water, sanitation, and hygiene infrastructure, and to improve healthcare delivery in underserved areas. Long-term interventions must also involve investments in the development of new antimicrobials and rapid diagnostic technologies, enabled through intersectoral collaboration involving policymakers, healthcare providers, researchers, and affected populations (Gupta et al., 2012). Such integrated interventions are pivotal to the preservation of the efficacy of current drugs and to the sustainability of NTD control in the face of increasingly accumulating AMR pressures.

1.11 EMERGING TECHNOLOGIES AND SYNTHETIC BIOLOGY

Gain of function is the process of generating new characteristics in organisms by altering their genes either naturally or in a laboratory setting. Gain of function research enhances understanding of viral processes such as transmissibility, replication, virulence, host range, immune evasion, and drug and vaccine resistance (Saalbach, 2022). Gain of function involves the deliberate alteration of viral genetic material, which can provide more insight into molecular processes such as viral attachment to cells, cell penetration, and taking over cellular machinery (Fischer & Smith, 2021).

Engineered probiotics are live microorganisms that confer beneficial traits to the host by modulating intestinal microbiota. Genetically modified probiotics have been developed to counter AMR and improve beneficial traits (Mazhar et al., 2020). Probiotics find extensive applications in health and possess functional parameters that affect their metabolic activity (Yadav & Shukla, 2020). Moreover, bacteriophages can eliminate antibiotic-sensitive and antibiotic-resistant bacteria and have been found to be effective in infection treatment (Lobocka et al., 2021).

AI has had a major influence on genomics, proteomics, and immunology, and integration of AI into the creation of biomedicine is more extensive. AI plays a major role in the creation of mRNA vaccines, particularly in the step of antigen selection, epitope prediction, and adjuvant identification (McCaffrey, 2022). Vaccine development is critical to public health, such that infectious diseases are cured and mortality and morbidity globally are reduced. AI-based approaches facilitate systematic immunogen design and the identification of non-adjuvant candidates with effective function (Olawade et al., 2024). AI facilitates algorithmic protein and genetic circuit design by employing deep learning models, e.g., deep neural networks and reinforcement learning, to predict protein structure (Oladele, 2024).

Dual-use research of concern (DURC) refers to scientific research that has the potential to advance knowledge to be used to damage national security, public health, or the environment (Patrone et al., 2012). Effective control of DURC should involve strict interdisciplinary methods that draw from history lessons and ethical implications (Gillum, 2024). Biosafety policies are necessary regulations aimed at reducing risks posed by the manipulation of biological agents in laboratories.

1.12 CASE STUDIES IN EMERGING INFECTIOUS DISEASES

EIDs case studies are valuable learning tools for understanding the complex dynamics of infectious disease outbreaks, as well as the success of response efforts. The 2014–2016 Ebola virus epidemic is a good case in point on the potential of case-based learning to increase student interest and foster higher-order thinking in the natural and health sciences. In US higher education institutions, the use of such approaches has been found to construct cognitive knowledge and dispel misconceptions regarding the disease and, therefore, enhance public health literacy (Dube et al., 2018).

From a clinical perspective, case studies play a vital role in the synthesis of interdisciplinary knowledge that consists of clinical medicine, microbiology, and epidemiology. The reference book *EIDs: Clinical Case Studies* provides an overview of infections like Middle East respiratory syndrome coronavirus (MERS-CoV) and severe fever with thrombocytopenia syndrome (SFTS). It provides healthcare workers with experiential knowledge in the form of hands-on suggestions on diagnosis, treatment, and prevention alongside identifying key knowledge gaps at the same time, thus making evidence-based decision-making feasible in real-life situations (Kynaston & Sinnott, 2015).

Aside from their importance in the clinical and academic settings, case studies play a pivotal role in global health policy-making, particularly disease surveillance. For instance, comparative case studies in Kenya, Peru, Thailand, and along the US-Mexico border identify the infrastructural and socio-political determinants that drive EID surveillance. The case studies highlight the need for active involvement of local specialists, efficient application of communication technologies, and application of centralized health networks. The case studies also emphasize the need for collaboration among civilian and military organizations to optimize global EID surveillance systems (Ear, 2012).

1.13 STRATEGIC FORESIGHT AND FUTURE PREPAREDNESS

Strategic foresight, as required to react suitably to EIDs, requires an overarching and foresight-based approach, solidly established in the synthesis of historical lessons from past outbreaks and forward-looking planning for future dangers. Key aspects of preparedness include rapid response systems, global collaboration, and infection control systems that are robust; these are required to enhance the resilience of both global and national health systems (Gama et al., 2024). Because of the inherent complexity and unpredictability associated with EIDs, sustained investment in prevention and control infrastructure is the order of the day. This includes the development of resilient surveillance systems and a "wartime-peacetime" medical transition system, the aim being to develop preparedness during emergent emergencies and inter-epidemic periods alike (Yang & Zhang, 2022). The escalating danger from AMR and ongoing emergence of new pathogens also operate to underscore the need for sustained global health vigilance and collaborative risk mitigation strategies (Gupta et al., 2012).

To predict and rank potential threats, tools such as environmental scanning, strategic foresight programs, and quantitative risk analysis models are required. The tools allow policymakers and public health practitioners to direct resources effectively and take specific mitigation measures prior to the outbreak (Brookes et al., 2015). Recent pandemic experience – most importantly COVID-19 – has served to emphasize the need for stronger surveillance infrastructure, improved real-time data collection, and supply chain integrity. Furthermore, the creation and refinement of outbreak prediction models and emergency response plans have become a priority area of concern in the building of resilience against future pandemics (Bloch et al., 2024).

1.14 CONCLUSION

The spread of infectious diseases is likely to continue as a persistent serious challenge to public health systems, particularly in a context of rising mobility, environmental deterioration, and microorganism evolution. This chapter contends that control of such threats needs more than reactive response; it needs anticipatory, data-guided, and integrative strategies. The epidemiological backdrop of EIDs is increasingly being characterized by drivers such as zoonotic spillover events, climate change, demographic stress, and antibiotic abuse. As with recent outbreaks like COVID-19, Ebola, and MERS, it is essential to detect these threats early and respond promptly to contain them effectively. Next-generation digital disease surveillance technologies, fueled by AI and real-time data fusion, have unprecedented potential for monitoring disease emergence, risk analysis, and health system response. In parallel with this, significant advances in therapeutic modalities – such as mRNA vaccines, monoclonal antibodies, and genome-edited antiviral drugs – have the potential for rapid response to newly emergent pathogens. Such technology, however, must be coupled with strong global cooperation, fair access to resources, and sustained investment in public health programs. To respond effectively to challenges on the horizon, a paradigm shift is needed toward a proactive, interdisciplinary approach that converges human, animal, and environmental health (One Health).

ABBREVIATIONS

AI	Artificial Intelligence
AMR	Antimicrobial Resistance
ARGs	Antibiotic Resistance Genes
DALYs	Disability-Adjusted Life Years
DETECTR	DNA Endonuclease Targeted CRISPR Trans Reporter
DHODH	Dihydroorotate Dehydrogenase
DURC	Dual-Use Research of Concern
EHRs	Electronic Health Records
EIDs	Emerging Infectious Diseases
EPS	Extracellular Polymeric Substance
HGT	Horizontal Gene Transfer
HIV	Human Immunodeficiency Virus
HLA	Human Leukocyte Antigen
HPV	Human Papillomavirus
HTAs	Host-Directed Antivirals
IHR	International Health Regulations
IL-1Ra	Interleukin-1 Receptor Antagonist
LAMP	Loop-Mediated Isothermal Amplification
LMICs	Low- and Middle-Income Countries
MDA	Mass Drug Administration
MERS-CoV	Middle East Respiratory Syndrome Coronavirus
ML	Machine Learning
NCDs	Non-communicable Diseases
NGS	Next-Generation Sequencing
NTDs	Neglected Tropical Diseases
PDMAEMA	Poly(Dimethylaminoethyl Methacrylate)
RSV	Respiratory Syncytial Virus
RV	Reverse Vaccinology
SARS-CoV	Severe Acute Respiratory Syndrome Coronavirus
SFTS	Severe Fever with Thrombocytopenia Syndrome
SHERLOCK	Specific High-sensitivity Enzymatic Reporter unLOCKing
VLPs	Virus-Like Particles

ACKNOWLEDGMENT

All the authors are thankful to their respective universities and institutions for their valuable support.

DATA AVAILABILITY

The authors confirm that the data supporting the findings of this study are available within the chapter.

REFERENCES

Abdallah, M. S. (2018). Review on emerging and reemerging infectious diseases and their origins. *Microbiology Research Journal International, 26*(1), 1–5.

Abdolmohammadi-Vahid, S., Baradaran, B., Adcock, I. M., & Mortaz, E. (2024). Immune checkpoint inhibitors and SARS-CoV2 infection. *International Immunopharmacology, 137*, 112419.

Adeoye, A., Onah, C. O., Orobator, E. T., Akintayo, E. A., Inuaeyen, J. U., Umoru, D. O., & Demola, M. B. (2025). AI and machine learning for early detection of infectious diseases in the US: Opportunities and challenges. *Journal of Medical Science, Biology, and Chemistry, 2*(1), 54–63.

Ahamad, S. K., Laxmi, R. V., Kumar, A. R., Kumar, M. N., Chakala, V., & Tadikonda, R. R. (2024). mRNA vaccines: Transforming disease prevention for the modern era. *International Journal of Health Sciences and Research, 14*(10), 505–515.

Ahsan, H. (2022). Monoplex and multiplex immunoassays: Approval, advancements, and alternatives. *Comparative Clinical Pathology, 31*(2), 333–345.

Akinsolu, F. T., Nemieboka, P. O., Njuguna, D. W., Ahadji, M. N., Dezso, D., & Varga, O. (2019). Emerging resistance of neglected tropical diseases: A scoping review of the literature. *International Journal of Environmental Research and Public Health, 16*(11), 1925.

Alagendran, S., Archunan, G., Prabhu, S. V., Orozco, B. E., & Guzman, R. G. (2010). Biochemical evaluation in human saliva with special reference to ovulation detection. *Indian Journal of Dental Research, 21*, 165–168.

Alamgir, W., & Sajjad, W. (2024). Strengthening health security through sustainable vaccine economics and equity. *Life Sciences, 5*(3), 02.

Alberca, L. N., & Talevi, A. (2020). The efficiency of multi-target drugs: A network approach. In Mariano Bizzarri (Ed.). *Approaching complex diseases: Network-based pharmacology and systems approach in bio-medicine* (pp. 63–75). Springer.

Alexander, E., & Leong, K. W. (2024). Discovery of nanobodies: A comprehensive review of their applications and potential over the past five years. *Journal of Nanobiotechnology, 22*(1), 661.

Al-Saeedi, M., & Al-Hajoj, S. (2017). Diversity and evolution of drug resistance mechanisms in Mycobacterium tuberculosis. *Infection and Drug Resistance, 10*, 333–342.

Alves, C., Souza, T., Meyer, I., Toralles, M. B. P., & Brites, C. (2006). Immunogenetics and infectious diseases: Special reference to the mayor histocompatibility complex. *Brazilian Journal of Infectious Diseases, 10*, 122–131.

Anelone, A. J., & Spurgeon, S. K. (2017). Prediction of the containment of HIV infection by antiretroviral therapy – A variable structure control approach. *IET Systems Biology, 11*(1), 44–53.

Artois, M., Bengis, R., Delahay, R. J., Duchêne, M. J., Duff, J. P., Ferroglio, E., & Smith, G. C. (2009). Wildlife disease surveillance and monitoring. In J. Delahay, Graham C. Smith, Michael R. Hutchings (Eds.). *Management of disease in wild mammals* (pp. 187–213). Tokyo: Springer.

Arumugam, V. A., Thangavelu, S., Fathah, Z., Ravindran, P., Sanjeev, A. M. A., Babu, S., Arun, M., Mohd. Iqbal, Y., Khan, S., Ruchi, T., Megha Katare, P., Ranjit, S., Chandra, R., & Dhama, K. (2020). COVID-19 and the world with co-morbidities of heart disease, hypertension and diabetes. *Journal of Pure and Applied Microbiology, 14*(3), 1623–1638.

Asaaga, F. A., Young, J. C., Oommen, M. A., Chandarana, R., August, J., Joshi, J., & Purse, B. V. (2021). Operationalising the "One Health" approach in India: Facilitators of and barriers to effective cross-sector convergence for zoonoses prevention and control. *BMC Public Health, 21*, 1–21.

Aulicino, P. C., & Kimata, J. T. (2024). Beyond surveillance: Leveraging the potential of next generation sequencing in clinical virology. *Frontiers in Tropical Diseases, 5*, 1512606.

Azegami, T., Yuki, Y., & Kiyono, H. (2014). Challenges in mucosal vaccines for the control of infectious diseases. *International Immunology, 26*(9), 517–528.

Babalola, O. E., & Ajayi, A. A. (2023). The place of ivermectin in the management of Covid-19: State of the evidence. *Medical Research Archives, 11*(4).

Bai, Y., Yang, B., Lin, L., Herrera, J. L., Du, Z., & Holme, P. (2017). Optimizing sentinel surveillance in temporal network epidemiology. *Scientific Reports, 7*(1), 4804.

Balakumar, A., Das, D., Datta, A., Mishra, A., Bryak, G., Ganesh, S. M., & Thangamani, S. (2024). Single-cell transcriptomics unveils skin cell specific antifungal immune responses and IL-1Ra-IL-1R immune evasion strategies of emerging fungal pathogen Candida auris. *PLoS Pathogens, 20*(11), e1012699.

Bartos, M. (2020). Modern pandemics and old methods: What AIDS, SARS, Ebola and the long history of quarantine tell us about Covid-19. *Asia-Pacific Journal, 18*(14), e2.

Bashetti, P., Kurhe, R., Khandare, R., Bansod, A., Namapalle, M., Kshirsagar, A. V., & Avhad, C. (2024). One health perspective for infectious diseases of animals. *International Journal of Advanced Biochemistry Research, 8*(2S), 450–456.

Becherer, L., Borst, N., Bakheit, M., Frischmann, S., Zengerle, R., & von Stetten, F. (2020). Loop-mediated isothermal amplification (LAMP) – Review and classification of methods for sequence-specific detection. *Analytical Methods, 12*(6), 717–746.

Belay, E. D., Kile, J. C., Hall, A. J., Barton-Behravesh, C., Parsons, M. B., Salyer, S., & Walke, H. (2017). Zoonotic disease programs for enhancing global health security. *Emerging Infectious Diseases*, 23(Suppl 1), S65.

Beshwari, F., Beshwari, M., & Beshwari, A. (2020). The role of artificial intelligence in mitigating unknown-unknown risks. *The Role of Artificial Intelligence in Mitigating Unknown-Unknown Risks*, 64(1), 13.

Bhakta, S., Paul, J., Bhattacharya, A., & Choudhury, S. (2021). Vaccine development through reverse vaccinology using artificial intelligence and machine learning approach. In Saptarshi Chatterjee (Ed.). *COVID-19: Tackling global pandemics through scientific and social tools* (pp. 33–49). Academic Press.

Bhure, S., & Shete, V. (2024). Revolutionizing healthcare: AI-powered imaging transforms medical diagnosis. In *2024 2nd DMIHER International Conference on Artificial Intelligence in Healthcare, Education and Industry (IDICAIEI)* (pp. 1–4). IEEE.

Bloch, E. M., Sullivan, D. J., Casadevall, A., Shoham, S., Tobian, A. A., & Gebo, K. (2024). Applying lessons of COVID-19 and other emerging infectious diseases to future outbreaks. *MBio*, 15(6), e01109–e01124.

Bojarska, J., Wang, X., & Skwarczynski, M. (2024). Peptides against infectious diseases: From antimicrobial peptides to vaccines. *Frontiers in Pharmacology*, 15, 1522148.

Borremans, B., Faust, C., Manlove, K. R., Sokolow, S. H., & Lloyd-Smith, J. O. (2019). Cross-species pathogen spillover across ecosystem boundaries: Mechanisms and theory. *Philosophical Transactions of the Royal Society B*, 374(1782), 20180344.

Branda, F., Ali, A. Y., Ceccarelli, G., Albanese, M., Binetti, E., Giovanetti, M., & Scarpa, F. (2024). Assessing the burden of neglected tropical diseases in low-income communities: Challenges and solutions. *Viruses*, 17(1), 29.

Brockmann, D. (2017). Global connectivity and the spread of infectious diseases. *Nova Acta Leopoldina*, 419, 129–136.

Brookes, V. J., Hernandez-Jover, M., Black, P. F., & Ward, M. P. (2015). Preparedness for emerging infectious diseases: Pathways from anticipation to action. *Epidemiology and Infection*, 143(10), 2043–2058.

Bucătariu, S., Cosman, B., Constantin, M., Ailiesei, G. L., Rusu, D., & Fundueanu, G. (2024). Thermally solvent-free cross-linked pH/thermosensitive hydrogels as smart drug delivery systems. *Gels*, 10(12), 834.

Burmeister, A. R. (2015). Horizontal gene transfer. *Evolution, Medicine, and Public Health*, 2015(1), 193–194.

Cancrini, G., & Iori, A. (2004). Traditional and innovative diagnostic tools: When and why they should be applied. *Parassitologia*, 46(1–2), 173–176.

Capitani, N., & Baldari, C. T. (2022). The immunological synapse: An emerging target for immune evasion by bacterial pathogens. *Frontiers in Immunology*, 13, 943344.

Carlson, C. J., Brookson, C. B., Becker, D. J., Cummings, C. A., Gibb, R., Halliday, F. W., & Poisot, T. (2025). Pathogens and planetary change. *Nature Reviews Biodiversity*, 1(1), 32–49.

Carrion, M., & Madoff, L. C. (2017). ProMED-mail: 22 years of digital surveillance of emerging infectious diseases. *International Health*, 9(3), 177–183.

Chakrabarti, S., Biswas, N., Jones, L. D., Kesari, S., & Ashili, S. (2022). Smart consumer wearables as digital diagnostic tools: A review. *Diagnostics*, 12(9), 2110.

Chen, D., & Kristensen, D. (2009). Opportunities and challenges of developing thermostable vaccines. *Expert Review of Vaccines*, 8(5), 547–557.

Chen, K. T. (2022). Emerging infectious diseases and one health: Implication for public health. *International Journal of Environmental Research and Public Health*, 19(15), 9081.

Colgrove, J. (2014). The Ebola outbreak: A historical perspective vis-à-vis past epidemics. *Journal of Communication in Healthcare*, 7(4), 250–251.

Coltart, C. E., & Behrens, R. H. (2012). The new health threats of exotic and global travel. *British Journal of General Practice*, 62(603), 512–513.

Danon, L., Ford, A. P., House, T., Jewell, C. P., Keeling, M. J., Roberts, G. O., & Vernon, M. C. (2011). Networks and the epidemiology of infectious disease. *Interdisciplinary Perspectives on Infectious Diseases*, 2011(1), 284909.

Das, S., Gajbhiye, R. L., Kumar, N., & Sarkar, D. (2023). Multi-targeted prediction of the antiviral effect of Momordica charantia extract based on network pharmacology. *Journal of Natural Remedies*, 23(1), 169–183.

Dawson-Hahn, E. E., Pidaparti, V., Hahn, W., & Stauffer, W. (2021). Global mobility, travel and migration health: Clinical and public health implications for children and families. *Paediatrics and International Child Health*, 41(1), 3–11.

DeVito, N., Morse, M. A., Hanks, B., & Clarke, J. M. (2018). Immune checkpoint combinations with inflammatory pathway modulators. In Sandip Pravin Patel, Razelle Kurzrock (Eds.). *Early Phase Cancer Immunotherapy* (pp. 219–241). Cham: Springer International Publishing.

di Bari, C., Venkateswaran, N., Pigott, D., Flastl, C., & Devleesschauwer, B. (2022). The global burden of neglected zoonotic diseases: Current state of evidence. *European Journal of Public Health, 32*(Supplement_3), ckac129–757.

Dovie, D. B., Miyittah, M. K., Dodor, D. E., Dzodzomenyo, M., Christian, A. K., Tete-Larbi, R., & Bawah, A. A. (2022). Earth system's gatekeeping of "One Health" approach to manage climate-sensitive infectious diseases. *GeoHealth, 6*(4), e2021GH000543.

Duan, C., Cao, H., Zhang, L. H., & Xu, Z. (2021). Harnessing the CRISPR-Cas systems to combat antimicrobial resistance. *Frontiers in Microbiology, 12*, 716064.

Dube, D., Addy, T. M., Teixeira, M. R., & Iadarola, L. M. (2018). Enhancing student learning on emerging infectious diseases: An Ebola exemplar. *The American Biology Teacher, 80*(7), 493–500.

Dunuweera, S. P., Rajapakse, R. M. S. I., Rajapakshe, R. B. S. D., Wijekoon, S. H. D. P., Nirodha Thilakarathna, M. G. G. S., & Rajapakse, R. M. (2019). Review on targeted drug delivery carriers used in nanobiomedical applications. *Current Nanoscience, 15*(4), 382–397.

Ear, S. (2012). Towards effective emerging infectious diseases surveillance: Evidence from Kenya, Peru, Thailand, and the US-Mexico border. In *Stanford Center for International Government Working Paper* (p. 464).

Edwards, H. M., & Rhodes, J. (2021). Accounting for the biological complexity of pathogenic fungi in phylogenetic dating. *Journal of Fungi, 7*(8), 661.

Ellwanger, J. H., Veiga, A. B. G. D., Kaminski, V. D. L., Valverde-Villegas, J. M., Freitas, A. W. Q. D., & Chies, J. A. B. (2021). Control and prevention of infectious diseases from a One Health perspective. *Genetics and Molecular Biology, 44*(1 Suppl 1), e20200256.

Engering, A., Hogerwerf, L., & Slingenbergh, J. (2013). Pathogen–host–environment interplay and disease emergence. *Emerging Microbes & Infections, 2*(1), 1–7.

Esser-Kahn, A. P. (2023). Innate immune memory for improved vaccine response. *The Journal of Immunology, 210*(Supplement_1), 223.10.

Eweas, A. F., Alhossary, A. A., & Abdel-Moneim, A. S. (2021). Molecular docking reveals ivermectin and remdesivir as potential repurposed drugs against SARS-CoV-2. *Frontiers in Microbiology, 11*, 592908.

Feng, Z., Dieckmann, U., & Levin, S. A. (Eds.). (2006). *Disease evolution: Models, concepts, and data analyses* (Vol. 71). American Mathematical Soc.

Fischer, T. H., & Smith, C. J. (2021). Gain-of-function and pathogenic viruses. *Ethics & Medicine, 46*(9), 1–2.

Flahault, A. (2007). Emerging infectious diseases: The example of the Indian Ocean chikungunya outbreak (2005–2006). *Bulletin de l'Académie Nationale de Médecine, 191*(1), 113–124.

Fortpied, J., Collignon, S., Moniotte, N., Renaud, F., Bayat, B., & Lemoine, D. (2020). The thermostability of the RTS,S/AS01 malaria vaccine can be increased by co-lyophilizing RTS,S and AS01. *Malaria Journal, 19*, 1–15.

Franc, S., Daoudi, A., Mounier, S., Boucherie, B., Laroye, H., Peschard, C., & Charpentier, G. (2011). Telemedicine: What more is needed for its integration in everyday life? *Diabetes & Metabolism, 37*, S71–S77.

Gama, K., Romeiro, E. T., da Silva, C. L., de Jesus, B. M. S., de Freitas, H. C., de Sousa, B. E., dos Santos, V. S., de Medeiros, G. A. A., Hackenhaar, T. S., & Lustosa, A. (2024). Strategies for control and prevention of emerging infectious diseases: Lessons learned and future directions. *Revista Ibero-Americana de Humanidades, Ciências e Educação 10*(3), 38–44.

Garcia, L. P., & da Silva, G. D. M. (2016). *Doenças transmissíveis e situação socioeconômica no Brasil: Análise espacial* (No. 2263). Texto para Discussão.

Georgiev, V. S. (2009). Emerging and re-emerging infectious diseases. In *National institute of allergy and infectious diseases, NIH: Volume 2: Impact on global health* (pp. 23–28). Brasília: Institute of Applied Economic Research (IPEA).

Ghatak, S., Milton, A. A. P., & Das, S. (2020). Drivers of emerging viral zoonoses. In Yashpal Singh Malik, Raj Kumar Singh, Kuldeep Dhama (Eds.). *Animal-origin viral zoonoses* (pp. 313–338). Singapore: Springer.

Ghosh, S., Chowdhury, S. R., Rahaman, M., Basu, B., & Prajapati, B. (2024). Revolutionizing influenza treatment: A deep dive into targeted drug delivery systems. *Current Pharmaceutical Biotechnology* https://doi.org/10.2174/0113892010326373241012061547

Gibb, R., Ryan, S. J., Pigott, D., Fernandez, M. D. P., Muylaert, R. L., Albery, G. F., & Carlson, C. J. (2024). The anthropogenic fingerprint on emerging infectious diseases. *MedRxiv*, 2024-05.

Gillum, D. R. (2024). Balancing innovation and safety: Frameworks and considerations for the governance of dual-use research of concern and potential pandemic pathogens. *Applied Biosafety*. https://doi.org/10.1089/apb.2024.0033

González-Candelas, F., Comas, I., Martínez, J. L., Galán, J. C., & Baquero, F. (2011). The evolution of antibiotic resistance. In Michel Tibayrenc (Ed.). *Genetics and evolution of infectious disease* (pp. 305–337). Elsevier.

Goswami, A. K., Gadela, V. R., & Venuganti, V. V. K. (2024). Transforming biologics delivery through microneedles. *Current Nanoscience*. https://doi.org/10.2174/0115734137317813241014113421

Gundappa, M., Arumugam, V. A., Hsieh, H. L., Balasubramanian, B., & Shanmugam, V. (2022). Expression of tissue factor and TF-mediated integrin regulation in HTR-8/SVneo trophoblast cells. *Journal of Reproductive Immunology, 150,* 103473.

Gupta, S. K., Gupta, P., Sharma, P., Shrivastava, A. K., & Soni, S. K. (2012). Emerging and re-emerging infectious diseases, future challenges and strategy. *Journal of Clinical and Diagnostic Research, 6*(6), 223.

Hajishengallis, G., & Chavakis, T. (2024). Central trained immunity and its impact on chronic inflammatory and autoimmune diseases. *Journal of Allergy and Clinical Immunology, 154*(5), 1113–1116.

Hajishengallis, G., Li, X., Mitroulis, I., & Chavakis, T. (2019). Trained innate immunity and its implications for mucosal immunity and inflammation. *Oral Mucosal Immunity and Microbiome, 1197,* 11–26.

Hauner, A., Onwuchekwa, C., & Ariën, K. K. (2024). Sample-to-result molecular diagnostic platforms and their suitability for infectious disease testing in low-and middle-income countries. *Expert Review of Molecular Diagnostics, 24*(5), 423–438.

Hayman, C. M., Hermans, I. F., & Painter, G. F. (2018). Increased efficacy of NKT cell-adjuvanted peptide vaccines through chemical conjugation. In Zbigniew J. Witczak, Roman Bielski (Eds.). *Coupling and decoupling of diverse molecular units in glycosciences* (pp. 309–335). Cham: Springer International Publishing.

He, Y., Zhou, J., Gao, H., Liu, C., Zhan, P., & Liu, X. (2024). Broad-spectrum antiviral strategy: Host-targeting antivirals against emerging and re-emerging viruses. *European Journal of Medicinal Chemistry, 265,* 116069.

Henning, K. J. (2004). What is syndromic surveillance? *MMWR: Morbidity & Mortality Weekly Report, 53.*

Holmes, K. K., Bertozzi, S., Bloom, B. R., & Jha, P. (2017). *Disease control priorities: Major infectious diseases* (Vol. 6). World Bank Publications.

Horsburgh, B. A., Walker, G. J., Kelleher, A., Lloyd, A. R., Bull, R. A., & Giallonardo, F. D. (2024). Next-generation sequencing methods for near-real-time molecular epidemiology of HIV and HCV. *Reviews in Medical Virology, 34*(6), e70001.

Husain, F., Akram, S., Al-Kubaisi, H. A. R., & Hameed, F. (2023). The COVID-19 pandemic exposes and exacerbates inequalities for vulnerable groups: A systematic review. *Pakistan Journal of Humanities and Social Sciences, 11*(3), 3755–3765.

Irfan, M., Almotiri, A., & AlZeyadi, Z. A. (2022). Antimicrobial resistance and its drivers–A review. *Antibiotics, 11*(10), 1362.

Jamaledin, R., Yiu, C. K., Zare, E. N., Niu, L. N., Vecchione, R., Chen, G., & Makvandi, P. (2020). Advances in antimicrobial microneedle patches for combating infections. *Advanced Materials, 32*(33), 2002129.

Jiao, Z., Ji, H., Yan, J., & Qi, X. (2023). Application of big data and artificial intelligence in epidemic surveillance and containment. *Intelligent Medicine, 3*(1), 36–43.

Jones, A., Dhanapala, L., Kankanamage, R. N., Kumar, C. V., & Rusling, J. F. (2019). Multiplexed immunosensors and immunoarrays. *Analytical Chemistry, 92*(1), 345–362.

Jones, B. A., Betson, M., & Pfeiffer, D. U. (2017). Eco-social processes influencing infectious disease emergence and spread. *Parasitology, 144*(1), 26–36.

Joung, J., Ladha, A., Saito, M., Segel, M., Bruneau, R., Huang, M. L. W., & Zhang, F. (2020). Point-of-care testing for COVID-19 using SHERLOCK diagnostics. *MedRxiv.*

Justino, C. I., Duarte, A. C., & Rocha-Santos, T. A. (2017). Recent progress in biosensors for environmental monitoring: A review. *Sensors, 17*(12), 2918.

Kamel, M., Aleya, S., Alsubih, M., & Aleya, L. (2024). Microbiome dynamics: A paradigm shift in combatting infectious diseases. *Journal of Personalized Medicine, 14*(2), 217.

Kellner, M. J., Koob, J. G., Gootenberg, J. S., Abudayyeh, O. O., & Zhang, F. (2019). SHERLOCK: Nucleic acid detection with CRISPR nucleases. *Nature Protocols, 14*(10), 2986–3012.

Khadka, S., Yuchi, A., Shrestha, D. B., Budhathoki, P., Al-Subari, S. M. M., Alhouzani, T. Z., & Butt, A. I. (2020). Repurposing drugs for COVID-19: An approach for treatment in the pandemic. *Alternative Therapies in Health and Medicine, 26*(S2), 100–107.

Khan, I., Gouxin, H., Khan, S., & Ali, A. (2024). The roles of pathogens in gut-related diseases and the strategies for inhibiting their growth. *Frontiers in Pharmacology, 15,* 1513694.

21

Kohli, R., & Tan, S. S. L. (2016). Electronic health records. *Mis Quarterly*, 40(3), 553–574.

Koo, B., Kim, M. G., Lee, K., Kim, J. Y., Lee, S., Kim, S. H., & Shin, Y. (2023). Automated sample-to-answer system for rapid and accurate diagnosis of emerging infectious diseases. *Sensors and Actuators B: Chemical*, 380, 133382.

Kozani, P. S., Kozani, P. S., & Rahbarizadeh, F. (2021). The potential applicability of single-domain antibodies (VHH): From checkpoint blockade to infectious disease therapy. *Trends in Medical Sciences*, 1(2), e114888.

Kuchi Bhotla, H., Balasubramanian, B., Meyyazhagan, A., Pushparaj, K., Easwaran, M., Pappusamy, Alwin Robert, A., Arumugam, V. A., Tsibizova, V., Msaad Alfalih, A., Aljowaie, R. M., Saravanan, M., & Di Renzo, G. C. (2021). Opportunistic mycoses in COVID-19 patients/survivors: Epidemic inside a pandemic. *Journal of Infection and Public Health*, 14(11), 1720–1726.

Kuchi Bhotla, H., Kaul, T., Balasubramanian, B., Easwaran, M., Arumugam, V. A., Pappusamy, M., Muthupandian, S., & Meyyazhagan, A. (2020). Platelets to surrogate lung inflammation in COVID-19 patients. *Medical Hypotheses*, 143, 110098.

Kwok, A. J., Mentzer, A., & Knight, J. C. (2021). Host genetics and infectious disease: New tools, insights and translational opportunities. *Nature Reviews Genetics*, 22(3), 137–153.

Kynaston, K., & Sinnott, J. (2015). *Emerging infectious diseases: Clinical case studies* In Önder Ergönül et al. (Eds.). *Clinical Infectious Diseases*, 61(3). https://doi.org/10.1093/cid/civ288

Le, T., Sun, C., Chang, J., Zhang, G., & Yin, X. (2022). mRNA vaccine development for emerging animal and zoonotic diseases. *Viruses*, 14(2), 401.

Lee, T. H., Liu, P. S., Wang, S. J., Tsai, M. M., Shanmugam, V., & Hsieh, H. L. (2021). Bradykinin, as a reprogramming factor, induces transdifferentiation of brain astrocytes into neuron-like cells. *Biomedicine*, 9(8), 923.

Lefrancq, N., Duret, L., Bouchez, V., Brisse, S., Parkhill, J., & Salje, H. (2025). Learning the fitness dynamics of pathogens from phylogenies. *Nature*, 1–8. https://doi.org/10.1101/2023.12.23.23300456

Lobocka, M., Dąbrowska, K., & Górski, A. (2021). Engineered bacteriophage therapeutics: Rationale, challenges and future. *BioDrugs*, 35(3), 255–280.

Lorenzo, M., & Picó, Y. (2019). Wastewater-based epidemiology: Current status and future prospects. *Current Opinion in Environmental Science & Health*, 9, 77–84.

Louten, J. (2016). Emerging and reemerging viral diseases. *Essential Human Virology*, 291.

Maarifi, G., Martin, M. F., Zebboudj, A., Boulay, A., Nouaux, P., Fernandez, J., & Nisole, S. (2022). Identifying enhancers of innate immune signaling as broad-spectrum antivirals active against emerging viruses. *Cell Chemical Biology*, 29(7), 1113–1125.

Machalaba, C. M., & Karesh, W. B. (2017). Emerging infectious disease risk: Shared drivers with environmental change. *Revue Scientifique et Technique [International Office of Epizootics]*, 36(2), 435–444.

MacIntyre, C. R., Chen, X., Kunasekaran, M., Quigley, A., Lim, S., Stone, H., & Gurdasani, D. (2023). Artificial intelligence in public health: The potential of epidemic early warning systems. *Journal of International Medical Research*, 51(3), https://doi.org/10.1177/03000605231159335

Mandl, K. D., Overhage, J. M., Wagner, M. M., Lober, W. B., Sebastiani, P., Mostashari, F., & Grannis, S. (2004). Implementing syndromic surveillance: A practical guide informed by the early experience. *Journal of the American Medical Informatics Association*, 11(2), 141–150.

Mao, K., Zhang, K., Du, W., Ali, W., Feng, X., & Zhang, H. (2020). The potential of wastewater-based epidemiology as surveillance and early warning of infectious disease outbreaks. *Current Opinion in Environmental Science & Health*, 17, 1–7.

Marcolin, L., Tonelli, A., & Di Marco, M. (2024). Early-stage loss of ecological integrity drives the risk of zoonotic disease emergence. *Journal of the Royal Society Interface*, 21(215), 20230733.

Martinus, K., Pauli, N., & Kragt, M. (2023). Key policy interventions to limit infectious disease emergence and spread. *Frontiers in Environmental Science*, 11, 1128831.

May, L., Chretien, J. P., & Pavlin, J. A. (2009). Beyond traditional surveillance: Applying syndromic surveillance to developing settings – Opportunities and challenges. *BMC Public Health*, 9, 1–11.

Mayorga-Ramos, A., Zúñiga-Miranda, J., Carrera-Pacheco, S. E., Barba-Ostria, C., & Guamán, L. P. (2023). CRISPR-Cas-based antimicrobials: Design, challenges, and bacterial mechanisms of resistance. *ACS Infectious Diseases*, 9(7), 1283–1302.

Mazhar, S. F., Afzal, M., Almatroudi, A., Munir, S., Ashfaq, U. A., Rasool, M., & Khurshid, M. (2020). The prospects for the therapeutic implications of genetically engineered probiotics. *Journal of Food Quality*, 2020(1), 9676452.

McCaffrey, P. (2022). Artificial intelligence for vaccine design. In Sunil Thomas (Ed.). *Vaccine design: Methods and protocols, Volume 3. Resources for vaccine development* (pp. 3–13). New York, NY: Humana.

McClure, N. S., & Day, T. (2014). A theoretical examination of the relative importance of evolution management and drug development for managing resistance. *Proceedings of the Royal Society B: Biological Sciences, 281*(1797), 20141861.

McFarlane, R. A., Sleigh, A. C., & McMichael, A. J. (2013). Land-use change and emerging infectious disease on an island continent. *International Journal of Environmental Research and Public Health, 10*(7), 2699–2719.

Medhasi, S., & Chantratita, N. (2022). Human leukocyte antigen (HLA) system: Genetics and association with bacterial and viral infections. *Journal of Immunology Research, 2022*(1), 9710376.

Meher, M. K., & Poluri, K. M. (2020). pH-sensitive nanomaterials for smart release of drugs. In Nabeel Ahmad and P. Gopinath (Eds.). *Intelligent nanomaterials for drug delivery applications* (pp. 17–41). Elsevier.

Mei, Y., Chen, Y., Sivaccumar, J. P., An, Z., Xia, N., & Luo, W. (2022). Research progress and applications of nanobody in human infectious diseases. *Frontiers in Pharmacology, 13*, 963978.

Menachemi, N., & Collum, T. H. (2011). Benefits and drawbacks of electronic health record systems. *Risk Management and Healthcare Policy, 4*, 47–55.

Merianos, A. (2007). Surveillance and response to disease emergence. In James E. Childs, John S. Mackenzie, Jürgen A. Richt (Eds.). *Wildlife and emerging zoonotic diseases: The biology, circumstances and consequences of cross-species transmission* (pp. 477–509). Berlin, Heidelberg: Springer.

Meurens, F., Dunoyer, C., Fourichon, C., Gerdts, V., Haddad, N., Kortekaas, J., & Zhu, J. (2021). Animal board invited review: Risks of zoonotic disease emergence at the interface of wildlife and livestock systems. *Animal, 15*(6), 100241.

Meyyazhagan, A., Balasubramanian, B., Easwaran, M., Alagamuthu, K. K., Shanmugam, S., Kuchi Bhotla, H., Pappusamy, M., Arumugam, V. A., Thangaraj, A., Kaul, T., Keshavarao, S., & Cacabelos, R. (2020). Biomarker study of the biological parameter and neurotransmitter levels in autistics. *Molecular and Cellular Biochemistry, 474*(1–2), 277–284.

Meyyazhagan, A., Pushparaj, K., Balasubramanian, B., Kuchi Bhotla, H., Pappusamy, M., Arumugam, V. A., Easwaran, M., Pottail, L., Mani, P., Tsibizova, V., & Di Renzo, G. C. (2022). COVID-19 in pregnant women and children: Insights on clinical manifestations, complexities, and pathogenesis. *International Journal of Gynecology & Obstetrics, 156*(2), 216–224.

Michaelis, C., & Grohmann, E. (2023). Horizontal gene transfer of antibiotic resistance genes in biofilms. *Antibiotics, 12*(2), 328.

Mohd, Y., Balasubramanian, B., Meyyazhagan, A., Kuchi Bhotla, H., Shanmugam, S. K., Ramesh Kumar, M. K., Pappusamy, M., Alagamuthu, K. K., Keshavarao, S., & Arumugam, V. A. (2021). Extricating the association between the prognostic factors of colorectal cancer. *Journal of Gastrointestinal Cancer, 52*(3), 022–1028.

Mohd, Y., Kumar, P., Kuchi Bhotla, H., Meyyazhagan, A., Balasubramanian, B., Ramesh Kumar, M. K., Pappusamy, M., Alagamuthu, K. K., Orlacchio, A., Keshavarao, S., Sampathkumar, P., & Arumugam, V. A. (2021). Transmission jeopardy of adenomatosis polyposis coli and methylenetetrahydrofolate reductase in colorectal cancer. *Journal of the Renin-Angiotensin-Aldosterone System, 2021*. 7010706.

Morse, S. S. (2014). Public health disease surveillance networks. In Ronald M. Atlas, Stanley Maloy (Eds.). *One health: People, animals, and the environment* (pp. 195–211). ASM Press.

Mustapha, A. Y., Ikhalea, N., Chianumba, E. C., & Forkuo, A. Y. (2023). A model for integrating AI and big data to predict epidemic outbreaks. *Journal of Frontiers in Multidisciplinary Research, 04*(01), 157–176.

Muylaert, R. L., Wilkinson, D. A., Kingston, T., D'Odorico, P., Rulli, M. C., Galli, N., & Hayman, D. T. (2023). Using drivers and transmission pathways to identify SARS-like coronavirus spillover risk hotspots. *Nature Communications, 14*(1), 6854.

Naghavi, M., Mestrovic, T., Gray, A., Hayoon, A. G., Swetschinski, L. R., Aguilar, G. R., & Murray, C. J. (2024). Global burden associated with 85 pathogens in 2019: A systematic analysis for the global burden of disease study 2019. *The Lancet Infectious Diseases, 24*(8), 868–895.

Nassimbwa Kabanda, D. (2024). *Nanotechnology in malaria treatment: Targeted drug delivery systems and future applications.* INOSR Experimental Sciences 13(2):56–61.

Nelemans, T., & Kikkert, M. (2019). Viral innate immune evasion and the pathogenesis of emerging RNA virus infections. *Viruses, 11*(10), 961.

Nevagi, R. J., Toth, I., & Skwarczynski, M. (2018). Peptide-based vaccines. In Sotirios Koutsopoulos (ed.). *Peptide applications in biomedicine, biotechnology and bioengineering* (pp. 327–358). Woodhead Publishing.

Nguyen-Thanh, L., Wernli, D., Målqvist, M., Graells, T., & Jørgensen, P. S. (2024). Characterising proximal and distal drivers of antimicrobial resistance: An umbrella review. *Journal of Global Antimicrobial Resistance, 36,* 50–58.

Nishamol, P. H., & Gopakumar, G. (2015). Multi-target drug discovery using system polypharmacology-state of the art. In *2015 IEEE International Conference on Signal Processing, Informatics, Communication and Energy Systems (SPICES)* (pp. 1–5). IEEE.

Notomi, T., Mori, Y., Tomita, N., & Kanda, H. (2015). Loop-mediated isothermal amplification (LAMP): Principle, features, and future prospects. *Journal of Microbiology, 53*(1), 1–5.

Oladele, O. K. (2024). AI-driven synthetic biology: Algorithmic design of proteins and genetic circuits. https://www.researchgate.net/profile/Oluwaseyi-Oladele-3/publication/390032994_AI-Driven_Synthetic_Biology_Algorithmic_Design_of_Proteins_and_Genetic_Circuits/links/67dc6c56fe0f5a760f5b3bfe/AI-Driven-Synthetic-Biology-Algorithmic-Design-of-Proteins-and-Genetic-Circuits.pdf

Olawade, D. B., Teke, J., Fapohunda, O., Weerasinghe, K., Usman, S. O., Ige, A. O., & David-Olawade, A. C. (2024). Leveraging artificial intelligence in vaccine development: A narrative review. *Journal of Microbiological Methods, 224,* 106998.

Ong, E., Wang, H., Wong, M. U., Seetharaman, M., Valdez, N., & He, Y. (2020). Vaxign-ML: Supervised machine learning reverse vaccinology model for improved prediction of bacterial protective antigens. *Bioinformatics, 36*(10), 3185–3191.

Oyeniyi, J., & Oluwaseyi, P. (2024). Emerging trends in AI-powered medical imaging: Enhancing diagnostic accuracy and treatment decisions. *International Journal of Enhanced Research in Science Technology & Engineering, 13,* 2319–7463.

Palacios Araya, D., Palmer, K. L., & Duerkop, B. A. (2021). CRISPR-based antimicrobials to obstruct antibiotic-resistant and pathogenic bacteria. *PLoS Pathogens, 17*(7), e1009672.

Paranitharan, N., Kataria, S., Arumugam, V. A., Hsieh, H., Muthukrishnan, S., & Velayuthaprabhu, S. (2025). Integrin α1 upregulation by tf:fviia complex promotes cervical cancer migration through par2-dependent mek1/2 activation. *Biochemical and Biophysical Research Communications, 742,* 151151.

Patil, N. K., Guo, Y., Luan, L., & Sherwood, E. R. (2017). Targeting immune cell checkpoints during sepsis. *International Journal of Molecular Sciences, 18*(11), 2413.

Patrone, D., Resnik, D., & Chin, L. (2012). Biosecurity and the review and publication of dual-use research of concern. *Biosecurity and Bioterrorism: Biodefense Strategy, Practice, and Science, 10*(3), 290–298.

Petri, K., & Pattanayak, V. (2018). SHERLOCK and DETECTR open a new frontier in molecular diagnostics. *The CRISPR Journal, 1*(3), 209–211.

Pisarski, K. (2019). The global burden of disease of zoonotic parasitic diseases: Top 5 contenders for priority consideration. *Tropical Medicine and Infectious Disease, 4*(1), 44.

Platt, R., Wilson, M., Chan, K. A., Benner, J. S., Marchibroda, J., & McClellan, M. (2009). The new Sentinel network-improving the evidence of medical-product safety. *The New England Journal of Medicine, 361*(7), 645–647.

Ponne, S., Kumar, R., Vanmathi, S. M., Brilhante, R. S. N., & Kumar, C. R. (2024). Reverse engineering protection: A comprehensive survey of reverse vaccinology-based vaccines targeting viral pathogens. *Vaccine, 42*(10), 2503–2518.

Pushkaran, A., Chattu, V. K., & Narayanan, P. (2023). A critical analysis of COVAX alliance and corresponding global health governance and policy issues: A scoping review. *BMJ Global Health, 8*(10), e012168.

Pushkaran, A., Chattu, V. K., & Narayanan, P. (2024). COVAX and COVID-19 vaccine inequity: A case study of G-20 and African Union. *Public Health Challenges, 3*(2), e185.

Pushparaj, K., Kuchi Bhotla, H., Arumugam, V. A., Pappusamy, M., Easwaran, M., Liu, W. C., Issara, U., Rengasamy, K. R. R., Meyyazhagan, A., & Balasubramanian, B. (2022). Mucormycosis (black fungus) ensuing COVID-19 and comorbidity meets – magnifying global pandemic grieve and catastrophe begins. *Science of the Total Environment, 805,* 150355.

Quinn, S. C., & Kumar, S. (2014). Health inequalities and infectious disease epidemics: A challenge for global health security. *Biosecurity and Bioterrorism: Biodefense Strategy, Practice, and Science, 12*(5), 263–273.

Rahman, S. Z., Senthil, R., Ramalingam, V., & Gopal, R. (2023). Predicting infectious disease outbreaks with machine learning and epidemiological data. *Journal of Advanced Zoology, 44*(S4), 110–121.

Rajendra Acharya, U., Paul Joseph, K., Kannathal, N., Lim, C. M., & Suri, J. S. (2006). Heart rate variability: A review. *Medical and Biological Engineering and Computing, 44,* 1031–1051.

Ramezanian, S., Moghaddas, J., Roghani-Mamaqani, H., & Rezamand, A. (2023). Dual pH-and temperature-responsive poly (dimethylaminoethyl methacrylate)-coated mesoporous silica nanoparticles as a smart drug delivery system. *Scientific Reports, 13*(1), 20194.

Ramon-Torrell, J. M. (2023). Perspective chapter: Emerging infectious diseases as a public health problem. In A. Michaud, S. P. Stawicki, & R. Izurieta (Eds.). *Global health security-contemporary considerations and developments*. Intechopen.

Ramya, S., Poornima, P., Jananisri, A., Geofferina, I. P., Bavyataa, V., Divya, M., Priyanga, P., Vadivukarasi, J., Sujitha, S., Elamathi, S., Anand, A. V., & Balamuralikrishnan, B. (2023). Role of hormones and the potential impact of multiple stresses on infertility. *Stress, 3*(2), 454–474.

Rana, S., Luo, W., Tran, T., Phung, D., Venkatesh, S., & Harvey, R. (2014). HealthMap: A visual platform for patient suicide risk review. In *Abstracts of the Scientific Stream at Big Data 2014, Melbourne*, Australia, April 3-4 (pp. 42–43). CEUR-WS Vol. 1149.

Restif, O., & Graham, A. L. (2015). Within-host dynamics of infection: From ecological insights to evolutionary predictions. *Philosophical Transactions of the Royal Society, B: Biological Sciences, 370*(1675), 20140304.

Rhee, J. H. (2020). Current and new approaches for mucosal vaccine delivery. In Hiroshi Kiyono and David W. Pascual (Eds.). *Mucosal vaccines* (pp. 325–356). Academic Press.

Richards, T. (2013). The health sector cannot mount an effective response on its own to global health threats. *BMJ 346*, f751.

Rizk, S. S., Moustafa, D. M., ElBanna, S. A., Nour El-Din, H. T., & Attia, A. S. (2024). Nanobodies in the fight against infectious diseases: Repurposing nature's tiny weapons. *World Journal of Microbiology and Biotechnology, 40*(7), 209.

Roche, K. L., Remiszewski, S., Todd, M. J., Kulp, J. L., Tang, L., Welsh, A. V., & Chiang, L. W. (2023). An allosteric inhibitor of sirtuin 2 deacetylase activity exhibits broad-spectrum antiviral activity. *The Journal of Clinical Investigation, 133*(12), e158978.

Rosa, B. G., Akingbade, O. E., Guo, X., Gonzalez-Macia, L., Crone, M. A., Cameron, L. P., & Li, B. (2022). Multiplexed immunosensors for point-of-care diagnostic applications. *Biosensors and Bioelectronics, 203*, 114050.

Rosendahl Huber, S., van Beek, J., de Jonge, J., Luytjes, W., & van Baarle, D. (2014). T cell responses to viral infections – Opportunities for peptide vaccination. *Frontiers in Immunology, 5*, 171.

Saalbach, K. P. (2022). Gain-of-function research. In Geoffrey Michael Gadd and Sima Sariaslani (Eds.). *Advances in applied microbiology* (Vol. 120, pp. 79–111). Academic Press.

Saini, G., Kumar, I., & Verma, K. K. (2024). The potential of drug repurposing as a rapid response strategy in COVID-19 therapeutics. *Journal of Advances in Medical and Pharmaceutical Sciences, 26*(12), 12–31.

Saldaña, F., Stollenwerk, N., Van Dierdonck, J. B., & Aguiar, M. (2024). Modelling spillover dynamics: Understanding emerging pathogens of public health concern. *Scientific Reports, 14*(1), 9823.

Sanaei, M., Setayesh, N., Sepehrizadeh, Z., Mahdavi, M., & Yazdi, M. H. (2020). Nanobodies in human infections: Prevention, detection, and treatment. *Immunological Investigations, 49*(8), 875–896.

Sangeetha, T., Anand, A. V., & Begum, T. N. (2022). Assessment of inter-relationship between anemia and COPD in accordance with altitude. *Open Respiratory Medicine Journal, 16*, e187430642206270.

Sangeetha, T., Nargis Begum, T., Balamuralikrishnan, B., Arun, M., Rengasamy, K. R. R., Senthilkumar, N., Velayuthaprabhu, S., Saradhadevi, M., Sampathkumar, P., & Vijaya Anand, A. (2022). Influence of SERPINA1 gene polymorphisms on anemia and chronic obstructive pulmonary disease. *Journal of the Renin-Angiotensin-Aldosterone System*, 2022.

Schnetzinger, F., Clénet, D., Gilbert, P. A., Guzzi, A., Paludi, M., Weusten, J., & Hesselink, R. (2024). Stability preparedness: The not-so-cold case for innovations in vaccine stability modelling and product release. *Vaccine, 12*(9), 1000.

Sedgley, C., & Dunny, G. (2015). Antimicrobial resistance in biofilm communities. In Luis E. Chávez de Paz, Christine M. Sedgley, Anil Kishen (Eds.). *The root canal biofilm* (pp. 55–84). Berlin, Heidelberg: Springer.

Seele, P. P. (2025). Structure-function relationships in the modification of liposomes for targeted drug delivery. In Benjamin S. Weeks (Ed.). *Liposomes as pharmaceutical and nutraceutical delivery systems* (pp. 91–112), IntechOpen.

Seprényi, K., Tamás, P., & Cservenák, Á. (2022). Logistics and cold chain relationship. *Advanced Logistic Systems-Theory and Practice, 16*(2), 47–53.

Shah, A., Malik, M. S., Khan, G. S., Nosheen, E., Iftikhar, F. J., Khan, F. A., & Aminabhavi, T. M. (2018). Stimuli-responsive peptide-based biomaterials as drug delivery systems. *Chemical Engineering Journal, 353*, 559–583.

Shah, R., Nadaf, S., Bhagwat, D., & Gurav, S. (2025). Trends in nano drug delivery system for infectious diseases. In Rajasekhar Chokkareddy, Suvardhan Kanchi, Gan G. Redhi (Eds.). *Sustainable nanomaterials for treatment and diagnosis of infectious diseases* (pp. 181–206). Scrivener Publishing LLC.

Shaikh, I. (2018). Critically analyse the different approaches of eco health, one health, planetary health and political economy and political ecology of global health to analyse current challenges in the anthropocene. *Journal of Ecosystem & Ecography, 8*(252), 2.

Shanks, S., van Schalkwyk, M. C., & Cunningham, A. A. (2022). A call to prioritise prevention: Action is needed to reduce the risk of zoonotic disease emergence. *The Lancet Regional Health–Europe, 23*, 100506.

Shanmugam, R., Thangavelu, S., Fathah, Z., Yatoo, M. I., Tiwari, R., Pandey, M. K., Dhama, J., Chandra, R., Malik, Y. S., Dhama, K., Sah, R., Chaicumpa, W., Shanmugam, V., & Arumugam, V. A. (2020). SARS-CoV-2/COVID-19 pandemic – An update. *Journal of Experimental Biology and Agricultural Sciences, 8*(Special Issue 1), S219–S245.

Shi, X., Tao, Y., & Lin, S. C. (2024). Deep neural network-based prediction of B-cell epitopes for SARS-CoV and SARS-CoV-2: Enhancing vaccine design through machine learning. In *2024 4th International signal processing, communications and engineering management conference (ISPCEM)* (pp. 259–263). IEEE.

Sims, N., & Kasprzyk-Hordern, B. (2020). Future perspectives of wastewater-based epidemiology: Monitoring infectious disease spread and resistance to the community level. *Environment International, 139*, 105689.

Sinha, S., Aggarwal, S., & Singh, D. V. (2024). Efflux pumps: Gatekeepers of antibiotic resistance in Staphylococcus aureus biofilms. *Microbial Cell, 11*, 368.

Smith, W. R., Atala, A. J., Terlecki, R. P., Kelly, E. E., & Matthews, C. A. (2020). Implementation guide for rapid integration of an outpatient telemedicine program during the COVID-19 pandemic. *Journal of the American College of Surgeons, 231*(2), 216–222.

Snowden, F. M. (2019). *Epidemics and society: From the black death to the present*. Yale University Press.

Stevens, E. J., Li, J. D., Hector, T. E., Drew, G. C., Hoang, K., Greenrod, S. T., & King, K. C. (2024). Within-host competition sparks pathogen molecular evolution and perpetual microbiota dysbiosis. *bioRxiv*, 2024-09.

Suster, C. J., Watt, A. E., Wang, Q., Chen, S. C. A., Kok, J., & Sintchenko, V. (2024). Combined visualization of genomic and epidemiological data for outbreaks. *Epidemiology and Infection, 152*, e110.

Swei, A., Couper, L. I., Coffey, L. L., Kapan, D., & Bennett, S. (2020). Patterns, drivers, and challenges of vector-borne disease emergence. *Vector Borne and Zoonotic Diseases, 20*(3), 159–170.

Theochari, I., Xenakis, A., & Papadimitriou, V. (2020). Nanocarriers for effective drug delivery. In Phuong Nguyen-Tri, Trong-On Do and Tuan Anh Nguyen (Eds.). *Smart nanocontainers* (pp. 315–341). Elsevier.

Thursina, T., Hasanbasri, M., & Mahendradhata, Y. (2024). Strengthening one health: Lessons learned of rabies response in Indonesia. In *BIO web of conferences* (Vol. 132, p. 01001). EDP Sciences.

Ulanova, M. (2025). Health inequalities in respiratory tract infections – beyond COVID-19. *Current Opinion in Infectious Diseases, 38*(2), 161–168.

Van der Ley, P., & Schijns, V. E. (2023). Outer membrane vesicle-based intranasal vaccines. *Current Opinion in Immunology, 84*, 102376.

Van Dorst, B., Mehta, J., Bekaert, K., Rouah-Martin, E., De Coen, W., Dubruel, P., & Robbens, J. (2010). Recent advances in recognition elements of food and environmental biosensors: A review. *Biosensors and Bioelectronics, 26*(4), 1178–1194.

Varghese, J. E., Shanmugam, V., Rengarajan, R. L., Meyyazhagan, A., Arumugam, V. A., Al-Misned, F. A., & El-Serehy, H. A., (2020). Role of vitamin D_3 on apoptosis and inflammatory-associated gene in colorectal cancer: An in vitro approach. *Journal of King Saud University – Science, 32*(6), 2786–2789.

Velayuthaprabhu, S., & Archunan, G. (2005). Evaluation of anticardiolipin antibodies and antiphosphatidylserine antibodies in women with recurrent abortion. *Indian Journal of Medical Sciences 59*, 347–352.

Velayuthaprabhu, S., Archunan, G., & Balakrishnan, K. (2007). Placental thrombosis in experimental anticardiolipin antibodies-mediated intrauterine fetal death. *American Journal of Reproductive Immunology, 57*, 270–276.

Velayuthaprabhu, S., Matsubayashi, H., & Archunan, G. (2016). Beta-2 GPI induced tissue factor and placental apoptosis for the pathophysiology of pregnancy loss in antiphospholipid syndrome. *International Journal of Research in Medical Sciences, 4*(8), 3109–3113.

Velayuthaprabhu, S., Matsubayashi, H., Sugi, T., Nakamura, M., Ohnishi, Y., Ogura, T., & Archunan, G. (2013). Expression of apoptosis in placenta of experimental antiphospholipid syndrome (APS) mouse. *American Journal of Reproductive Immunology, 69*(5), 486–494.

Velayuthaprabhu, S., Matsubayashi, H., Sugi, T., Nakamura, M., Ohnishi, Y., Ogura, T., Tomiyama, T., & Archunan, G. (2011). A unique preliminary study on fetal resorption and placental apoptosis in mice with passive immunization of anti-phosphatidylethanolamine antibodies and anti-Factor XII antibodies. *American Journal of Reproductive Immunology, 66,* 373–384.

Verma, R., Raj, S., Berry, U., Ranjith-Kumar, C. T., & Surjit, M. (2023). Drug repurposing for COVID-19 therapy: Pipeline, current status and challenges. In RC Sobti, SK Lal, RK Goyal (Eds.). In *Drug repurposing for emerging infectious diseases and cancer*, Springer Nature, 451–478, Springer.

Verweij, M., & Dawson, A. (2011). Infectious disease control. In Angus Dawson (Ed.). *Public health ethics: Key concepts and issues in policy and practice* (pp. 100–117). Cambridge University Press.

Wanakule, P., & Roy, K. (2012). Disease-responsive drug delivery: The next generation of smart delivery devices. *Current Drug Metabolism, 13*(1), 42–49.

Wang, F. Y., Chen, Y., Huang, Y. Y., & Cheng, C. M. (2021). Transdermal drug delivery systems for fighting common viral infectious diseases. *Drug Delivery and Translational Research, 11*(4), 1498–1508.

Wang, S., Liu, H., Zhang, X., & Qian, F. (2015). Intranasal and oral vaccination with protein-based antigens: Advantages, challenges and formulation strategies. *Protein & Cell, 6*(7), 480–503.

Wang, Y., Yang, J., Zhuang, X., Ling, Y., Cao, R., Xu, Q., & Zhang, G. (2022). Linking genomic and epidemiologic information to advance the study of COVID-19. *Scientific Data, 9*(1), 121.

Washburne, A. D., Crowley, D. E., Becker, D. J., Manlove, K. R., Childs, M. L., & Plowright, R. K. (2019). Percolation models of pathogen spillover. *Philosophical Transactions of the Royal Society B, 374*(1782), 20180331.

Wykes, M. N., & Lewin, S. R. (2018). Immune checkpoint blockade in infectious diseases. *Nature Reviews Immunology, 18*(2), 91–104.

Yadav, M., & Shukla, P. (2020). Efficient engineered probiotics using synthetic biology approaches: A review. *Biotechnology and Applied Biochemistry, 67*(1), 22–29.

Yang, R., Yao, T., Xu, J., Liu, X., Yang, Y., Ding, J., & Gao, X. (2024). Peptide-TLR7/8a-coordinated DNA vaccines elicit enhanced immune responses against infectious diseases. *ACS Biomaterials Science & Engineering, 10*(7), 4374–4387.

Yang, W. Z., & Zhang, T. (2022). Strategy and measures in response to highly uncertain emerging infectious disease. *Zhonghua liu xing bing xue za zhi= Zhonghua liuxingbingxue zazhi, 43*(5), 627–633.

Yon, L., Duff, J. P., Ågren, E. O., Erdélyi, K., Ferroglio, E., Godfroid, J., & Gavier-Widén, D. (2019). Recent changes in infectious diseases in European wildlife. *Journal of Wildlife Diseases, 55*(1), 3–43.

Yow, H. Y., Govindaraju, K., Lim, A. H., & Abdul Rahim, N. (2022). Optimizing antimicrobial therapy by integrating multi-omics with pharmacokinetic/pharmacodynamic models and precision dosing. *Frontiers in Pharmacology, 13,* 915355.

Zahra, A., Shahid, A., Shamim, A., Khan, S. H., & Arshad, M. I. (2023). The SHERLOCK platform: An insight into advances in viral disease diagnosis. *Molecular Biotechnology, 65*(5), 699–714.

Zhang, F., & Cheng, W. (2022). The mechanism of bacterial resistance and potential bacteriostatic strategies. *Antibiotics, 11*(9), 1215.

Zhang, H. (2024). Therapeutic applications of CRISPR-Cas system in infectious diseases. In *Third International Conference on Biological Engineering and Medical Science (ICBioMed2023)* (Vol. 12924, pp. 337–341). SPIE.

Zhang, M., Hussain, A., Yang, H., Zhang, J., Liang, X. J., & Huang, Y. (2023). mRNA-based modalities for infectious disease management. *Nano Research, 16*(1), 672–691.

Zheng, Y., Li, S., Song, K., Ye, J., Li, W., Zhong, Y., & Xu, K. (2022). A broad antiviral strategy: Inhibitors of human DHODH pave the way for host-targeting antivirals against emerging and re-emerging viruses. *Viruses, 14*(5), 928.

Zur Wiesch, P. A., Kouyos, R., Engelstädter, J., Regoes, R. R., & Bonhoeffer, S. (2011). Population biological principles of drug-resistance evolution in infectious diseases. *The Lancet Infectious Diseases, 11*(3), 236–247.

2 Integrating AI Technology into Biomedical Research and Health Innovation

Edge Tools and Emerging Trends

Loganathan Murugesan, Mani Manoj, Jothi Dheivasikamani Abidharini,
Gunasekaran Arthi, Sivaramakrishnan Sharmili, Srikamali Sundar,
Asirvatham Alwin Robert, and Arumugam Vijaya Anand

2.1 INTRODUCTION: THE ARTIFICIAL INTELLIGENCE (AI) REVOLUTION IN BIOMEDICINE

In our industrialized world, the necessity for accurate and timely detection of many diseases is crucial for managing patient health and effectively halting their spread. Furthermore, diseases are spreading faster. The factors contributing to the quick spread of diseases include increased international travel, population growth, and environmental changes (Velayuthaprabhu and Archunan, 2005; Velayuthaprabhu et al. 2007; Alagendran et al., 2010; Velayuthaprabhu et al. 2011; Velayuthaprabhu et al., 2013; Velayuthaprabhu et al., 2016; Varghese et al., 2020; Lee et al., 2021; Mohd, Balasubramanian, et al., 2021; Mohd, Kumar, et al., 2021; Gundappa et al., 2022; Sangeetha, Anand, and Begum, 2022; Sangeetha, Nargis Begum, et al., 2022; Meyyazhagan et al., 2022; Ramya et al., 2023; Paranitharan et al., 2025).

The application of artificial intelligence (AI) in the areas of health innovation and biomedical research has seen tremendous evolution since its initial conceptualization in the 1950s. From its initial stages as rudimentary expert systems and fuzzy logic models in the 1960s, it laid the foundational basis for the development of more advanced paradigms like artificial neural networks (ANNs), machine learning (ML), deep learning (DL), and natural language processing (NLP). There has been a phenomenal explosion of AI-directed research in the healthcare sector over the last three decades, with both the United States and China emerging as publication leaders, as well as other technologically advanced nations (Xie et al., 2025). The critical juncture of the application of AI in the health sciences has emerged in tandem with the advent and acceptance of ML and DL paradigms. These technologies have revolutionized the redescription of clinical prediction models, improved diagnostic capabilities, and hastened drug discovery pathways (Tran, Vu, et al., 2019). AI systems today enable the identification of biomedically significant patterns in large and intricate datasets like genomics, proteomics, and high-resolution medical imaging, thereby revealing insights that would likely remain hidden through traditional analytical methods (Chaudhari & Telrandhe, 2024).

The COVID-19 pandemic (Arumugam et al., 2020; Meyyazhagan et al., 2020; Shanmugam et al., 2020; Kuchi Bhotla et al., 2020; Kuchi Bhotla et al., 2021; Pushparaj et al., 2022) served as one of the key drivers for the speedy adoption of AI throughout the healthcare industry. AI-enabled solutions were instrumental in augmenting telemedicine capabilities, streamlining healthcare processes, and employing chatbots for patient outreach and triage. These solutions not only demonstrated the effectiveness of AI in managing public health crises but also underscored its potential to enhance healthcare access and enable digital medical training (Hirani et al., 2024). The pandemic scenario reaffirmed the necessity of AI integration into healthcare infrastructure, with a greater emphasis on virtual delivery of medicine and real-time data monitoring. At the same time, AI technologies in the domain of biomedical imaging have seen unprecedented progress. AI-enabled tools, particularly those employing DL algorithms, have greatly improved the recognition, characterization, and monitoring of diseases. Through computer-aided image interpretation and enhanced resolution, these tools are enhancing treatment planning and enabling earlier and more accurate diagnosis (Altara et al., 2024). Furthermore, AI is being utilized increasingly across a range of applications, from drug design, toxicology, and biomaterial identification, enabling automated hypothesis generation, study design optimization, and high-throughput data analysis.

In domains like precision oncology and pharmacogenomics, autonomous experimentation (AE) platforms enable rapid target discovery of therapeutics while optimizing the reproducibility of biomedical experiments (da Silva, 2024). AI models like AlphaFold have revolutionized the field of structural biology by providing unprecedented accuracy in protein structure prediction, thus transforming the scope and pace of biological research (Pandian & Murugan, 2024). Additionally, AI is radically changing the way healthcare education is taught. AI literacy must be incorporated into

DOI: 10.1201/9781003615699-2

biomedical and medical education to provide aspiring professionals with the skills they need to understand and utilize AI-generated inputs. Education in AI ethics, data privacy, and algorithmic bias is critical to enable responsible innovation and safeguard patient rights (Sharma et al., 2024). Ethical considerations are particularly critical in situations involving patient data, where informed consent, data anonymization, and interpretability of AI algorithms are critical to enable transparency and accountability (Haque et al., 2024).

Notwithstanding the vast transformative power of AI, its adoption in clinical practice remains less than optimal due to a constellation of barriers. Major impediments are the non-availability of large, high-quality annotated datasets; model interpretability and generalizability problems; and the lack of robust scientific validation procedures. In addition, ethical concerns related to data property rights, algorithmic bias, and healthcare disparities reinforce the need for the establishment of holistic governance frameworks (Dzobo et al., 2020; Palvadi & Kadiravan, 2025). Such frameworks must steer the ethical deployment of AI technologies while enhancing inclusivity and reducing health disparities. Decentralization of AI infrastructure and support for local AI research environments can fill technology gaps and enhance access to AI-based healthcare solutions in developing economies (da Silva, 2024). Such intervention would ensure equitable sharing of the benefits of AI and customized solutions to region-specific healthcare needs. In the years to come, the trajectory of AI in biomedical research and healthcare combines promise with complexity. Advances in precision medicine, robotic-assisted surgical procedures, and explainable AI (XAI) will transform diagnostics and therapeutics. However, the actualization of the full benefits of AI requires collaborative synergy between AI and human intelligence (HI), protected by ethical governance, cross-disciplinary education, and policy-oriented frameworks. Only through such integrative and inclusive strategies can AI be a transformational agent for the promotion of global health and biomedical innovation. Figure 2.1 presents a landscape of key AI technologies that are shaping biomedical research and clinical innovation.

2.2 CORE CONCEPTS AND FOUNDATIONS OF AI IN BIOMEDICAL SCIENCES

2.2.1 Core AI Technologies in Biomedical Research

Key AI technologies like ML, DL, NLP, and computer vision are the pillars of contemporary biomedical innovation. ML algorithms are employed to detect intricate patterns in patient data, thereby facilitating disease prediction, diagnosis, and individualized treatment planning (Subasi, 2024). DL, leveraging neural networks, demonstrates impressive capability in handling heterogeneous datasets collected from electronic health records (EHRs), medical imaging, and genomics, thereby serving the aims of precision medicine (De Freitas et al., 2021). Convolutional neural networks (CNNs) have revolutionized the domain of medical imaging by facilitating precise image-based diagnostics and risk determinations (Al Gharrawi & Al-Joda, 2023). NLP is essential in extracting data from unstructured clinical text and facilitating the mining of large volumes of patient records and biomedical literature. Computer vision, conversely, makes visual medical data analysis easier, increasing workflow efficiency and diminishing diagnostic errors. All of these AI technologies combined empower healthcare practitioners by enriching decision-support systems, streamlining routine procedures, and enhancing patient outcomes through real-time disease tracking and risk assessment (Subasi, 2024).

2.2.2 Multimodal Biomedical Data and AI-Driven Analysis

Biomedical data are multidimensional, covering a range of modalities such as multi-omics (e.g., mRNA, DNA methylation, miRNA), EHRs, medical imaging, and wearable device-generated physiological signals. Each modality provides unique information, jointly improving understanding of disease mechanisms and patient health. Multi-omics data are particularly crucial in defining complex biological interactions and molecular biomarker discovery. Graph convolutional network-based sophisticated AI frameworks, e.g., MORONET, can effectively integrate these datasets to obtain accurate disease classification and biomarker discovery (Wang, Shao, et al., 2020). Integration with imaging data from EHRs leads to enhanced diagnostic accuracy, especially in complex domains such as neurology (Mohsen et al., 2022). Additionally, the incorporation of wearable device data allows continuous monitoring of physiological signals, therefore enabling real-time intervention and personalized healthcare delivery (Tong et al., 2023). Upcoming examples such as the MultiMed benchmark illustrate the importance of multimodal AI systems, showing enhanced performance in varied clinical applications (Mo & Liang, 2024). However, the integration of heterogenous data types

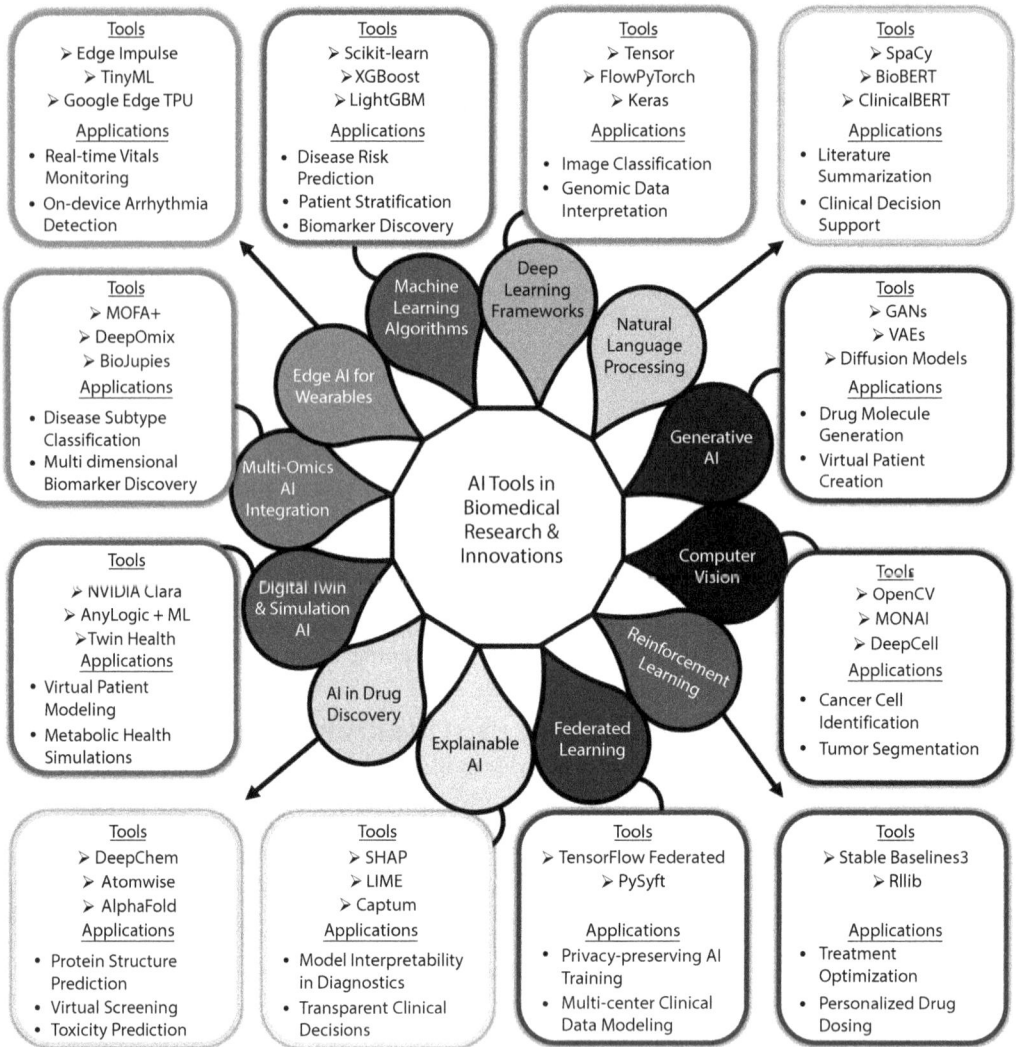

Figure 2.1   AI technology landscape in biomedical research and innovations.

still encounters problems of standardization, interoperability, and the need for interdisciplinary expertise (Wang & Li, 2023).

2.2.3 Cloud, Edge Computing, and Big Data Analytics in Healthcare

The intersection of AI with big data analytics and distributed computing platforms, i.e., cloud and edge computing, has transformed biomedical research and healthcare delivery. Predictive modeling, real-time decision-making, and individualized treatment planning are made possible by big data analytics, which enables the quick processing and interpretation of massive amounts of health data. Cloud computing offers a scalable and secure data storage platform and shared access, hence interoperability and operational efficiency. Problems with data privacy, latency, and regulatory compliance, however, remain contentious. Edge computing addresses the issue through the decentralization of the data analysis process, hence allowing sensitive health information to be processed close to the point of origin. The approach significantly reduces latency and enhances real-time clinical applications, particularly in resource-constrained or privacy-sensitive settings (Barik et al., 2017). The convergence of edge-cloud solutions advances the hybrid model where computational jobs are optimally distributed across networks, hence enhancing the reliability, scalability, and responsiveness of healthcare systems. Under the Internet of Medical Things (IoMT) paradigm, the technologies converge to promote smart resource allocation and data management, hence driving remote patient monitoring, telemedicine, and emergency response innovations, while their revolutionary potential, however, has challenges such as interoperability, energy efficiency, and infrastructure development, which call for further research and enabling policy interventions (Belcastro et al., 2024).

2.3 CUTTING-EDGE AI TOOLS AND PLATFORMS FOR BIOMEDICAL RESEARCH

The use of AI and ML in drug discovery and molecular modeling has revolutionized the biomedical research practice and healthcare innovation. These technologies are revolutionizing conventional practices by making it to become more convenient and economical to handle big data, speeding up the process of drug development. One of the breakthroughs in the field is AlphaFold, developed by DeepMind, which has shaken the world of protein structure prediction. AlphaFold makes highly accurate predictions of protein three-dimensional structures from amino acid sequences based on the use of DL models. This breakthrough has far-reaching implications in biomedical science, specifically in protein function identification, understanding molecular interactions, and uncovering likely treatment targets. Interestingly, AlphaFold's accurate structure of the SARS-CoV-2 spike protein enabled rapid manufacture of the COVID-19 vaccines, highlighting its value for pandemic preparedness, vaccine design, and personalized medicine.

ML and AI are applied to speed up different aspects of drug discovery, not only structural biology. They are employed to apply techniques such as structure-based virtual screening and ligand design, designing drugs de novo, and drug repurposing. AI narrows down the drug candidates by rapid screening of large chemical libraries so that AI identifies drug candidates that bind tightly and produce positive effects on health. AI also enhances the way we predict the behavior of drugs in the body, their effects, safety, and efficacy, making it more convenient to test drugs early and perform clinical trials. This assists in reducing the time and cost incurred in the development of drugs (Shah et al., 2024). AI platforms have also assisted in identifying allosteric modulators, which play a crucial role in drug discovery since they act on specific regions and cause fewer side effects. It is still difficult to predict allosteric binding sites with high accuracy since protein structures are dynamic and context-dependent. New advances also involve the application of AI in nanomedicine, where it assists in designing and deploying nanorobots for targeted drug delivery. These nano systems, aided by AI, enhance treatment by making it more precise and efficient through delivering it exactly where it is needed while restricting undesirable side effects. Alhough they hold enormous potential, there are yet problems such as variability in the data, understanding the algorithms, and ethical issues regarding applying AI in the healthcare sector that must be addressed (Blanco et al., 2024).

2.3.1 Multi-omics Integration and AI-Driven Biomarker Discovery

Combining multi-omics and AI is revolutionizing biomedical research by delivering end-to-end insights into intricate biological systems. Multi-omics approaches facilitate in-depth analysis of disease pathogenesis and therapeutic targets. Multi-omics, through the integration of information from other layers of molecules, facilitates the identification of pivotal biomarkers that can be used to predict disease outcomes, enhancing diagnostic accuracy and treatment modalities. For instance, in cancer research, multi-omics analysis has been particularly crucial in categorizing and forecasting other cancers. Variational autoencoders (VAEs) and graph convolutional neural networks (GCNNs) have been great assets to cancer type prediction, averaging 93.90% accuracy in pan-cancer tasks (Ivanisevic & Sewduth, 2023). Furthermore, AI-driven multi-omics data analysis has also identified novel biomarkers like SLK, which is associated with cancer prognosis, and TK1, whose concentration can be used for glioma diagnosis, with robust evidence coming from receiver operating characteristic (ROC) curve analysis (Zhao et al., 2020; Shao et al., 2023). These findings show how multi-omics can individualize treatment modalities and predict patient outcomes with precision. The use of AI streamlines, speeds up, and enhances the interpretation of multi-omics data, thus playing a critical role in the development of precision medicine (Pasupuleti et al., 2024).

2.3.2 AI-Enhanced Medical Imaging and Radiomics in Oncology

AI-facilitated radiomics and medical imaging are breakthroughs in non-invasive diagnosis, particularly in cancer treatment. Radiomics is the process of extracting many detailed features from medical images – images impossible for humans to visually recognize but very informative about tissue variability, tumor morphology, and tumor context. By leveraging AI, specifically methods like ML and DL (and some types of neural networks), radiomics work has become more automatic, accurate, and reproducible. These AI tools help develop computer-aided diagnosis (CAD) software that aids radiologists in detecting, categorizing, and assessing disease (Maniaci et al., 2024; Huang & Gao,

2024). All stages of the radiomics process are aided by AI-from image capture and image preparation to feature extraction, data abstraction, and prediction. Such end-to-end unification results in stable systems that help doctors work more effectively, individualize treatment plans, and improve patient outcomes. For example, radiomic features have been used to predict the behavior of treatments and whether cancers like lung, breast, and glioblastoma will recur (Vrettos et al., 2024). However, issues persist in making image-acquisition processes equal, harmonizing feature extraction methods, and cross-validating AI models across different clinical contexts.

2.3.3 Robotic Laboratory Automation and AI-Guided Experimental Design

Robotics and AI-based experimental design are revolutionizing laboratory science by accelerating experiments, enhancing efficiency, and improving reproducibility. Among the advantages here is the advent of AE platforms, also referred to as self-driving laboratories. These platforms carry out each step, ranging from hypothesis generation to experiment execution, data analysis, and optimization, accelerating scientific discovery. In precision oncology and drug discovery, AE systems are revolutionaries. They can run thousands of experiments automatically, rapidly determining the optimal conditions for synthesizing molecules, screening their activity, or developing scales (da Silva, 2024). AI algorithms in the platforms assist in selecting experiments from previous outcomes, reducing trial-and-error experiments and enabling better decision-making (Schneider et al., 2024).

New technological innovations in robotics, such as robotic arms for sample manipulation, liquid handling robots, and AI algorithms, have enhanced the speed and accuracy of the laboratory. For instance, intelligent robotic assistants can perform advanced operations such as flexible liquid dispensing and microfluidic manipulation with greater accuracy than humans, making biomedical experiments safer and more reproducible (Knobbe et al., 2022). Moreover, AI-enabled lab platforms support researchers in adapting their experiments using up-to-date data. This capacity to adapt is crucial to research in large and intricate fields such as synthetic biology, chemical biology, and materials science (Pandy et al., 2025). Nevertheless, robotic automation comes with its drawbacks. High cost, complex systems, and the requirement of special training are gigantic hindrances, particularly in low- and middle-income countries. Surmounting these challenges with decentralized and modular state-of-the-art equipment could render new lab technologies more affordable and enhance global research collaboration (Omair et al., 2023).

2.4 EDGE AI AND TINYML: REAL-TIME, LOW-POWER BIOMEDICAL INNOVATIONS

The intersection of AI with biomedical research and healthcare innovation has given birth to two complementary yet distinct computational frameworks: Edge AI and Cloud AI. Each framework has its distinct functionalities based on the specific healthcare context, data processing requirement, and real-time response need. Edge AI, as termed by Jain et al., refers to the implementation of AI algorithms on wearable devices like wearables, implantable devices, and Internet of Things (IoT) sensors, deployed at or near the point of origin of the data (Jain et al., 2023). This configuration facilitates real-time processing of data with minimal latency, a requirement that is necessary for time-sensitive applications like continuous monitoring of health, detection of early diseases, and management of chronic diseases (Badidi, 2023). Moreover, by keeping sensitive information local and not sending it to far-off servers, Edge AI ensures data privacy and security, a factor that is of strategic importance in healthcare settings where preserving patient confidentiality is paramount. Moreover, by virtue of its distributed nature, Edge AI opens up space for scalability and system failure tolerance by facilitating decentralized analysis of big and heterogeneous datasets like EHRs, demographic information, and genomic data (Chandrasekaran et al., 2024). The integration of federated learning into the Edge AI framework further makes the former applicable in the clinical setting because it facilitates collaborative training of AI models across institutions without compromising individual data privacy.

Alternatively, Cloud AI demonstrates outstanding capability in high-performance computing centralization, enabling computationally intensive tasks requiring high computational and storage capability to be performed. It covers a range of applications, such as biomedical image analysis, high-throughput genomics, and personalized medicine, which will greatly benefit from Cloud AI, as it enables training and deployment of sophisticated DL models for conditions like detection, prognosis modeling, and therapeutic planning (Chaudhari & Telrandhe, 2024). This centralization, however, incurs greater latency and greater potential for privacy breach, especially when patient information must be moved to remote servers. These are addressed by Edge AI, which is based on local analysis and real-time processing. Although the past constraints in energy efficiency and computational capability of edge devices were limiting, progress in AI hardware and algorithmic efficiency is opening up the possibility of developing more resilient, energy-efficient, and responsive Edge AI systems. These developments are likely to transform digital health by enabling health monitoring and intervention processes to be smart, quick, and secure (Jain et al., 2023).

2.4.1 Wearables, Implantable, and IoT: Revolutionizing Health Monitoring and Patient Care

The convergence of wearable technologies, implantable devices, and IoT technologies has catalyzed a paradigm revolution in biomedical research and healthcare provision through continuous and real-time tracking of health status and remote patient monitoring. Wearable and implantable medical devices with the integration of tiny sensors, low-power microcontrollers, and wireless communication capabilities can track vital physiological parameters, such as heart rate, blood pressure, and body temperature, with remarkable accuracy (Anasica, 2023). The devices send data to cloud storage servers, where it is processed to detect anomalies and initiate immediate medical intervention. The UbiMon project is a classic example of the use of wearable and implantable sensors in continuous monitoring of physiological status, thus playing a crucial role in the early detection of predictors of adverse medical events illustrated in Figure 2.2 (Van Laerhoven et al., 2004). The technology not only improves patient outcomes but also yields massive amounts of real-world data, which is valuable in clinical trials, population health studies, and biomedical informatics. The impact of IoT on the health sector is increasing at an alarming rate. According to Algfari et al., the healthcare IoT

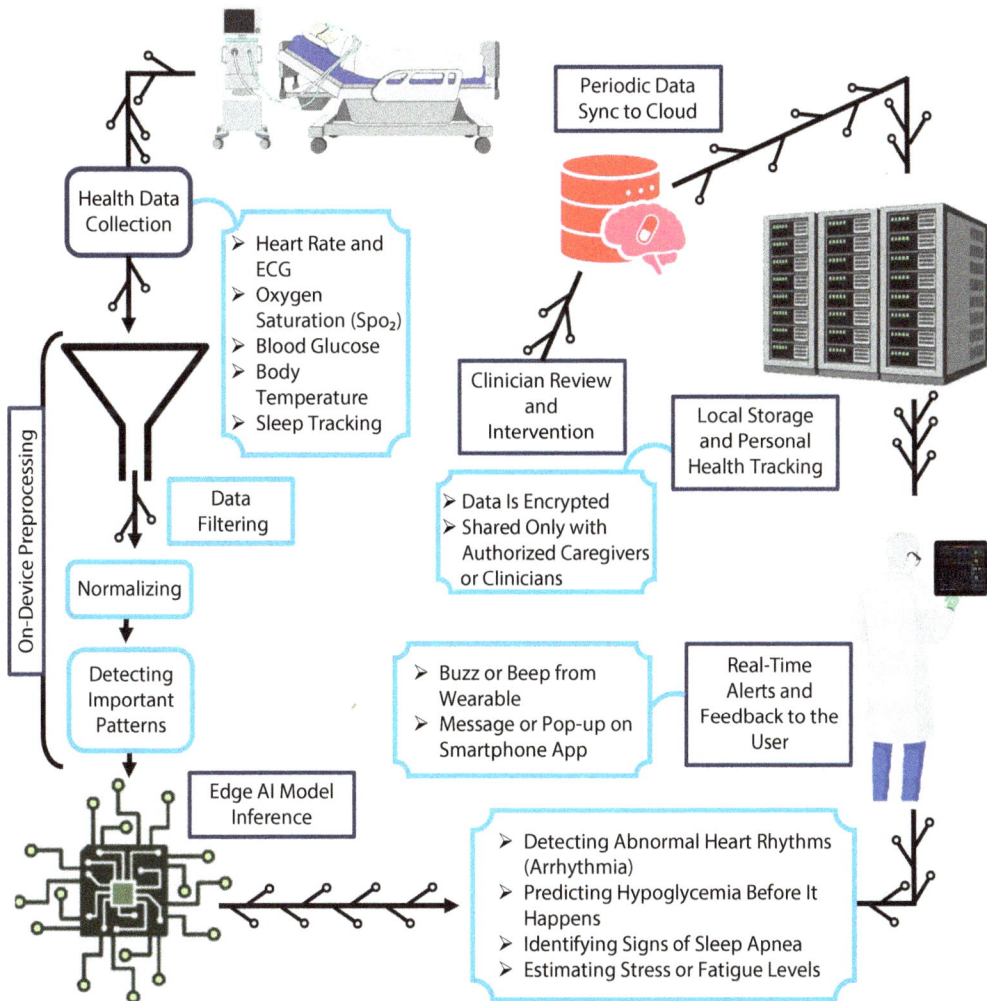

Figure 2.2 Edge AI in healthcare devices – step-by-step flow.

market was estimated to be around $409.9 billion in 2022, owing mostly to its use in chronic disease management, emergency response systems, and customized healthcare plans. The invention of self-sustaining sensors and near-field communication technologies further enhances IoT performance in healthcare applications through improved energy efficiency and seamless data transmission. They not only alleviate the pressure on healthcare infrastructures via remote and preventive care but also make healthcare accessible to the masses through the democratization of healthcare by providing personalized, real-time monitoring access to more people to (Algfari et al., 2020).

2.4.2 Ultra-Low Latency Inference and On-Device Learning in Biomedical AI

The potential advantage of AI breakthroughs is positioned within the domains of ultra-low latency inference and on-device learning, both of which are paramount in real-time biomedical applications. The breakthroughs are encapsulated within the 3U-EdgeAI framework, which seeks to address the edge device hardware limitations through three primary strategies: ultra-low memory training, ultra-low bit width quantization, and ultra-low latency acceleration. In total, these breakthroughs dramatically reduce the memory footprint and computational complexity of deep neural networks (DNNs), enabling them to be deployable on resource-constrained edge devices (Chen et al., 2021). Supplementing this framework is the BioAIP processor, a configurable architecture optimized for wearable health monitoring applications. The processor incorporates an event-driven architecture to reduce energy consumption and an adaptive learning component to enhance classification accuracy by permitting inter-patient variability. It has demonstrated astounding performance in real-time biomedical applications, including ECG-based arrhythmia classification, EEG-based seizure detection, and EMG-based gesture recognition, with the lowest energy-per-inference among competing platforms (Liu et al., 2023). These edge-centric AI processors not only increase the accuracy and reliability of real-time health monitoring but also mitigate the inherent variability and complexity of biomedical signals, which are often misclassified by conventional AI models. Consequently, these breakthroughs are the foundation of personalized healthcare, enabling AI systems to learn from and respond to individual patient characteristics and evolving physiological states.

2.4.3 Ethical, Privacy, and Security Considerations in AI-Driven Healthcare

AI use in the biomedical sector holds enormous transformational power, but at the same time, it initiates strong ethical, privacy, and security issues. The privacy issue is especially vital, given that AI models trained on divulging sensitive biomedical data might inadvertently disclose identifiable information, even if seemingly anonymized summary statistics are used (Torkzadehmahani et al., 2022). Such weaknesses have raised regulatory concerns and made sharing data vital for collaborative biomedicine research more challenging. To address such challenges, privacy-preserving technology like federated learning and differential privacy has been advocated. Federated learning would then be capable of enabling AI model training on numerous decentralized devices or institutions without divulging raw data, thereby ensuring confidentiality while enabling collaborative breakthroughs (Yekaterina, 2024). Such methods do have some compromises, such as the added computational overhead and network latency, which may degrade real-time performance and energy efficiency.

The rapidly changing nature of AI in the healthcare industry is generating important ethical concerns of consent for the use of data, algorithmic bias, and equitable access to AI-based services. Federico and Trotsyuk emphasize the importance of a common worldwide regulatory framework to realize transparency, equity, and accountability in AI-based healthcare solutions (Federico & Trotsyuk, 2024). Further, the sharing of publicly accessible medical databases, providing unprecedented scope for research and innovation, necessitates the development of robust data protection mechanisms to ensure patient confidentiality (Yang, Pan, et al., 2024). To this end, an interdisciplinary and collaborative process synthesizing the efforts of clinicians, AI researchers, ethicists, and policymakers is crucial in developing robust and effective AI systems. Such activities are intended to achieve the maximum benefit of AI in healthcare while maintaining ethical considerations and patient trust.

2.5 ADVANCED AI METHODOLOGIES IN HEALTH INNOVATION

A client-server architecture is used in federated learning (FL), where multiple clients collaborate to train a global model without disclosing their raw data. Li et al. claim that this architecture is designed to reduce communication overhead, safeguard data privacy, and enable scalability across a range of hardware configurations (Li et al., 2020). Collins & Wang provide a comprehensive examination of FL, covering its lifetime, architecture, and privacy characteristics such as differential privacy and secure aggregation. Their work highlights problems, including non-independent and identically distributed data (IID) data, system heterogeneity, and communication efficiency while examining integration with quantum computing and reinforcement learning (Collins & Wang, 2025).

2.5.1 Privacy Leakage in Federated Learning

Even while sharing knowledge in the form of model gradients or coefficients is significantly more secure than sharing the underlying data, there is a small but controllable risk of malicious actors committing what are known as privacy attacks. "Black box" attacks, in which the model (architecture and parameters) and training data are unknown, and "white box" attacks, in which the adversary has total access to the model and/or dataset from which the training data was derived, are two examples of the various types of privacy attacks. Although most AI models are vulnerable to these types of attacks, FL increases the likelihood of them since gradients between local institutions and the coordinating center can be intercepted, and several centers have white box access to the model. Patient-level privacy breaches can be caused by membership inference attacks (Melis et al., 2019). The more privacy protection a change provides, the worse the models perform, even when the models can internally adjust for the noise. Another well-liked method for privacy protection is binning input data to generate k-anonymity and/or l-diversity. One way to help protect privacy is to divide age into deciles, such as 0–9 years, 10–19 years, and 20–29 years. Depending on the bin width and the link between the binned variables, this approach might provide a lower compromise between privacy and performance than rounding or introducing noise. Finally, encrypting changes between the local and central models is a strategy to reduce the vulnerability of such attacks without compromising efficiency. It is important to remember that these types of attacks require a large amount of external data to be successful, including information about specific patients to test against, access to the data source from which the training data was gathered, knowledge of the framework and parameters of the model, and the ability to query the model thousands to millions of times. The information that an adversary could extract from such an attack is limited to whatever information was fed into the model, which is highly unlikely to contain hard identifiers like birthdate, social security number, dates of service, medical record number, full name, etc. Lastly, even with a wealth of external information about individuals, it may be nearly impossible to re-identify a specific patient or group of patients. Data privacy can be protected during data gathering by employing clustering-based anonymization, which lowers the possibility of hostile attacks (Onesimu et al., 2021). Healthcare professionals are increasingly using clinical decision-support systems (CDSSs), which are ML-driven AI tools, to predict patient outcomes. These cutting-edge CDSSs can provide recommendations based on a multitude of patient data much more rapidly than healthcare professionals (HCPs) (Kubben et al., 2019).

2.5.2 Brief Explainability for CDSSs

Explainability approaches are quite promising for CDSSs. Explaining the workings of AI systems can facilitate the identification of potential flaws and the identification of the root causes of errors (Bhatt et al., 2020). However, explainability allows HCPs to evaluate the output of the system's dependability. According to Tonekaboni et al. the AI system's recommendation of a specific course of action may also promote trust in the HCP-patient relationship and allow them to interact with patients more successfully (Tonekaboni et al., 2019). According to Doshi-Velez & Kim, explanations can also give us fresh perspectives on what the AI system learned from data and help us better understand what the algorithm optimizes for and the related trade-offs (Doshi-Velez & Kim, 2017). Furthermore, compared to either the AI system or the HCP alone, it has been shown that providing a qualified second opinion on an AI system can increase diagnostic accuracy (Tschandl et al., 2020). Some writers even argue that explainability is a normative need for medical applications (Amann et al., 2020).

2.5.3 Challenges and Drawbacks

In opposition to the general concept of explainability, Sendak et al. contended that HCPs prioritize the effective application of knowledge from trustworthy sources over a comprehensive understanding of the information's acquisition (Sendak et al., 2020). Experimental results suggest that explainability may not actually enhance users' propensity to follow an AI system's predictions and may even make it more difficult to spot inaccurate predictions (Poursabzi-Sangdeh et al., 2021). These circumstances impose a significant and potentially unfeasible load on the level of explanation.

Users can misread likely links that AI systems have found. It is crucial to keep in mind that while an AI-based CDSS may be able to justify its predictions, physicians may erroneously assume that a causal relationship exists. However, because explanations rely on correlation, they may be distorted by chance elements such as overfitting and spurious correlation; hence, clinical inference is required to determine the reasons (London, 2019). It also highlights the need for an interdisciplinary approach, like the Z-Inspection process, to study and assess AI-powered CDSS because of the diverse range of stakeholders affected by these solutions and the interdependencies that must be considered. A detailed flow of this synthetic data generation process using generative AI is shown in Figure 2.3.

2.6 AI INTEGRATION WITH CUTTING-EDGE BIOMEDICAL TECHNOLOGIES

Advances in CRISPR gene editing, pharmacogenomics, and AI are revolutionizing precision medicine by enabling the implementation of highly tailored therapy regimens. While pharmacogenomics helps identify genetic differences that affect the toxicity, efficacy, and metabolism of drugs, computational techniques driven by AI improve drug optimization and biomarker identification. According to Srivastav et al., CRISPR-based genome editing is an accurate method for modifying gene expression and correcting detrimental mutations. The significance of precision medicine is particularly evident in treatment outcomes for metabolic illnesses and cancer, where genetic heterogeneity and molecular complexity result in widely differing outcomes (Srivastav et al., 2025).

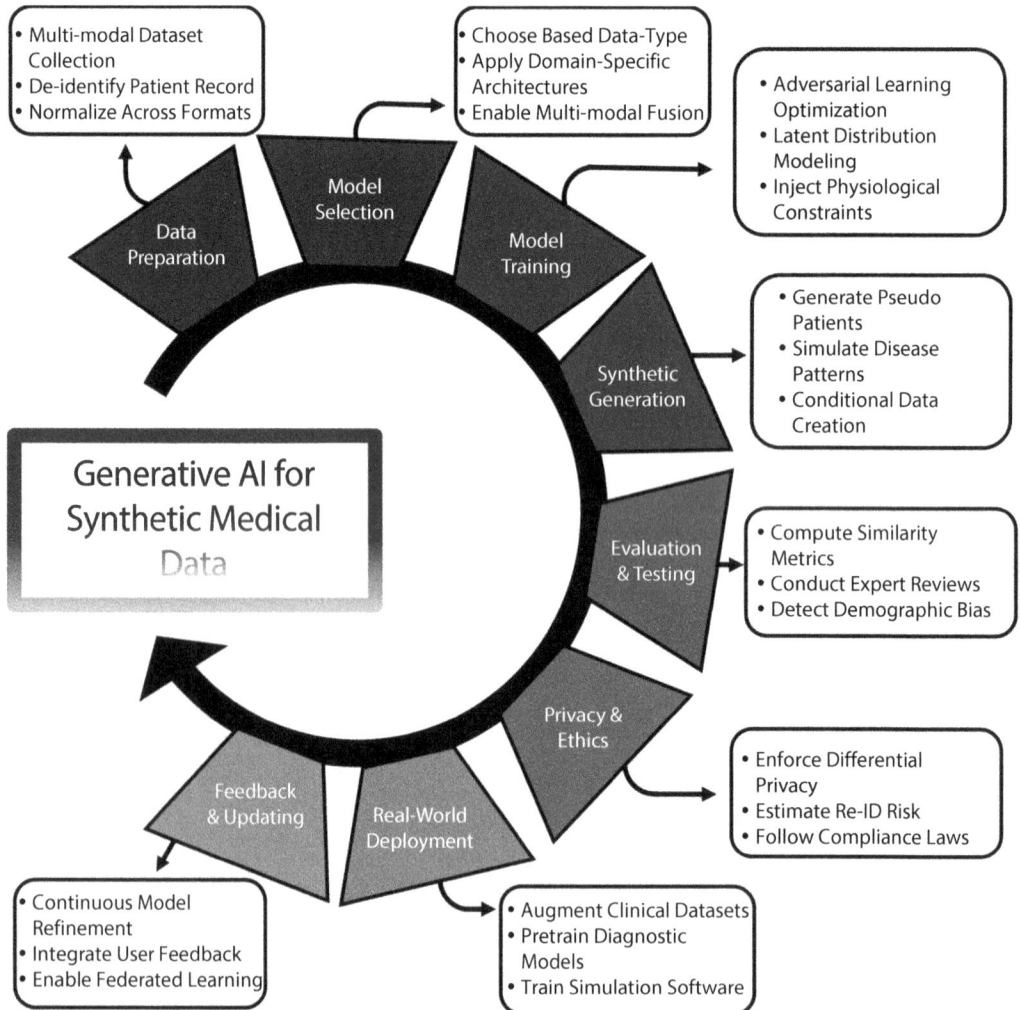

Figure 2.3 Generative AI creating synthetic medical data.

Oncology's tumor heterogeneity necessitates the identification of genetic markers that predict therapeutic outcomes to support pharmacogenomics-driven cancer patient stratification for tailored chemotherapy, immunotherapy, and targeted treatments. This customized approach improves therapy efficacy, decreases toxicity, and increases patient survival. Precision-based medicines are necessary to enhance therapeutic responses and halt the progression of metabolic illnesses, including diabetes and obesity, as genetic predispositions also impact insulin sensitivity, lipid control, and medication metabolism (Krzyszczyk et al., 2018).

2.6.1 CRISPR Genome Editing

The CRISPR-Cas9 system is a revolutionary genome-editing technique because of its ability to precisely modify genes at specific genomic locations. Unlike other genome-editing methods that rely on complex protein engineering, CRISPR uses a programmable guide RNA (gRNA) to direct the Cas9 endonuclease to a complementary target DNA sequence. Aljabali et al. claim that because CRISPR can target nearly any gene by changing the gRNA sequence, it is an essential tool in functional genomics, precision medicine, and gene therapy. In addition to conventional CRISPR-Cas9 editing, other techniques such as base editing and prime editing have been developed to increase the accuracy and efficacy of genomic modifications. These techniques reduce undesirable mutations and off-target effects by preventing the formation of double-strand breaks (DSBs) (Aljabali et al., 2024; Xue & Greene, 2021).

2.6.2 Molecular Mechanism of CRISPR-Cas9

The CRISPR-Cas9 gene-editing system is derived from the adaptive immune systems of bacteria and archaea, which work as a defense mechanism against viral infections. Bacteria can recognize and target invasive viruses when they are reexposed to them because they naturally incorporate spacers, which are pieces of viral DNA, into their CRISPR locus. The Cas9 endonuclease is guided by a complementary gRNA, binds to the target DNA, and creates a DSB at a particular genomic location. This sequence-specific cleavage is the foundation for genome editing in mammalian cells (Loureiro & da Silva, 2019). The CRISPR-Cas9 system consists of two primary components: gRNA is a synthetic, programmable RNA sequence that uses a protospacer-adjacent motif (PAM) to target a certain DNA region. The gRNA's two regions are: The spacer sequence, which has roughly 20 nucleotides, is complementary to the target DNA. The scaffold sequence is required for Cas9 to bind and stabilize. The second is Cas9 endonuclease, an RNA-guided enzyme that locates and binds to the target DNA region immediately upstream of a PAM sequence (typically 5′-NGG-3′ for *Streptococcus pyogenes* Cas9; Asmamaw & Zawdie 2021).

2.6.3 Genetic Variability: Its Role in Drug Discovery

Pharmacogenomics is crucial to precision medicine because it elucidates how genetic variability influences drug metabolism, efficacy, and toxicity. Inter-individual differences in medicine response may be due to genetic variations that affect pharmacokinetics (drug absorption, distribution, metabolism, and excretion) and pharmacodynamics (interaction with target receptors, enzymes, or transporters). These genetic changes, which mostly appear as single-nucleotide polymorphisms (SNPs), copy number variations (CNVs), and structural rearrangements, have a significant impact on drug metabolism by altering the function of crucial enzymes and transporters. By tailoring treatment regimens to a patient's genetic profile through the analysis of certain genetic markers, clinicians can enhance therapeutic outcomes and reduce adverse pharmaceutical reactions (Ahmed et al., 2016). The pharmacogenomics of drug transporters also has a major impact on the range of pharmacological reactions. Drug efflux may be impacted by changes in ABCB1 (P-glycoprotein), which may lead to changes in plasma drug concentrations. The correlation between increased statin-induced myopathy and mutations in SLCO1B1, which encodes the organic anion-transporting polypeptide 1 B1 (OATP1 B1), demonstrates how transporter polymorphisms might impact drug disposition and toxicity (Sissung et al., 2014).

2.6.4 Challenges, Barriers, and Solutions

Data quality, curation, and standardization are foundational to the success of data-driven initiatives across various sectors. Poor data quality, manifested as missing values, inconsistencies, duplicates, or outdated records, can severely impair the reliability of analytical models, leading to flawed insights and decision-making. One of the major challenges lies in integrating data from heterogeneous sources, each potentially having its structure, format, and standards. For instance, when merging datasets from different healthcare providers or financial institutions, the lack of uniform data entry protocols can lead to discrepancies in key attributes, such as dates, terminologies, or measurement units (Mushtaq et al., 2020). Furthermore, the increasing volume and velocity of data generated in real time demand scalable and automated quality assurance processes, which many organizations still struggle to implement effectively (Wang & Strong, 1996).

Curation and standardization are equally complex, particularly in domains requiring semantic interoperability, such as biomedical research or supply chain logistics. Data curation involves continuous monitoring and enrichment to ensure relevance and usability, while standardization requires adherence to established formats, taxonomies, and ontologies. A lack of universally accepted standards can hinder data sharing and collaboration across institutions or countries (Whetzel et al., 2011). Moreover, manual curation efforts are labor-intensive and susceptible to human error, while automated solutions often face difficulties interpreting unstructured data or resolving context-sensitive ambiguities (Aroyo & Welty, 2015). Addressing these challenges necessitates a balanced approach combining robust governance frameworks, advanced AI-driven tools, and stakeholder collaboration to ensure that data remains clean, consistent, and contextually meaningful throughout its lifecycle.

2.7 EMERGING TRENDS AND FUTURE DIRECTIONS

The application of various types of AI for medical analysis and medicine is transforming the way we apply precision medicine and bioinformatics. This new paradigm unites various types of data, such as medical images, genomic data, health records, and behavior data, to support the understanding of intricate biological systems and disease processes (Dankan Gowda et al., 2025). Multimodal AI operates with DL and intelligent data fusion techniques to extract meaning from these heterogeneous sources, enabling clinicians to make more informed decisions and resulting in improved health outcomes. Multimodal methods, as opposed to models of one type, are better, with a mean gain of 6.2 percentage points in prediction accuracy, and hence effective in actual medicine (Schouten et al., 2025). Multimodal AI is applied in numerous key applications, such as personalized medicine, decentralized and online clinical trials, telemedicine, remote healthcare monitoring, and real-time pandemic tracking systems (Acosta et al., 2022). In personalized medicine, multimodal AI customizes the treatment by employing genetic, physical, and lifestyle data, enhancing diagnosis and treatment precision. In clinical trials, integrating behavior and health data facilitates trial designs that are more patient-focused and lighter. Multimodal AI also has some noteworthy challenges, such as variations in data forms, system cooperation problems, and the need for standard procedures for mixing dissimilar types of data. Furthermore, extensive clinical validation and transparent model explanations should be performed to win the hearts of clinicians and patients.

Areas for future work include applying XAI methodologies to untangle decision-making processes. This will be complemented by privacy-preserving mechanisms like FL for data protection and compliance with ethical standards. New architectures to enable data to be federated seamlessly and accommodate larger models to be trained are also necessary to realize the true potential of multimodal AI in medicine. As technology progresses, it can re-engineer healthcare delivery, making it efficient, fair, and personalized, yet humane in the values that medicine cherishes (Chodvadiya, 2025).

2.7.1 AI in Precision Medicine: Transforming Personalized Healthcare

AI is increasingly employed in precision medicine. It is capable of examining and reading vast amounts of varied biomedical information with a very high degree of accuracy. By merging the prints of genetic profiles, molecular markers, and long patient clinical histories, AI allows one to identify complex patterns and risk factors that describe how various patients perceive and react to disease and therapy (Fatima et al., 2024). This ability is necessary to achieve the objectives of precision medicine, which is to move away from one-size-fits-all to highly tailored diagnosis and treatment approaches. One of the most wide-ranging uses of AI in medicine is revolutionizing the way clinical trials are designed and carried out. Virtual control groups based on AI are used to generate placebo or control groups based on historical data or real-world data and thereby reducing the requirement for recruitment of control group patients. The technology is also making trials more ethical, particularly for orphan diseases with smaller patient numbers (Haider, 2024). Drug discovery is also being accelerated by AI, with target identification, repurposing of drugs, and prediction of toxicity being automated.

In cancer, AI algorithms are transforming the way we diagnose and treat patients by enabling us to classify patients according to their tumor genes and response to certain markers. This aids in developing personalized treatments for the specific shape of every patient's cancer, and it is more effective with fewer toxic side effects (Akter, 2024). AI also assists with the optimization of the type of patients by identifying patient cohorts with different clinical and biological profiles by integrating gene data with other data (Johnson et al., 2021). The wider application of AI to precision medicine is confronted by some challenges, such as data privacy issues, the extent to which algorithms can be explained, and the degree of cost of mirroring bias in training data. To combat these challenges, cooperation among doctors, bioinformaticians, and data ethicists is necessary. These partnerships are crucial in the development of robust, understandable, and moral AI systems that can be applied in full to medicine. As the field expands, the overlap of AI and precision medicine can usher in massive breakthroughs in customized treatment and drug discovery, ultimately transforming healthcare for the better (Akter, 2024).

2.7.2 AI in Mental Health: Personalization, Access, and Ethical Considerations

The use of AI in mental health is growing very fast, providing innovative diagnostic, treatment, and care technologies. AI systems are being developed to analyze complex behavior data like speech, social media usage, facial expressions, and mobile sensor data to detect early warning signs of mental illness like depression, anxiety, and post-traumatic stress disorder (Pandey, 2024). Such technologies provide earlier detection of vulnerable people and enable the implementation of early intervention strategies, which are critical for enhancing long-term mental health outcomes. AI also makes personalized mental health treatments possible. Predictive models of analytics allow for people to be anticipated based on how they react to medications and treatments, which allows doctors to create customized treatment plans based on the person's needs (Alhuwaydi, 2024). Virtual chatbots and cognitive-behavioral therapy (CBT) websites can be another possible use, providing accessible, real-time mental healthcare. These web pages facilitate the dismantling of the past barriers to mental healthcare, including unavailability, stigma, and provider shortages (Olawade et al., 2024).

The use of AI in mental health therapy is of immense practical and ethical importance. Patient secrecy and data security are of the highest priority, particularly since mental health data is confidential. The risk of bias in the algorithms is also present, which would most likely result in faulty evaluation or rejection of certain groups. This risk can be eliminated by ongoing verification of models and inputting data that is representative of all groups (Pandey, 2024). Special care should also be taken to document that therapeutic relationships are human-centered; although AI can be useful, it cannot substitute for the empathy and compassion of well-trained mental health professionals (Alhuwaydi, 2024). To deploy AI in mental health responsibly, future research must be aimed at developing open and validated models, setting standards, and undertaking cross-domain research. This conservative stance will make AI a strong tool to enhance mental health outcomes without compromising ethical principles and human dignity (Beg et al., 2024).

2.7.3 Hybrid Quantum-Classical AI in Biomedicine: A New Frontier for Complex Problem-Solving

The emergence of hybrid quantum-classical AI systems is a breakthrough in the field of biology and medical technology. Such systems take advantage of the computational power of quantum mechanics and the adaptability of conventional ML models to address complex problems that conventional solutions are not capable of addressing (Pulicharla, 2023). With the assistance of quantum concepts such as entanglement and superposition, such hybrid models can explore many complex solutions at top speed, and this is priceless in applications such as protein folding, drug discovery, and systems biology. Hybrid quantum-classical CNNs have been demonstrated through experimentation to predict intricate protein-ligand interactions over the past few years, and this is crucial in smart drug discovery from an AI perspective. Such platforms have achieved 20% computational complexity reduction and 40% training time reduction, accelerating preliminary research (Martín-Cuevas & Calleja, 2025). Moreover, bio-inspired computing techniques such as artificial immune systems (AIS) are being enhanced with quantum algorithms to process intricate biological data, yielding improved solutions to mimic systems in immunogenomics and disease networks (Myakala et al., 2025).

Design of these systems is informed by guidelines that seek to use data efficiently, keep algorithms stable, and scale toward quantum hardware development. With better quantum computing using more accurate qubits and error correction, hybrid AI systems will operate much more effectively, and applications to biomedical research will be more common. The technology, however, has its constraints, like insufficient hardware, algorithm noise, and the need for experienced individuals able to link quantum theory with biology. Investment in research infrastructure, learning, and interdisciplinary integration will be key in balancing these constraints. With the evolution in technology, hybrid quantum-classical AI has the potential to transform the biomedical discovery of the future by eliminating problems previously thought to be unsolvable (Hadap & Patil, 2024).

2.8 ETHICAL, LEGAL, AND REGULATORY CONSIDERATIONS

Health innovation and biomedical research are being revolutionized today through the application of AI to improve health. AI can be utilized to enhance the way physicians diagnose illness and develop tailored treatment regimes, offering new means of treating patients and enhancing healthcare. Applications of AI in healthcare are conducted with a careful deployment considering its advantages against ethical concerns, privacy policy, and legislation such as the Health Insurance Portability and Accountability Act (HIPAA) in the US and the General Data Protection Regulation (GDPR) in the EU. The rapid application of AI to medicine has raised concerns about patient privacy, data security, and the ethics of using data, as per Momani. Compliance with HIPAA and GDPR is not only a legal issue; it is a requirement in maintaining the public's trust in new digital health technology (Momani, 2025; Utomi et al., 2024). All this relies heavily on having robust data protection mechanisms such as encryption, data anonymization, and robust access controls that safeguard patient-sensitive data.

To further enhance privacy and compliance, newer privacy-preserving technologies such as FL, where AI models can be trained on decentralized data without its movement, and differential privacy, where noise is injected into data to safeguard identities, are becoming promising tools (Singhal, 2024). These techniques enable the continued utilization of real-world data within AI systems without compromising the privacy standards. Also, the marriage of AI systems with blockchain technology has been extremely promising in terms of data integrity and protection from unauthorized access. The immutable and decentralized ledger system of blockchain provides transparent tracking and verification of data, thus ensuring regulatory compliance on a global scale. Application of privacy-by-design principles, embedding data protection into the design and architecture of AI systems from the outset, also maximizes transparency, accountability, and user control over personal health data. Bringing ethically strong and privacy-friendly AI in healthcare into being needs a collaborative approach involving healthcare professionals, AI developers, ethicists, lawyers, and policymakers. There should be joint governing mechanisms put in place to ensure that the development and use of AI technologies are executed with a well-defined commitment to ethical principles, regulatory compliance, and patient-centered practice (Yekaterina, 2024).

2.8.1 Bias, Fairness, and Equity in Biomedical AI Models

Even though AI promises immense potential to drive biomedical discovery and healthcare, there exist enormous concerns of bias, fairness, and health equity to make their responsible use challenging. If the AI systems are created or applied without consideration of social factors and group differences, then they can exacerbate current health issues. This is particularly relevant with the use of AI in biomedical NLP and computer vision, where bias could be due to training data that does not depict all humans or due to decisions while developing algorithms (Yang, Lin, et al., 2024). AI model bias can occur at numerous points throughout the development cycle, anywhere from data collection and processing to model training and model deployment. For instance, if some racial, ethnic, or gender populations are underrepresented in training data, then the model would not work correctly for those populations and might even exacerbate healthcare disparities. Due to this, approaches to mitigate bias, including algorithmic reweighting, adversarial training, and domain adaptation, are being researched to make possible the correction of such problems (Griffin et al., 2023).

To make AI systems equitable, we need to test them using transparent and open methods. XAI enables individuals to comprehend how models make decisions. This enables physicians to determine the fairness and reliability of AI suggestions. We also require ethical guidelines to encourage equitable AI system design. This involves the use of a vast set of data, ongoing testing of the algorithms, and incorporation of social and legal factors in AI testing (Mienye et al., 2024). Policy measures such as the Fairness of AI Recommendations (FAIR) framework mention the involvement of diverse groups to eliminate bias and make healthcare solutions with AI equitable (Ueda et al., 2024). The guidelines facilitate ease in incorporating voices from marginalized communities, provide frequent fairness testing, and deploy AI systems designed with community outreach for groups with healthcare inequalities.

2.8.2 Algorithmic Governance and Adaptive Regulation of AI in Healthcare

Successful AI healthcare governance is essential to close the gap between what the technology can accomplish and how it operates within real-world medical settings. Traditional regulation, for example, like that used by the U.S. Food and Drug Administration (FDA), has the effect of "locking" AI algorithms once they are certified. They cannot then change in response to further data and evolving medical conditions, which is a fundamental nature of medical ML systems. This strictness is making it difficult to keep models accurate and functional within fast-changing clinical settings (Mashar et al., 2023). To address this challenge, elastic rules are being created to allow AI models to learn safely and incrementally. The Healthcare AI Governance Readiness Assessment (HAIRA) model features a multi-layered approach to determine whether the organizations are ready to deploy and govern AI, with a focus on resource differences between the large academic hospitals and small healthcare facilities (Hussein et al., 2024).

Governance programs must incorporate ethical practices such as non-harm to others, transparency, openness, and accountability. Prosperi et al. contend that algorithmic bias can lead to unfairness, referring to the need for governance mechanisms that affirm fairness and that provide equal access to the benefits of AI (Prosperi et al., 2023). The necessity for a national AI health strategy based on the Quintuple Aim, better patient experience, better population health, reduced costs, empowered care teams, and enhanced health equity (Hurd et al., 2024), is increasingly being accepted. To make AI governance effective, there needs to be supporting organizations or sandboxes for regulation to aid organizations with AI validation, integration, and monitoring once they are deployed. The platforms should be capable of supporting innovation while ensuring compliance and ethical responsibility (Hassan et al., 2025).

2.8.3 Ethical Implications of Autonomous AI in Clinical Decision-Making

The increasing use of autonomous AI systems in making medical decisions raises serious ethical concerns. These systems can assist doctors in making more accurate decisions, reducing diagnostic mistakes, and providing patients with more control and more accurate predictions, but they are also a source of responsibility, justice, and moral duty concerns. One of the main issues is that AI decision-making is opaque, or the "black box" issue. Transparent AI would prevent doctors from knowing and being able to trust AI recommendations, possibly compromising patient safety and informed consent (Azad, 2018). Furthermore, the implementation of AI systems that translate human judgment into statistical information might not reflect the ethical nuances required in complex clinical cases, particularly in end-of-life care or psychiatric diagnoses (Benzinger et al., 2023).

The second core question is who to hold accountable for ethical actions. Where autonomous AI systems are determining or creating clinical decisions, there is no clear understanding of who bears the ultimate responsibility: developers, clinicians, or the organization. Traditional ethics, under which there are assumed-to-be clear lines of responsibility, are not highly suited to shared decision-making. This entails the application of relational ethics and models of collective responsibility, taking into account the interrelated roles of all the parties involved in the AI system (Marques et al., 2024). To address these moral problems, the development of AI systems has to conform to moral design principles, be regulated, and keep stakeholders engaged. Ethical auditing, algorithm effect checks, and open debate among clinicians, developers, and patients are required to instill confidence and ensure that AI systems uphold human dignity and social values.

2.9 INTEROPERABILITY AND INTEGRATION ACROSS PLATFORMS AND DATA MODALITIES

Interoperability and integration across platforms and data modalities are critical for enabling seamless data exchange, unified analytics, and collaborative innovation in today's digital ecosystems. Interoperability refers to the ability of diverse systems and applications to communicate and operate effectively together, while integration involves combining data from different sources into a cohesive and usable format. A major challenge in achieving interoperability is the heterogeneity of data formats, communication protocols, and metadata standards used across platforms. This issue is especially pronounced in healthcare, smart cities, and industrial IoT, where systems often rely on legacy infrastructure or proprietary technologies that resist standardization (Raghupathi & Raghupathi, 2014). Without a common framework, valuable data remains siloed, limiting its potential for cross-domain insights and comprehensive analysis.

Integration across multiple data modalities, text, images, time-series, and genomic data, introduces further complexity. Multimodal integration requires sophisticated alignment methods that preserve semantic meaning across different data types (Ngiam et al., 2011). For instance, in precision medicine, integrating clinical notes, imaging, and genetic data is essential but technically demanding due to differences in data granularity, structure, and scale. Middleware solutions, APIs, and ontology-based mapping are among the strategies used to bridge these gaps, yet they often require significant customization and domain expertise (Kass-Hout & Alhinnawi, 2013). Achieving true interoperability and integration not only requires technological solutions but also institutional collaboration, regulatory support, and adherence to open standards such as HL7 FHIR, RDF, or ISO standards for data interchange.

2.9.1 Workforce Training and Preparing Clinicians for AI Collaboration

As AI becomes increasingly integrated into clinical practice, preparing the healthcare workforce, especially clinicians, for effective collaboration with AI tools is paramount. One of the key challenges is bridging the knowledge gap between clinical expertise and technical literacy. Many clinicians lack formal training in data science, ML, or the interpretability of AI algorithms, which can hinder trust and adoption (Topol, 2019a). Educational institutions and healthcare organizations must therefore update curricula and professional development programs to include AI fundamentals, data ethics, and practical case studies demonstrating AI's application in diagnostics, treatment planning, and patient management (Kolachalama & Garg, 2018). These efforts should emphasize not just technical competence, but also the ability to critically evaluate AI outputs in a clinical context. Moreover, effective human-AI collaboration requires fostering a culture of interdisciplinary teamwork, where clinicians, data scientists, and engineers can communicate seamlessly. Without this, AI systems may be misused, underutilized, or even resisted altogether. Concerns about liability, workflow integration, and potential deskilling further complicate adoption (Jiang et al., 2017). Therefore, workforce training must also address ethical considerations, accountability frameworks, and the importance of clinician oversight to ensure that AI augments rather than replaces human judgment. Simulation-based training environments, participatory design approaches, and continuous learning systems are promising strategies to build confidence and competence in using AI responsibly and effectively in clinical care (Rajkomar et al., 2019).

2.10 REAL-WORLD CASE STUDIES AND APPLICATIONS

2.10.1 AI in COVID-19 Diagnostics, Prediction, and Epidemiology

AI played a pivotal role in combating the COVID-19 pandemic, particularly in diagnostics, prediction, and epidemiological modeling. In diagnostics, AI-enabled tools, especially those leveraging DL, were rapidly deployed to analyze medical imaging data such as chest X-rays and CT scans for detecting COVID-19-related abnormalities. These AI systems not only helped reduce the burden on radiologists but also enabled faster and scalable screening in regions with limited healthcare infrastructure (Ardakani et al., 2020). Additionally, NLP was used to analyze unstructured clinical notes and EHRs to assist in triaging patients and identifying symptom patterns indicative of COVID-19 (Wang, Lo, et al., 2020). In prediction and epidemiology, AI was instrumental in forecasting case trends, modeling the spread of infection, and evaluating the impact of public health interventions. ML algorithms used real-time data, such as mobility patterns, social behavior, and climate variables, to predict outbreak hotspots and assess risk levels (Chimmula & Zhang, 2020). Tools like BlueDot and HealthMap employed AI to provide early warnings about the spread of the virus, even before official confirmations were made (Kraemer et al., 2020). Moreover, AI-supported contact tracing apps and surveillance systems were critical in containment strategies, although they raised ethical and privacy concerns that highlighted the importance of transparent and equitable deployment (Naudé, 2020). Overall, while AI significantly augmented the global response to COVID-19, the crisis also underscored the need for robust data governance and interdisciplinary collaboration to maximize its effectiveness.

2.10.2 AI for Rare Disease Diagnosis and Treatment Personalization

AI has emerged as a transformative tool in the diagnosis and treatment personalization of rare diseases, which often present unique clinical challenges due to limited data, heterogeneous symptoms, and a lack of standardized treatment protocols. Traditional diagnostic processes for rare diseases can take years, involving multiple misdiagnoses and clinical visits. AI helps overcome these barriers by analyzing large-scale, multimodal data, such as genomic sequences, EHRs, and medical imaging, to detect patterns that may be too subtle or complex for human interpretation (Greene et al., 2016). For example, ML algorithms can match patient symptoms with known rare disease phenotypes or gene mutations, significantly accelerating time to diagnosis (Liu et al., 2019).

In treatment personalization, AI facilitates precision medicine by tailoring therapies to an individual's unique genetic, environmental, and clinical profile. This is especially valuable in rare diseases, where one-size-fits-all treatments are often ineffective. AI-driven models can identify potential therapeutic targets, repurpose existing drugs, and predict patient-specific treatment responses using data from omics technologies and clinical trials (Topol, 2019b). Moreover, NLP enables the mining of biomedical literature and case reports to uncover insights about rare disease mechanisms and management strategies. Despite these advances, challenges remain, including the need for high-quality annotated datasets, ethical considerations around data use, and the integration of AI tools into real-world clinical workflows. Continued interdisciplinary collaboration and open data sharing will be essential to fully harness AI's potential in this critical area of medicine.

2.10.3 AI Optimization in Reproductive Medicine and IVF Success

AI is revolutionizing reproductive medicine, particularly in optimizing in vitro fertilization (IVF) procedures and improving their success rates. Traditional IVF is a complex, costly, and emotionally demanding process with relatively low success rates, especially for older patients. AI applications are increasingly being used to enhance various stages of the IVF process, from selecting the most viable embryos to predicting treatment outcomes, through advanced data analytics and ML algorithms (Khosravi et al., 2019). For example, AI models can analyze thousands of time-lapse images to assess embryo morphology and development, providing more objective and consistent embryo grading than manual evaluation by embryologists (Tran, Cooke, et al., 2019). Moreover, AI enables the personalization of reproductive treatments by integrating data from patients' hormonal profiles, genetic markers, age, and prior fertility history. These models can help predict optimal stimulation protocols, reduce the likelihood of ovarian hyperstimulation syndrome (OHSS), and improve implantation and live birth rates (Blank et al., 2022). In addition, predictive analytics powered by AI can assist in counseling patients about their chances of success, enabling more informed decision-making and reducing emotional distress. However, the use of AI in reproductive medicine also raises ethical and regulatory questions, including concerns about algorithmic transparency, bias in training datasets, and equitable access to AI-enhanced fertility services (Esteva et al., 2019). As the technology continues to mature, robust validation and clinical integration strategies will be key to ensuring AI's safe and effective use in reproductive healthcare.

2.11 CONCLUSION

The use of AI in biomedical research is a paradigm change in data analysis, interpretation, and application strategies used in the context of health advancement. AI-based tools and methods, ranging from traditional ML algorithms to advanced generative models, are now unavoidable in investigating intricate biological systems, speeding up drug discovery, improving diagnostic accuracy, and facilitating customized medical approaches. As the biomedical sector relies more and more on data, AI offers the computing capacity and analytical models required to handle large-scale genomic, proteomic, imaging, and clinical data sets. Progressive deployments, such as Edge AI for real-time monitoring, digital twins for simulating physiological processes, and XAI for explaining clinical decision-making mechanisms, are shaping the future of patient-centric healthcare. Moreover, FL and synthetic data generation address important ethical and privacy concerns, enabling collaborative research while keeping sensitive data confidential. Despite the vast potential offered by these technologies, their ethical application continues to be paramount. Maintaining data integrity, algorithmic fairness, model interpretability, and regulatory concordance is crucial to the biomedical application of AI. AI is not a substitute for HI but a potent tool for augmentation, enabling researchers and clinicians to make more accurate, informed, and predictive choices. Continued integration of AI and biomedical science has the potential to transform healthcare, making it predictive, preventive, and personalized.

ACKNOWLEDGMENT

All the authors are thankful to their respective universities and institutions for their valuable supports.

DATA AVAILABILITY

The authors confirm that the data supporting the findings of this study are available within the book/chapter.

ABBREVIATIONS

AE	Autonomous Experimentation
AI	Artificial Intelligence
AIS	Artificial Immune Systems
ANNs	Artificial Neural Networks
CAD	Computer-Aided Diagnosis
CDSSs	Clinical Decision-Support Systems
CNNs	Convolutional Neural Networks
CNVs	Copy Number Variations
DL	Deep Learning
DNNs	Deep Neural Networks
DSB	Double-Strand Break
EHRs	Electronic Health Records
FAIR	Fairness of AI Recommendations
FDA	Food and Drug Administration
FL	Federated Learning
GCNNs	Graph Convolutional Neural Networks
GDPR	General Data Protection Regulation
gRNA	Guide RNA
HAIRA	Healthcare AI Governance Readiness Assessment
HI	Human Intelligence
HIPAA	Health Insurance Portability and Accountability Act
IoMT	Internet of Medical Things
IoT	Internet of Things
IVF	In Vitro Fertilization
ML	Machine Learning
NLP	Natural Language Processing
OATP1 B1	Organic Anion-Transporting Polypeptide 1 B1
OHSS	Ovarian Hyperstimulation Syndrome
PAM	Protospacer-Adjacent Motif
ROC	Receiver Operating Characteristic
SNPs	Single-Nucleotide Polymorphisms
VAEs	Variational Autoencoders
XAI	Explainable AI

REFERENCES

Acosta, J. N., Falcone, G. J., Rajpurkar, P., & Topol, E. J. (2022). Multimodal biomedical AI. *Nature Medicine, 28*(9), 1773–1784.

Ahmed, S., Zhou, Z., Zhou, J., & Chen, S. Q. (2016). Pharmacogenomics of drug metabolizing enzymes and transporters: Relevance to precision medicine. *Genomics, Proteomics & Bioinformatics, 14*(5), 298–313.

Akter, S. (2024). AI-driven precision medicine: Transforming personalized cancer treatment. *JAPMI, 2*, 10–21.

Al Gharrawi, R. J., & Al-Joda, A. A. (2023). A survey of medical image analysis based on machine learning techniques. *Journal of Al-Qadisiyah for Computer Science and Mathematics, 15*(1), 48.

Alagendran, S., Archunan, G., Prabhu, S.V., Orozco, B. E., & Guzman, R. G. (2010). Biochemical evaluation in human saliva with special reference to ovulation detection. *Indian Journal of Dental Research, 21*, 165–168.

Algfari, S. M., Alghamdi, A. S., Ali, E. I. A., Almuhaylib, A. M., Alzaher, M. A., Alshammry, M. M., & Alumtairi, A. Z. (2020). Real-time health monitoring using IoT devices for patients with chronic conditions. *International Journal of Health Sciences, 4*(S1), 214–228.

Alhuwaydi, A. M. (2024). Exploring the role of AI in mental healthcare: Current trends and future directions—A narrative review for a comprehensive insight. *Risk Management and Healthcare Policy, 17*, 1339–1348.

Aljabali, A. A., El-Tanani, M., & Tambuwala, M. M. (2024). Principles of CRISPR-Cas9 technology: Advancements in genome editing and emerging trends in drug delivery. *Journal of Drug Delivery Science and Technology, 92*, 105338.

Altara, R., Basson, C. J., Biondi-Zoccai, G., & Booz, G. W. (2024). Exploring the promise and challenges of artificial intelligence in biomedical research and clinical practice. *Journal of Cardiovascular Pharmacology, 83*(5), 403–409.

Amann, J., Blasimme, A., Vayena, E., Frey, D., Madai, V. I., & Precise4Q Consortium. (2020). Explainability for AI in healthcare: A multidisciplinary perspective. *BMC Medical Informatics and Decision Making, 20,* 1–9.

Anasica, S. (2023). Implementation of wearable IoT devices for continuous physiological monitoring and analysis. *Research Journal of Computer Systems and Engineering, 4*(2), 188–200.

Ardakani, A. A., Kanafi, A. R., Acharya, U. R., Khadem, N., & Mohammadi, A. (2020). Application of deep learning technique to manage COVID-19 in routine clinical practice using CT images: Results of 10 convolutional neural networks. *Computers in Biology and Medicine, 121,* 103795.

Aroyo, L., & Welty, C. (2015). Truth is a lie: Crowd truth and the seven myths of human annotation. *AI Magazine, 36*(1), 15–24.

Arumugam, V. A., Thangavelu, S., Fathah, Z., Ravindran, P., Sanjeev, A. M. A., Babu, S., Arun, M., Yatoo, M. I., Khan, S., Ruchi, T., Megha Katare, P., Ranjit, S., Chandra, R., & Dhama, K. (2020). COVID-19 and the world with co-morbidities of heart disease, hypertension and diabetes. *Journal of Pure and Applied Microbiology, 14*(3), 1623–1638.

Asmamaw, M., & Zawdie, B. (2021). Mechanism and applications of CRISPR/Cas-9-mediated genome editing. *Biologics: Targets and Therapy, 15,* 353–361.

Azad, Y. (2018). Ethical implications of AI in decision making. *International Journal of AI and Machine Learning, 1*(2), 631–639.

Badidi, E. (2023). Edge AI for early detection of chronic diseases and the spread of infectious diseases: Opportunities, challenges, and future directions. *Future Internet, 15*(11), 370.

Barik, R. K., Dubey, H., & Mankodiya, K. (2017). SOA-FOG: Secure service-oriented edge computing architecture for smart health big data analytics. In *2017 IEEE global conference on signal and information processing (GlobalSIP),* Montreal (pp. 477–481). IEEE *Xplore.*

Beg, M. J., Verma, M., & Verma, M. K. (2024). AI for psychotherapy: A review of the current state and future directions. *Indian Journal of Psychological Medicine* https://doi.org/10.1177/02537176241260819

Belcastro, L., Carretero, J., & Talia, D. (2024). Edge-cloud solutions for big data analysis and distributed machine learning-1. *Future Generation Computer Systems, 159,* 323–326.

Benzinger, L., Ursin, F., Balke, W. T., Kacprowski, T., & Salloch, S. (2023). Should AI be used to support clinical ethical decision-making? A systematic review of reasons. *BMC Medical Ethics, 24*(1), 48.

Bhatt, U., Xiang, A., Sharma, S., Weller, A., Taly, A., Jia, Y., & Eckersley, P. (2020). Explainable machine learning in deployment. In *Proceedings of the 2020 conference on fairness, accountability, and transparency,* Barcelona (pp. 648–657) Association for Computing Machinery.

Bhotla, H. K., Balasubramanian, B., Meyyazhagan, A., Pushparaj, K., Easwaran, M., Pappusamy, A. R., Arumugam, V. A., Tsibizova, V., Msaad Alfalih, A., Aljowaie, R. M., Saravanan, M., & Di Renzo, G. C. (2021). Opportunistic mycoses in COVID-19 patients/survivors: Epidemic inside a pandemic. *Journal of Infection and Public Health, 14*(11), 1720–1726.

Blanco, M. J., Buskes, M. J., Govindaraj, R. G., Ipsaro, J. J., Prescott-Roy, J. E., & Padyana, A. K. (2024). Allostery illuminated: Harnessing AI and machine learning for drug discovery. *ACS Medicinal Chemistry Letters, 15*(9), 1449–1455.

Blank, C., Heusner, C., Arck, P., & Pils, S. (2022). AI and machine learning in reproductive medicine: Current applications and future possibilities. *Journal of Assisted Reproduction and Genetics, 39*(4), 851–862.

Chandrasekaran, S., Athinarayanan, S., Masthan, M., Kakkar, A., Bhatnagar, P., & Samad, A. (2024). Edge intelligence paradigm shift on optimizing the edge intelligence using AI state-of-the-art models. In Suman Rajest, Salvatore Moccia, Bhopendra Singh, R. Regin & Joseph Jeganathan (Eds.). *Advancing intelligent networks through distributed optimization* (pp. 1–18). IGI Global.

Chaudhari, G. R., & Telrandhe, U. B. (2024). The role of AI (AI) in transforming biomedical imaging: A comprehensive overview. In *2024 2nd DMIHER international conference on AI in healthcare, education and industry (IDICAIEI),* Wardha (pp. 1–5). IEEE *Xplore.*

Chen, Y., Hawkins, C., Zhang, K., Zhang, Z., & Hao, C. (2021). 3U-EdgeAI: Ultra-low memory training, ultra-low bitwidth quantization, and ultra-low latency acceleration. In *Proceedings of the 2021 Great Lakes symposium on VLSI* (pp. 157–162), Association for Computing Machinery.

Chimmula, V. K. R., & Zhang, L. (2020). Time series forecasting of COVID-19 transmission in Canada using LSTM networks. *Chaos, Solitons & Fractals, 135,* 109864.

Chodvadiya, K. (2025). The role of multimodal AI in revolutionizing healthcare: A perspective. *National Journal of Medical Research, 15*(1), 75–76.

Collins, E., & Wang, M. (2025). Federated learning: A survey on privacy-preserving collaborative intelligence. *arXiv preprint arXiv:2504.17703*.

da Silva, R. G. L. (2024). The advancement of AI in biomedical research and health innovation: Challenges and opportunities in emerging economies. *Globalization and Health, 20*(1), 44.

Dankan Gowda, V., Palanikkumar, D., Prasad, K. D. V., Kaur, M., & Singh, S. (2025). Future directions and emerging trends in multimodal data fusion for bioinformatics. In Umesh Kumar Lilhore, Abhishek Kumar, Narayan Vyas, Sarita Simaiya, & Vishal Dutt (Eds.). *Multimodal data fusion for bioinformatics AI* (pp. 247–282). Wiley Online Library.

De Freitas, J., Glicksberg, B. S., Johnson, K. W., & Miotto, R. (2021). Deep learning for biomedical applications. In Subhi J. Al'Aref, Gurpreet Singh, Lohendran Baskaran & Dimitris Metaxas (Eds.). *Machine learning in cardiovascular medicine* (pp. 71–94). Academic Press.

Doshi-Velez, F., & Kim, B. (2017). Towards a rigorous science of interpretable machine learning. *arXiv Preprint arXiv:1702.08608*.

Dzobo, K., Adotey, S., Thomford, N. E., & Dzobo, W. (2020). Integrating artificial and human intelligence: A partnership for responsible innovation in biomedical engineering and medicine. *Omics: A Journal of Integrative Biology, 24*(5), 247–263.

Esteva, A., Robicquet, A., Ramsundar, B., Kuleshov, V., DePristo, M., Chou, K., & Dean, J. (2019). A guide to deep learning in healthcare. *Nature Medicine, 25*(1), 24–29.

Fatima, G., Siddiqui, Z., & Parvez, S. (2024). *AI and precision medicine: Paving the way for future treatment*. Preprints.

Federico, C. A., & Trotsyuk, A. A. (2024). Biomedical data science, AI, and ethics: Navigating challenges in the face of explosive growth. *Annual Review of Biomedical Data Science, 7*, 1–14.

Greene, D., Richardson, S., & Turro, E. (2016). Phenotype similarity regression for identifying the genetic determinants of rare diseases. *The American Journal of Human Genetics, 98*(3), 490–499.

Griffin, A., Wang, K. H., Leung, T. I., & Facelli, J. (2023). Fairness and inclusion in biomedical AI research and clinical use: Technical and social perspectives. https://doi.org/10.1016/j.jbi.2024.104693

Gundappa, M., Arumugam, V. A., Hsieh, H. L., Balasubramanian, B., & Shanmugam, V. (2022). Expression of tissue factor and TF-mediated integrin regulation in HTR-8/SVneo trophoblast cells. *Journal of Reproductive Immunology, 150*, 103473.

Hadap, S., & Patil, M. (2024). Quantum computing in AI: A paradigm shift. *International Journal of Advanced Research in Science, Communication and Technology*, 530–534.

Haider, S. (2024). Reshaping healthcare: Advancements in AI across clinical trials, precision therapies, and diagnostics. *Journal of Applied Sciences and Clinical Practice, 5*(2), 120–121.

Haque, Z., Hanif, F., & Owais, M. (2024). AI in biomedical research: Unleashing the potential of a transformative partnership. *JPMA. The Journal of the Pakistan Medical Association, 74*(1 (Supple-2)), S1.

Hassan, M., Borycki, E. M., & Kushniruk, A. W. (2025). AI governance framework for healthcare. In *Healthcare management forum* (Vol. 38, pp. 125–130). SAGE Publications.

Hirani, R., Noruzi, K., Khuram, H., Hussaini, A. S., Aifuwa, E. I., Ely, K. E., & Etienne, M. (2024). AI and healthcare: A journey through history, present innovations, and future possibilities. *Lifestyles, 14*(5), 557.

Huang, B., & Gao, B. (2024). AI in Medical Imaging. *iRadiology, 2*(6), 525–526.

Hurd, T. C., Payton, F. C., & Hood, D. B. (2024). Targeting machine learning and AI algorithms in health care to reduce bias and improve population health. *The Milbank Quarterly, 102*(3), 577–604.

Hussein, R., Zink, A., Ramadan, B., Howard, F. M., Hightower, M., Shah, S., & Beaulieu-Jones, B. K. (2024). Advancing healthcare AI governance: A comprehensive maturity model based on systematic review. *medRxiv*, 2024-12.

Ivanisevic, T., & Sewduth, R. N. (2023). Multi-omics integration for the design of novel therapies and the identification of novel biomarkers. *Proteomes, 11*(4), 34.

Jain, P., Pateria, N., Anjum, G., Tiwari, A., & Tiwari, A. (2023). Edge AI and on-device machine learning for real time processing. *International Journal of Innovative Research in Computer and Communication Engineering, 12*, 8137–8146.

Jiang, F., Jiang, Y., Zhi, H., Dong, Y., Li, H., Ma, S., & Wang, Y. (2017). AI in healthcare: Past, present and future. *Stroke and Vascular Neurology, 2*(4).

Johnson, K. B., Wei, W. Q., Weeraratne, D., Frisse, M. E., Misulis, K., Rhee, K., & Snowdon, J. L. (2021). Precision medicine, AI, and the future of personalized health care. *Clinical and Translational Science, 14*(1), 86–93.

Kass-Hout, T. A., & Alhinnawi, H. (2013). Social media in public health. *British Medical Bulletin, 108*(1), 5–24.

Khosravi, P., Kazemi, E., Zhan, Q., Malmsten, J. E., Toschi, M., Zisimopoulos, P., & Hajirasouliha, I. (2019). Deep learning enables robust assessment and selection of human blastocysts after in vitro fertilization. *NPJ Digital Medicine, 2*(1), 21.

Knobbe, D., Zwirnmann, H., Eckhoff, M., & Haddadin, S. (2022). Core processes in intelligent robotic lab assistants: Flexible liquid handling. In *2022 IEEE/RSJ international conference on intelligent robots and systems (IROS)* (pp. 2335–2342), Kyoto. IEEE *Explore*.

Kolachalama, V. B., & Garg, P. S. (2018). Machine learning and medical education. *NPJ Digital Medicine, 1*(1), 54.

Kraemer, M. U., Yang, C. H., Gutierrez, B., Wu, C. H., Klein, B., Pigott, D. M., & Scarpino, S. V. (2020). The effect of human mobility and control measures on the COVID-19 epidemic in China. *Science, 368*(6490), 493–497.

Krzyszczyk, P., Acevedo, A., Davidoff, E. J., Timmins, L. M., Marrero-Berrios, I., Patel, M., & Yarmush, M. L. (2018). The growing role of precision and personalized medicine for cancer treatment. *Technology, 6*(03n04), 79–100.

Kubben, P., Dumontier, M., & Dekker, A. (2019). *Fundamentals of clinical data science*. Springer Nature.

Kuchi Bhotla, H., Kaul, T., Balasubramanian, B., Easwaran, M., Arumugam, V. A., Pappusamy, M., Muthupandian, S., & Meyyazhagan, A. (2020). Platelets to surrogate lung inflammation in COVID-19 patients. *Medical Hypotheses, 143*, 110098.

Lee, T. H., Liu, P. S., Wang, S. J., Tsai, M. M., Shanmugam, V., & Hsieh, H. L. (2021). Bradykinin, as a reprogramming factor, induces transdifferentiation of brain astrocytes into neuron-like cells. *Biomedicine, 9*(8), 923.

Li, T., Sahu, A. K., Talwalkar, A., & Smith, V. (2020). Federated learning: Challenges, methods, and future directions. *IEEE Signal Processing Magazine, 37*(3), 50–60.

Liu, J., Fan, J., Zhong, Z., Qiu, H., Xiao, J., Zhou, Y., & Zhou, J. (2023). An ultra-low power reconfigurable biomedical ai processor with adaptive learning for versatile wearable intelligent health monitoring. *IEEE Transactions on Biomedical Circuits and Systems, 17*(5), 952–967.

Liu, X., Li, Y. I., & Pritchard, J. K. (2019). Trans effects on gene expression can drive omnigenic inheritance. *Cell, 177*(4), 1022–1034.

London, A. J. (2019). AI and black-box medical decisions: Accuracy versus explainability. *Hastings Center Report, 49*(1), 15–21.

Loureiro, A., & da Silva, G. J. (2019). CRISPR-Cas: Converting a bacterial defence mechanism into a state-of-the-art genetic manipulation tool. *Antibiotics, 8*(1), 18.

Maniaci, A., Lavalle, S., Gagliano, C., Lentini, M., Masiello, E., Parisi, F., & La Via, L. (2024). The integration of radiomics and AI in modern medicine. *Lifestyles, 14*(10), 1248.

Marques, M., Almeida, A., & Pereira, H. (2024). The medicine revolution through AI: Ethical challenges of machine learning algorithms in decision-making. *Cureus, 16*(9), e69405.

Martín-Cuevas, R., & Calleja, G. (2025). Hybrid quantum-classical computing architectures. In Mohammad Hammoudeh, Clinton M. Firth, Harbaksh Singh, Christoph Capellaro, Mohamed Al Kuwaiti (Eds.). *Quantum technology applications, impact, and future challenges* (pp. 97–106). CRC Press.

Mashar, M., Chawla, S., Chen, F., Lubwama, B., Patel, K., Kelshiker, M. A., & Peters, N. S. (2023). AI algorithms in health care: Is the current food and drug administration regulation sufficient? *JMIR AI, 2*(1), e42940.

Melis, L., Song, C., De Cristofaro, E., & Shmatikov, V. (2019). Exploiting unintended feature leakage in collaborative learning. In *2019 IEEE symposium on security and privacy (SP)* (pp. 691–706) San Francisco. IEEE *Xplore*.

Meyyazhagan, A., Balasubramanian, B., Easwaran, M., Alagamuthu, K. K., Shanmugam, S., Kuchi Bhotla, H., Pappusamy, M., Arumugam, V. A., Thangaraj, A., Kaul, T., Keshavarao, S., & Cacabelos, R. (2020). Biomarker study of the biological parameter and neurotransmitter levels in autistics. *Molecular and Cellular Biochemistry, 474*(1–2), 277–284.

Meyyazhagan, A., Pushparaj, K., Balasubramanian, B., Kuchi Bhotla, H., Pappusamy, M., Arumugam, V. A., Easwaran, M., Pottail, L., Mani, P., Tsibizova, V., & Di Renzo, G. C. (2022). COVID-19 in pregnant women and children: Insights on clinical manifestations, complexities, and pathogenesis. *International Journal of Gynecology & Obstetrics, 156*(2), 216–224.

Mienye, I. D., Obaido, G., Jere, N., Mienye, E., Aruleba, K., & Emmanuel, I. D., & Ogbuokiri, B. (2024). A survey of explainable artificial intelligence in healthcare: Concepts, applications, and challenges. *Informatics in Medicine Unlocked, 51*, 101587.

Mo, S., & Liang, P. P. (2024). Multimed: Massively multimodal and multitask medical understanding. *arXiv preprint arXiv:2408.12682*.

Mohd, Y., Balasubramanian, B., Meyyazhagan, A., Kuchi Bhotla, H., Shanmugam, S. K., Ramesh Kumar, M. K., Pappusamy, M., Alagamuthu, K. K., Keshavarao, S., & Arumugam, V. A. (2021). Extricating the association between the prognostic factors of colorectal cancer. *Journal of Gastrointestinal Cancer*, 52(3), 022–1028.

Mohd, Y., Kumar, P., Kuchi Bhotla, H., Meyyazhagan, A., Balasubramanian, B., Ramesh Kumar, M. K., Pappusamy, M., Alagamuthu, K. K., Orlacchio, A., Keshavarao, S., Sampathkumar, P., & Arumugam, V. A. (2021). Transmission jeopardy of adenomatosis polyposis coli and methylenetetrahydrofolate reductase in colorectal cancer. *Journal of the Renin-Angiotensin-Aldosterone System*, 2021, 7010706.

Mohsen, F., Ali, H., El Hajj, N., & Shah, Z. (2022). AI-based methods for fusion of electronic health records and imaging data. *Scientific Reports*, 12(1), 17981.

Momani, A. (2025). Artificial intelligent implications on health data privacy and confidentiality. *arXiv preprint arXiv:2501.01639*.

Mushtaq, M., Mukhtar, M. A., Lapotre, V., Bhatti, M. K., & Gogniat, G. (2020). Winter is here! A decade of cache-based side-channel attacks, detection & mitigation for RSA. *Information Systems*, 92, 101524.

Myakala, P. K., Bura, C., & Jonnalagadda, A. K. (2025). Artificial immune systems: A bio-inspired paradigm for computational intelligence. *Journal of AI and Big Data*, 5(1), 10–31586.

Naudé, W. (2020). AI vs COVID-19: Limitations, constraints and pitfalls. *AI & Society*, 35(3), 761–765.

Ngiam, J., Khosla, A., Kim, M., Nam, J., Lee, H., & Ng, A. Y. (2011). Multimodal deep learning. *ICML*, 11, 689–696.

Olawade, D. B., Wada, O. Z., Odetayo, A., David-Olawade, A. C., Asaolu, F., & Eberhardt, J. (2024). Enhancing mental health with AI: Current trends and future prospects. *Journal of Medicine, Surgery, and Public Health*, 3, 100099.

Omair, A. O. M., Jabbar, A. M. A., & Albulushi, M. O. (2023). Recent advancements in laboratory automation technology and their impact on scientific research and laboratory procedures. *International Journal of Health Sciences*, 7(S1), 3043–3052.

Onesimu, J. A., Karthikeyan, J., & Sei, Y. (2021). An efficient clustering-based anonymization scheme for privacy-preserving data collection in IoT based healthcare services. *Peer-to-Peer Networking and Applications*, 14(3), 1629–1649.

Palvadi, S. K., & Kadiravan, G. (2025). Future trends in bioinformatics AI integration. In Umesh Kumar Lilhore, Abhishek Kumar, Narayan Vyas, Sarita Simaiya, & Vishal Dutt (Eds.). *Multimodal data fusion for bioinformatics AI* (pp. 283–303). Wiley Online Library.

Pandey, H. M. (2024). AI in 0: Evolution, current applications, future challenges, and emerging evidence. *arXiv Preprint arXiv:2501.10374*.

Pandian, S., & Murugan, R. (2024). Unveiling the recent trends in biomedical AI research. *i-manager's Journal on Software Engineering*, 18(4).

Pandy, G., Pugazhenthi, V. J., Murugan, A., & Jeyarajan, B. (2025). AI-powered robotics and automation: Innovations, challenges, and pathways to the future. *European Journal of Computer Science and Information Technology*, 13(1), 33–44.

Paranitharan, N., Kataria, S., Arumugam, V. A., Hsieh, H., Muthukrishnan, S., & Velayuthaprabhu, S. (2025). Integrin α1 upregulation by tf:fviia complex promotes cervical cancer migration through par2-dependent mek1/2 activation. *Biochemical and Biophysical Research Communications*, 742, 151151.

Pasupuleti, V., Thuraka, B., & Kodete, C. S. (2024). Machine learning and big data analytics in smart cities and healthcare: Applications, challenges, and future directions. In *2024 international conference on computer and applications (ICCA)*, Cairo (pp. 01–06). IEEE *Xplore*.

Poursabzi-Sangdeh, F., Goldstein, D. G., Hofman, J. M., Wortman Vaughan, J. W., & Wallach, H. (2021, May). Manipulating and measuring model interpretability. In *Proceedings of the 2021 CHI conference on human factors in computing systems*, Yokohama (pp. 1–52). Association for Computing Machinery.

Prosperi, M., Bian, J., Harle, C. A., Wang, M., Lee, G., Li, Y., & Veltri, P. (2023). Challenges for AI regulation in health and for healthcare organizations: Notes from the University of Florida's NSF-sponsored workshop on AI governance. *ACM SIGBioinformatics Record*, 12(1), 1–3.

Pulicharla, M. R. (2023). Hybrid quantum-classical machine learning models: Powering the future of AI. *Journal of Science and Technology*, 4(1), 40–65.

Pushparaj, K., Kuchi Bhotla, H., Arumugam, V. A., Pappusamy, M., Easwaran, M., Liu, W. C., Issara, U., Rengasamy, K. R. R., Meyyazhagan, A., & Balasubramanian, B. (2022). Mucormycosis (black fungus) ensuing COVID-19 and comorbidity meets – Magnifying global pandemic grieve and catastrophe begins. *Science of the Total Environment*, 805, 150355.

Raghupathi, W., & Raghupathi, V. (2014). Big data analytics in healthcare: Promise and potential. *Health Information Science and Systems*, 2, 1–10.

Rajkomar, A., Dean, J., & Kohane, I. (2019). Machine learning in medicine. Reply. *The New England Journal of Medicine*, 380(26), 2589–2590.

Ramya, S., Poornima, P., Jananisri, A., Geofferina, I. P., Bavyataa, V., Divya, M., Priyanga, P., Vadivukarasi, J., Sujitha, S., Elamathi, S., Anand, A. V., & Balamuralikrishnan, B. (2023). Role of hormones and the potential impact of multiple stresses on infertility. *Stress*, 3(2), 454–474.

Sangeetha, T., Anand, A. V., & Begum, T. N. (2022). Assessment of inter-relationship between anemia and COPD in accordance with altitude. *Open Respiratory Medicine Journal*, 16, e187430642206270.

Sangeetha, T., Nargis Begum, T., Balamuralikrishnan, B., Arun, M., Rengasamy, K. R. R., Senthilkumar, N., Velayuthaprabhu, S., Saradhadevi, M., Sampathkumar, P., & Vijaya Anand, A. (2022). Influence of SERPINA1 gene polymorphisms on anemia and chronic obstructive pulmonary disease. *Journal of the Renin-Angiotensin-Aldosterone System*, 2022, 2238320.

Schneider, T., Altintas, I., & Atkins, D. (2024). Accelerating scientific discovery with AI-aided automation. *Computing in Science & Engineering*, 25(5), 27–30.

Schouten, D., Nicoletti, G., Dille, B., Chia, C., Vendittelli, P., Schuurmans, M., & Khalili, N. (2025). Navigating the landscape of multimodal AI in medicine: A scoping review on technical challenges and clinical applications. *Medical Image Analysis*, 105, 103621.

Sendak, M., Elish, M. C., Gao, M., Futoma, J., Ratliff, W., Nichols, M., & O'Brien, C. (2020). "The human body is a black box": Supporting clinical decision-making with deep learning. In *Proceedings of the 2020 conference on fairness, accountability, and transparency*, Barcelona (pp. 99–109). Association for Computing Machinery.

Shah, P., Thakkar, D., Panchal, N., & Jha, R. (2024). AI and machine learning in drug discovery. In Rati Kailash Prasad Tripathi and Shrikant Tiwari (Eds.). *Converging pharmacy science and engineering in computational drug discovery* (pp. 54–75). IGI Global.

Shanmugam, R., Thangavelu, S., Fathah, Z., Yatoo, M. I., Tiwari, R., Pandey, M. K., Dhama, J., Chandra, R., Malik, Y. S., Dhama, K., Sah, R., Chaicumpa, W., Shanmugam, V., & Arumugam, V. A. (2020). SARS-CoV-2/COVID-19 pandemic – An update. *Journal of Experimental Biology and Agricultural Sciences*, 8(Special Issue 1), S219–S245.

Shao, C., Wang, P., Liao, B., Gong, S., & Wu, N. (2023). Multi-omics integration analysis of TK1 in glioma: A potential biomarker for predictive, preventive, and personalized medical approaches. *Brain Sciences*, 13(2), 230.

Sharma, A., Al-Haidose, A., Al-Asmakh, M., & Abdallah, A. M. (2024). Integrating AI into biomedical science curricula: Advancing healthcare education. *Clinics and Practice*, 14(4), 1391–1403.

Singhal, S. (2024). Data privacy, compliance, and security including AI ML: Healthcare. In Pawan Whig, Sachinn Sharma, Seema Sharma, Anupriya Jain, & Nikhitha Yathiraju (Eds.). *Practical applications of data processing, algorithms, and modelling* (pp. 111–126). IGI Global.

Sissung, T. M., Goey, A. K., Ley, A. M., Strope, J. D., & Figg, W. D. (2014). Pharmacogenetics of membrane transporters: A review of current approaches. *Pharmacogenomics in Drug Discovery and Development*, 1175, 91–120.

Srivastav, A. K., Mishra, M. K., Lillard Jr, J. W., & Singh, R. (2025). Transforming pharmacogenomics and CRISPR gene editing with the power of AI for precision medicine. *Pharmaceutics*, 17(5), 555.

Subasi, A. (2024). AI techniques for healthcare and biomedicine. In Abdulhamit Subasi (Ed.). *Applications of AI healthcare and biomedicine* (pp. 1–35). Academic Press.

Tonekaboni, S., Joshi, S., McCradden, M. D., & Goldenberg, A. (2019). What clinicians want: Contextualizing explainable machine learning for clinical end use. In *Machine learning for healthcare conference*, Ann Arbor (pp. 359–380). PMLR.

Tong, L., Shi, W., Isgut, M., Zhong, Y., Lais, P., Gloster, L., & Wang, M. D. (2023). Integrating multi-omics data with EHR for precision medicine using advanced AI. *IEEE Reviews in Biomedical Engineering*, 17, 80–97.

Topol, E. (2019a). *Deep medicine: How AI can make healthcare human again*. Hachette UK.

Topol, E. J. (2019b). High-performance medicine: The convergence of human and AI. *Nature Medicine*, 25(1), 44–56.

Torkzadehmahani, R., Nasirigerdeh, R., Blumenthal, D. B., Kacprowski, T., List, M., Matschinske, J., & Baumbach, J. (2022). Privacy-preserving AI techniques in biomedicine. *Methods of Information in Medicine*, 61(S 01), e12–e27.

Tran, B. X., Vu, G. T., Ha, G. H., Vuong, Q. H., Ho, M. T., Vuong, T. T., & Ho, R. C. (2019). Global evolution of research in AI in health and medicine: A bibliometric study. *Journal of Clinical Medicine*, 8(3), 360.

Tran, D., Cooke, S., Illingworth, P. J., & Gardner, D. K. (2019). Deep learning as a predictive tool for fetal heart pregnancy following time-lapse incubation and blastocyst transfer. *Human Reproduction*, 34(6), 1011–1018.

Tschandl, P., Rinner, C., Apalla, Z., Argenziano, G., Codella, N., Halpern, A., & Kittler, H. (2020). Human–computer collaboration for skin cancer recognition. *Nature Medicine*, 26(8), 1229–1234.

Ueda, D., Kakinuma, T., Fujita, S., Kamagata, K., Fushimi, Y., Ito, R., & Naganawa, S. (2024). Fairness of AI in healthcare: Review and recommendations. *Japanese Journal of Radiology*, 42(1), 3–15.

Utomi, E., Osifowokan, A. S., Donkor, A. A., & Yowetu, I. A. (2024). Evaluating the impact of data protection compliance on AI development and deployment in the US health sector. *World Journal of Advanced Research and Reviews*, 24(2), 1100–1110.

Van Laerhoven, K., Lo, B. P., Ng, J. W., Thiemjarus, S., King, R., Kwan, S., & Yang, G.-Z. (2004). Medical healthcare monitoring with wearable and implantable sensors. In *Proc. of the 3rd International Workshop on Ubiquitous Computing for Pervasive Healthcare Applications*, Paphos, 2006 (pp. 1–6) IWUC.

Varghese, J. E., Shanmugam, V., Rengarajan, R. L., Meyyazhagan, A., Arumugam, V. A., Al-Misned, F. A., & El-Serehy, H. A., (2020). Role of vitamin D3 on apoptosis and inflammatory-associated gene in colorectal cancer: An in vitro approach. *Journal of King Saud University – Science*, 32(6), 2786–2789.

Velayuthaprabhu, S., & Archunan, G. (2005). Evaluation of anticardiolipin antibodies and antiphosphatidylserine antibodies in women with recurrent abortion. *Indian Journal of Medical Sciences* 59, 347–352.

Velayuthaprabhu, S., Archunan, G., & Balakrishnan, K. (2007). Placental thrombosis in experimental anticardiolipin antibodies-mediated intrauterine fetal death. *American Journal of Reproductive Immunology*, 57, 270–276.

Velayuthaprabhu, S., Matsubayashi, H., Archunan, G. (2016). Beta-2 GPI induced tissue factor and placental apoptosis for the pathophysiology of pregnancy loss in antiphospholipid syndrome. *International Journal of Research in Medical Sciences*, 4(8), 3109–3113.

Velayuthaprabhu, S., Matsubayashi, H., Sugi, T., Nakamura, M., Ohnishi, Y., Ogura, T., Tomiyama, T., & Archunan, G. (2011). A unique preliminary study on fetal resorption and placental apoptosis in mice with passive immunization of anti-phosphatidylethanolamine antibodies and anti-Factor XII antibodies. *American Journal of Reproductive Immunology*, 66, 373–384.

Velayuthaprabhu, S., Matsubayashi, H., Sugi, T., Nakamura, M., Ohnishi, Y., Ogura, T., & Archunan, G. (2013). Expression of apoptosis in placenta of experimental antiphospholipid syndrome (APS) mouse. *American Journal of Reproductive Immunology*, 69(5), 486–494.

Vrettos, K., Triantafyllou, M., Marias, K., Karantanas, A. H., & Klontzas, M. E. (2024). AI-driven radiomics: Developing valuable radiomics signatures with the use of AI. *BJR | Artificial Intelligence*, 1(1), ubae011.

Wang, B., & Li, L. (2023). Research progress in biomedical big data. In Dong-Qing Ye (Ed.). *Progress in China epidemiology*, Vol. 1 (pp. 391–400). Singapore: Springer.

Wang, L. L., Lo, K., Chandrasekhar, Y., Reas, R., Yang, J., Burdick, D., & Kohlmeier, S. (2020). Cord-19: The covid-19 open research dataset. *ArXiv*, arXiv-2004.

Wang, R. Y., & Strong, D. M. (1996). Beyond accuracy: What data quality means to data consumers. *Journal of Management Information Systems*, 12(4), 5–33.

Wang, T., Shao, W., Huang, Z., Tang, H., Zhang, J., Ding, Z., & Huang, K. (2020). MORONET: Multi-omics integration via graph convolutional networks for biomedical data classification. *bioRxiv*, 2020-07.

Whetzel, P. L., Noy, N. F., Shah, N. H., Alexander, P. R., Nyulas, C., Tudorache, T., & Musen, M. A. (2011). BioPortal: Enhanced functionality via new web services from the National Center for Biomedical Ontology to access and use ontologies in software applications. *Nucleic Acids Research*, 39(suppl_2), W541–W545.

Xie, Y., Zhai, Y., & Lu, G. (2025). Evolution of AI in healthcare: A 30-year bibliometric study. *Frontiers in Medicine*, 11, 1505692.

Xue, C., & Greene, E.C. (2021). DNA repair pathway choices in CRISPR-Cas9-mediated genome editing. *Trends in Genetics* 2021, 37, 639–656.

Yang, A., Pan, M. L., Lu, H. H. S., Lien, C. Y., Wang, D. W., & Chen, C. H. (2024). Assessing the evolution and influence of medical open databases on biomedical research and healthcare innovation: A 25-year perspective with a focus on privacy and privacy-enhancing technologies. *JMIR, Canada*, 194.

Yang, Y., Lin, M., Zhao, H., Peng, Y., Huang, F., & Lu, Z. (2024). A survey of recent methods for addressing AI fairness and bias in biomedicine. *Journal of Biomedical Informatics*, 104646.

Yekaterina, K. (2024). Challenges and opportunities for AI in healthcare. *International Journal of Public Law and Policy*, 2, 11–15.

Zhao, N., Guo, M., Wang, K., Zhang, C., & Liu, X. (2020). Identification of pan-cancer prognostic biomarkers through integration of multi-omics data. *Frontiers in Bioengineering and Biotechnology*, 8, 268.

3 AI Technology-Powered Precision Medicine and Smart Healthcare

Technical Advancements and Prospects

Kazi Asraf Ali, Sabyasachi Choudhuri, Rideb Chakraborty, Anindya Pradhan, Amit Kotal, and Md Adil Shaharyar

3.1 INTRODUCTION

The advent of precision medicine and smart healthcare has revolutionized the traditional paradigms of disease diagnosis, treatment, and prevention. Precision medicine, an innovative approach that tailors medical interventions to individual patient characteristics, leverages genetic, environmental, and lifestyle data to optimize therapeutic outcomes. Unlike the conventional "one-size-fits-all" model, precision medicine emphasizes personalized care,[1] enabling clinicians to predict disease susceptibility, customize treatment regimens, and improve patient prognosis with unprecedented accuracy. Smart healthcare, on the other hand, integrates advanced digital technologies such as artificial intelligence (AI), the Internet of Medical Things (IoMT), big data analytics, and wearable devices[2,3] to create a seamless, data-driven healthcare ecosystem. This paradigm shift enhances real-time patient monitoring, automated diagnostics, and predictive analytics, fostering proactive rather than reactive medical interventions. The convergence of precision medicine and smart healthcare has thus set the stage for a transformative era in medicine, where data-driven insights and AI-powered tools synergize to deliver superior patient care.[4,5]

AI has become a fundamental component of contemporary healthcare, providing exceptional capabilities in analyzing data, identifying patterns, and supporting clinical decisions. Machine learning (ML) algorithms, deep learning (DL) models, and natural language processing (NLP) methods are progressively utilized to examine large-scale datasets, including electronic health records (EHRs), medical images, genomic data, and information from wearable sensors.[6] AI's capacity to derive valuable insights from intricate and diverse datasets has made it indispensable in precision medicine, where individualized treatment strategies combine multi-omics information (such as genomics, proteomics, and metabolomics) with clinical data. In infectious disease management, AI has demonstrated remarkable potential in early pathogen detection, drug repurposing, and outbreak prediction. For instance, AI-driven diagnostic tools can rapidly identify microbial pathogens from sequencing data, while predictive models can forecast disease spread based on epidemiological trends.

Furthermore, AI-powered decision-support systems assist clinicians in selecting optimal antimicrobial therapies by analyzing resistance patterns and patient-specific factors, thereby mitigating the global threat of antimicrobial resistance (AMR). Beyond diagnostics and therapeutics, AI enhances smart healthcare through intelligent automation, robotic-assisted surgeries, and virtual health assistants. These advancements boost clinical efficiency, reduce errors and costs, and improve care access, especially in low-resource areas. As AI advances, its fusion with precision medicine is set to transform healthcare into a more predictive, preventive, and personalized system.[7,8]

This chapter examines the revolutionary role of AI-driven precision medicine and intelligent healthcare, emphasizing their application in managing infectious diseases as illustrated in Figure 3.1.

The primary objectives are to elucidate the technological advancements driving AI in precision medicine, including ML algorithms, deep neural networks, and federated learning for secure data sharing, to examine AI applications in infectious disease diagnostics and treatment, covering next-generation sequencing (NGS), AI-based imaging analysis, and computational drug discovery. This chapter highlights smart healthcare's role in real-time disease tracking, remote monitoring, and AI-powered telemedicine. It also addresses challenges like data privacy, bias, and regulation while exploring future directions such as quantum computing, blockchain for data security, and AI-assisted vaccine development. The chapter also offers a concise overview of how AI is transforming precision medicine and smart healthcare, focusing on improving infectious disease management. It explores emerging trends, highlights AI's potential to enhance diagnostics, treatment, and public health response, and addresses key technical, ethical, and regulatory challenges. The goal is to guide researchers, clinicians, and policymakers in leveraging AI responsibly and effectively.

3.2 FUNDAMENTALS OF AI IN HEALTHCARE

AI has become a game-changer in healthcare, reshaping diagnostics, personalized treatment, and overall patient care. By leveraging computational techniques that mimic human cognition, AI

DOI: 10.1201/9781003615699-3

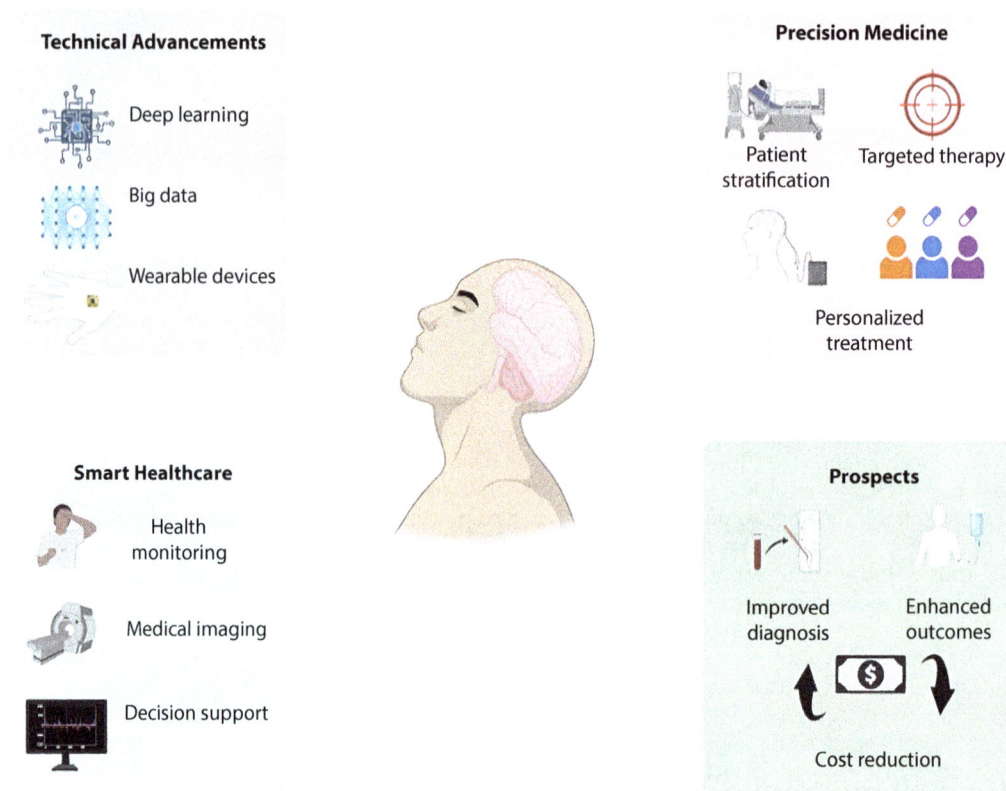

Figure 3.1 Advancements and prospects of AI technology-powered precision medicine and smart healthcare.

enables the analysis of vast and complex medical datasets with unprecedented accuracy and efficiency. The core AI technologies driving this paradigm shift include ML, DL, NLP, and big data analytics. These tools facilitate precision medicine by extracting actionable insights from heterogeneous healthcare data, ultimately improving clinical decision-making and patient outcomes.

3.3 MACHINE LEARNING AND DEEP LEARNING IN HEALTHCARE

ML is a branch of AI that utilizes algorithms to detect patterns within data without explicit programming. In healthcare, ML models are trained on structured (e.g., EHRs, genomic data) and unstructured (e.g., medical images, clinician notes) datasets to perform tasks such as disease diagnosis, supervised learning models, including support vector machines (SVMs) and random forests, classify diseases based on clinical and imaging data. Predictive analytics and time-series forecasting models predict disease progression, hospital readmissions, and patient deterioration. Personalized treatment and reinforcement learning methods optimize therapeutic regimens by analyzing patient responses to interventions.[8–12]

3.4 DEEP LEARNING FOR ADVANCED MEDICAL IMAGING AND DIAGNOSTICS

DL has revolutionized medical imaging and diagnostics by enabling automated, high-precision analysis of complex biomedical data. Convolutional neural networks (CNNs) and transformer-based architectures have demonstrated exceptional performance in detecting, segmenting, and classifying anomalies in radiology, pathology, and ophthalmology. For instance, DL models trained on large-scale datasets (e.g., ChestX-ray14, MIMIC-CXR) have achieved radiologist-level accuracy in diagnosing pneumonia, tuberculosis, and COVID-19 from X-rays and CT scans. In infectious disease management, DL enhances early detection by identifying subtle imaging biomarkers that often overlooked in manual assessments. Techniques like U-Net and Mask R-CNN facilitate precise lesion segmentation, aiding in disease progression monitoring. Moreover, generative adversarial networks

(GANs) can synthesize high-resolution medical images for data augmentation, addressing the challenge of limited annotated datasets in low-resource settings. Beyond structural imaging, DL integrates multi-modal data (e.g., genomics, EHRs) for holistic diagnostics. Recurrent neural networks (RNNs) and attention mechanisms enable temporal analysis of longitudinal imaging data, improving prognostic predictions. Explainable AI (XAI) methods, such as Grad-CAM and SHAP, provide interpretable insights into model decisions, fostering clinician trust and regulatory compliance.[13–15]

3.5 NATURAL LANGUAGE PROCESSING FOR MEDICAL DATA ANALYSIS

NLP has emerged as a transformative tool in medical data analysis, enabling the extraction of meaningful insights from unstructured clinical texts, such as EHRs, medical literature, and physician notes. Traditional healthcare data analysis often struggles with the heterogeneity and complexity of unstructured text. However, NLP techniques, powered by advanced ML and DL models, facilitate efficient data interpretation, decision support, and predictive analytics. One of the primary applications of NLP in medicine is clinical text mining, where algorithms parse and categorize information from EHRs to identify diagnoses, treatments, and patient outcomes. Techniques like named entity recognition (NER) and relation extraction help structure free-text data into standardized formats, enhancing interoperability and enabling large-scale analytics. For instance, NLP models can automatically extract symptoms, lab results, and medication histories, aiding in early disease detection and personalized treatment planning. Another critical use case is automated medical coding, where NLP converts physician notes into standardized billing codes (e.g., ICD-10),[16–18] reducing administrative burdens and minimizing errors. Furthermore, NLP-powered sentiment analysis can assess patient feedback from surveys or social media, offering insights into patient experiences and healthcare service quality.

3.6 CHALLENGES IN MEDICAL NLP

Medical NLP transforms unstructured clinical text into actionable insights for precision medicine and smart healthcare. However, several technical and domain-specific challenges hinder its widespread adoption and reliability in clinical settings. Clinical narratives exhibit vast structure, terminology, and linguistic style variability across different healthcare systems, specialties, and languages. Variations in abbreviations, acronyms, and colloquial expressions further complicate automated interpretation, necessitating robust normalization techniques. High-quality labeled medical corpora are scarce due to the need for expert annotation, patient privacy concerns, and data siloing across institutions. Supervised learning models, which rely on large, annotated datasets, often underperform in low-resource scenarios, necessitating the use of semi-supervised or few-shot learning approaches. Medical text contains complex biomedical jargon, temporal expressions (e.g., "two weeks post-admission"), and implicit contextual dependencies. Pre-trained language models (e.g., BERT, GPT) often struggle to understand rare medical terms, requiring specialized fine-tuning on domain-specific corpora like PubMed or MIMIC-III. NLP systems processing EHRs must comply with stringent regulations (e.g., the Health Insurance Portability and Accountability Act (HIPAA) and the General Data Protection Regulation (GDPR)). De-identification is challenging, as re-identification risks persist even after anonymization, necessitating advanced privacy-preserving NLP techniques such as federated learning. Many medical NLP models are trained on data from specific demographics or institutions, resulting in biases related to race, gender, or socioeconomic status. Ensuring fairness and achieving generalizability across diverse populations remain critical challenges.[16,19,20] Clinical decision-support systems require real-time NLP capabilities, however latency issues often emerge due to computational complexity. Seamless integration with EHRs and interoperability with existing healthcare IT infrastructure pose additional hurdles.

3.7 BIG DATA IN HEALTHCARE: OPPORTUNITIES AND CHALLENGES

Integrating big data analytics into healthcare has revolutionized precision medicine by enabling data-driven decision-making, predictive modeling, and personalized treatment strategies. The exponential growth of healthcare data – from EHRs, genomic sequences, wearable device outputs, and medical imaging – provides unprecedented opportunities for AI-powered diagnostics and therapeutic interventions. Advanced ML algorithms can analyze vast, heterogeneous datasets to uncover hidden patterns, predict disease progression, and optimize treatment protocols, particularly in infectious disease management. For instance, big data analytics facilitates real-time outbreak surveillance, pathogen genomics, and drug-resistance profiling, enhancing early intervention strategies. However, leveraging big data in healthcare presents significant challenges. Data heterogeneity, inconsistent formats, and interoperability issues hinder seamless integration across healthcare systems. Ensuring data quality, accuracy, and completeness remains a critical barrier, as biased or incomplete datasets may lead to erroneous AI predictions.

Additionally, the sheer volume of data demands robust computational infrastructure and scalable storage solutions, posing logistical and financial constraints.[19,21–23] Privacy and security concerns further complicate data sharing, necessitating stringent compliance with regulations such as the GDPR and HIPAA. Ethical considerations, including algorithmic bias and equitable access to AI-driven healthcare, must also be addressed to prevent disparities in treatment outcomes.

3.8 AI IN PRECISION MEDICINE

3.8.1 Genomic and Multi-omics Data Integration

The complexity and volume of omics data – encompassing genomics, transcriptomics, proteomics, metabolomics, and epigenomics – necessitate advanced analytical tools capable of multi-dimensional data processing.[24] For example, deep autoencoders can learn hierarchical representations from high-dimensional gene expression data, facilitating the identification of biomarkers and therapeutic targets. Bayesian networks and multi-view ML models enable researchers to integrate heterogeneous omics layers, enabling a more holistic understanding of disease biology.

In cancer genomics, AI tools such as DeepVariant by Google have demonstrated high accuracy in variant calling from sequencing data, which is crucial for identifying somatic mutations relevant to personalized cancer therapy.[25] Integration platforms like PANDAOmics[26] and tranSMART[27] are also gaining traction for multi-omics data analysis in clinical settings.

3.8.2 Predictive Analytics for Disease Risk and Diagnosis

Predictive modeling is central to the AI-enabled precision medicine paradigm. These models are trained on diverse datasets, including EHRs, medical imaging, lifestyle factors, and genetic data, to identify individuals at risk of developing diseases before clinical symptoms emerge.

ML classifiers such as random forests, XGBoost, and SVM have been widely used in predictive analytics for conditions such as cardiovascular disease, Alzheimer's disease, and cancer.[28] For instance, AI models analyzing retinal images in combination with EHR data can predict the five-year risk of cardiovascular events with performance comparable to that of traditional scoring systems.[29]

In oncology, AI-based radiomics extract features from medical images (CT, MRI, PET) to predict tumor aggressiveness and therapeutic response. PathAI and Tempus are AI-driven platforms that analyze histopathology and molecular data thereby assisting oncologists in early and accurate diagnosis.[30]

3.9 AI IN PERSONALIZED TREATMENT PLANNING AND DRUG DEVELOPMENT

AI systems are increasingly being employed to develop personalized treatment strategies. Reinforcement learning and genetic algorithms help optimize treatment pathways by considering patient-specific parameters such as genomic profiles, comorbidities, and prior treatment outcomes.[31] Furthermore, AI has revolutionized drug discovery pipelines by predicting drug–target interactions, bioavailability, toxicity, and efficacy.[32] For instance, AlphaFold2 by DeepMind[33] predicts protein structures with unprecedented accuracy, thereby accelerating target identification. Insilico Medicine,[34] Atomwise,[35] and BenevolentAI[36] leverage deep generative models to design novel molecules, thereby reducing both time and costs in early-phase drug development. Applications of AI technologies across various domains using different tools are summarized in Table 3.1.

3.10 SMART HEALTHCARE TECHNOLOGIES

3.10.1 AI-Powered Wearables and Remote Monitoring

Advanced AI models such as long short-term memory (LSTMs) and temporal convolutional networks (TCNs) detect patterns over time, enabling real-time anomaly detection.[43,44] For example, Apple Watch and Fitbit Sense utilize photoplethysmography (PPG) signals and AI algorithms to detect atrial fibrillation, stress levels, and blood-oxygen saturation variations.[45] In chronic disease management, AI-powered wearables like FreeStyle Libre (for diabetes) and Zio Patch (for arrhythmia) allow physicians to monitor patients remotely, thereby facilitating timely intervention and reducing hospital readmissions.[46]

3.10.2 Virtual Health Assistants and Chatbots

Virtual health assistants (VHAs) are AI-driven software tools that communicate with users via voice- or text-based interfaces.[47] These systems use NLP, sentiment analysis, and knowledge graphs to interpret and respond to health-related queries.

TABLE 3.1 AI Applications in Precision Medicine

Domain	AI Technique	Application	Example Tools/ Platforms	Reference
Genomic analysis	DL, CNNs	Variant calling, mutation detection	DeepVariant, DNAscope	37
Multi-omics integration	Multi-view ML, Bayesian networks	Biomarker identification, disease subtyping	PANDAOmics, tranSMART	38
Disease prediction	Random forest, XGBoost	Risk modeling, early detection	Framingham AI model, HealthMap	39
Imaging and pathology	Radiomics, CNNs	Tumor classification, therapy-response prediction	PathAI, Tempus	40
Drug discovery	GANs, reinforcement learning	Molecule generation, virtual screening	AlphaFold2, Atomwise	41
Personalized treatment planning	Genetic algorithms, NLP	Treatment optimization, patient stratification	IBM Watson for Oncology	42

For instance, Babylon Health integrates a chatbot with symptom triage-tools to deliver initial consultations and health recommendations. Ada Health employs a medical reasoning engine to suggest diagnoses based on user input. These tools enhance healthcare accessibility and reduce the burden on clinical staff, particularly in underserved regions. Additionally, in the field of mental health, AI chatbots like Woebot provide cognitive behavioral therapy (CBT)-based interventions, offering support for individuals with anxiety and depression.[48]

3.10.3 Robotics and Automation in Patient Care

Robotics, underpinned by AI, is increasingly being integrated into surgical, rehabilitative, and elderly care. In surgery, robotic systems like da Vinci enhance the surgeon's capabilities with high-precision instrumentation, tremor reduction, and 3D visualization.[49] AI augments these systems by enabling motion planning, tissue differentiation, and real-time feedback.

In rehabilitation, socially assistive robots (SARs) like Paro and Pepper engage patients both cognitively and emotionally, which is particularly beneficial for dementia and autism care.[50] AI-powered exoskeletons support mobility-impaired individuals in walking and therapy exercises, adapting dynamically to user feedback.[51] AI also enables automation in medication dispensing (e.g., PillPack), clinical documentation, patient triaging, and administrative workflows, thereby contributing to operational efficiency in healthcare systems.[52] Recent AI-driven smart healthcare technologies are summarized in Table 3.2.

3.11 CASE STUDIES AND REAL-WORLD APPLICATIONS

3.11.1 Early Disease Detection and Diagnosis

CNNs, a type of DL model, have demonstrated exceptional performance in medical imaging applications.[55,56] A 3D CNN developed by Ardila et al. (2019) detected lung nodules from low-dose CT scans with an AUC of 94.4%, matching the performance of professional radiologists and reducing false positives by 11%.[57] A DL system trained on more than 14,000 optical coherence tomography (OCT) images achieved diagnostic accuracy above 94% in identifying over 50 retinal abnormalities, as reported by De Fauw et al. (2018) in ophthalmology.[58] AI has improved prognostic accuracy in critical care by integrating clinical and physiological data in multiple ways. For example, Shashikumar et al. (2017) proposed DeepSOFA, an ICU-based AI model capable of diagnosing sepsis up to six hours in advance, thereby exceeding standard scoring methods in mortality prediction.

AI has demonstrated significant value in mobile diagnostics, mental health, and genetics in addition to traditional imaging.[59] DeepGestalt, a facial-analysis model developed by Gurovich et al. (2019), supported early rare illness diagnosis by identifying over 200 rare genetic disorders with 91% top-ten accuracy (10.3390/s21196595). Cummins et al. (2018) demonstrated that multi-modal AI models that use physiological, facial, and acoustic inputs could diagnose depression with over 80% classification accuracy in mental health.[60-62] Furthermore, even with few resources, high-sensitivity diagnostic testing for illnesses like COVID-19 and malaria is now possible thanks to AI-powered smartphone apps. These applications highlight how AI is increasingly enhancing diagnostic reach and accuracy, especially in early intervention settings and among underrepresented populations.[63,64] Applications of AI in early disease detection are discussed in Table 3.3.

TABLE 3.2 AI-Driven Smart Healthcare Technologies

Technology	AI Functionality	Application	Example Systems	Reference
Wearables & biosensors	Time-series analysis, anomaly detection	Vital-sign monitoring, chronic-disease tracking	Apple Watch, FreeStyle Libre	45
Virtual health assistants	NLP, Chatbot algorithms, Sentiment analysis	Symptom triage, mental health support	Babylon Health, Ada, Woebot	48
Surgical robotics	Image-guided AI, Precision-motion control	Minimally invasive surgery	da Vinci Surgical System, ROSA	53
Rehabilitation & elder care	Reinforcement learning, emotion AI	Assistive care, cognitive stimulation	Paro, Pepper, ReWalk	54
Automation in workflow	Computer vision, NLP, RPA	Clinical documentation, healthcare logistics	Suki, Olive, Amazon PillPack	52

TABLE 3.3 AI in Early Disease Detection

Medical Domain	Disease(s) Detected	AI Techniques Used	Key Benefits	References
Oncology	Breast cancer, lung cancer, colorectal cancer	DL (CNNs), SVM, Random Forest	Early tumor detection, reduced false positives/ negatives, non-invasive screening	65–67
Neurology	Alzheimer's disease, Parkinson's disease, brain tumors	CNN, ANN, DaTscan analysis, MRI pattern recognition	Detection of subtle biomarkers, early-stage intervention	68, 69
Cardiology	Arrhythmia, cardiomyopathy, coronary artery disease	AI-ECG (Aire), SVM, ANN, echo analysis	Prediction of cardiac events, structural diagnosis, risk stratification	70
Respiratory Medicine	Pneumonia, COPD, COVID-19, asthma	Chest X-ray/CT with CNNs, federated learning	Fast diagnosis, triage support, outbreak prediction	71–73
Endocrinology	Type 1 and Type 2 diabetes, thyroid dysfunction	Logistic regression, decision trees, ANN	Risk prediction, continuous glucose monitoring, lifestyle pattern analysis	74
Ophthalmology	Diabetic retinopathy, macular degeneration	DeepMind CNN, fundus image analysis	High diagnostic accuracy, ophthalmologist-level diagnosis, rural healthcare accessibility	75–77
Genetic Disease	Inherited metabolic and chromosomal disorders	Genomic ML models, feature-selection algorithms	Early carrier detection, prenatal risk assessment	78, 79
Mental Health	Depression, anxiety, PTSD	NLP, Voice, and facial recognition, behavioral AI	Early warning systems, digital therapy guidance	80
Dermatology	Skin cancers (melanoma, carcinoma), eczema	Computer vision, CNNs, mobile diagnostic tools	Rapid screening, high diagnostic accuracy, accessible via smartphones	81, 82
Infectious Disease	Tuberculosis, hepatitis, COVID-19	X-ray AI tools, cough analysis, federated DL	Early detection, contactless diagnostics, global pandemic support	83, 84
Multi-modal Diagnostics	Multisystem disease (e.g., sepsis, cancer + comorbidities)	Data fusion, ensemble learning, multi-modal networks	Holistic patient assessment, precision medicine	85, 86

3.11.2 Personalized Cancer Therapy

AI is changing personalized cancer therapy by leveraging genomic, proteomic, and phenotypic tumor data to enable targeted and individualized treatment approaches.[87] DL algorithms have been successfully used for large-scale datasets such as The Cancer Genome Atlas (TCGA), efficiently stratifying breast cancer subtypes and guiding precision medicine selection.[88] Kourou et al.

(2015) showed that ML algorithms, including SVMs and random forests, could predict cancer prognosis and treatment outcomes with accuracies ranging from 85% to 95%, indicating the therapeutic value of AI-based stratification.[89] Obermeyer et al. (2016) further validated AI's predictive powers by training a model using oncology patient records, which improved six-month mortality prediction by 20% above traditional approaches.[90] AI also optimizes medication response prediction by integrating multi-omics and clinical data, offering patient-specific guidance for chemotherapy, targeted therapy, and immunotherapy.[91]

AI-powered breakthroughs such as liquid biopsy have dramatically increased non-invasive cancer monitoring. Wan et al. (2019) demonstrated that ML models could classify circulating tumor DNA (ctDNA) with over 90% sensitivity, allowing early recurrence diagnosis in colorectal and breast cancer patients.[92] In radiomics, Aerts et al. (2014) retrieved over 400 features from CT images to predict lung cancer survival noninvasively, emphasizing AI's importance in imaging biomarkers.[93] AI also accelerates medication discovery; for example, Chen et al. (2021) employed knowledge graph embeddings to identify repurposable Food and Drug Administration (FDA)-approved drugs for prostate cancer.[94] Additionally, AI is increasingly used to optimize immunotherapy by predicting biomarkers such as PD-L1 expression and tumor mutational burden, and it enhances computational pathology by detecting tumor micro-environmental features in histopathology slides, thereby refining both diagnosis and treatment planning.[63,64,95]

3.11.3 Chronic Disease Management and Rehabilitation

AI has revolutionized chronic disease management by providing early identification, real-time monitoring, and personalized self-care. AI-integrated wearable devices continuously track vital signs such as heart rate and glucose levels, boosting early diagnosis of critical conditions.[96] For example, Steinhubl et al. (2015) revealed that wearable ECG monitors linked with AI significantly improved early diagnosis of atrial fibrillation, enabling earlier intervention.[97] Similarly, AI-driven mobile health (mHealth) applications improve disease management through individualized coaching and reminders. Almirall et al. (2020) found a 1.2% reduction in HbA1c levels and an 89% improvement in adherence among diabetic patients using an AI-powered app.[98] VHAs, including AI chatbots, have expanded access to remote chronic care, proving particularly effective during the COVID-19 pandemic in reducing hospital admissions and maintaining continuity of care.[99]

In rehabilitation, AI technologies are accelerating recovery processes and improving therapy outcomes. Radhakrishnan et al. (2016) found that AI-powered robotic gait trainers enhanced mobility and decreased rehabilitation time by up to 20% in stroke patients.[100] AI also aids individualized nutrition interventions; Zeevi et al. (2015) demonstrated that integrating microbiome data with ML allowed accurate prediction of individual glycemic responses, leading to optimal dietary recommendations.[101] Remote monitoring systems and AI-guided video tools now offer home-based physical therapy and chronic disease management. Suresh et al. (2022) created an AI-guided motion capture system that improved patient compliance by 30% and substantially boosted recovery results.[102] These achievements illustrate AI's rising importance in allowing patient-centered, continuous care beyond the clinical setting.

3.12 TECHNICAL ADVANCEMENTS AND TOOLS IN AI-DRIVEN HEALTHCARE

3.12.1 Advanced Algorithms and Models

Recent breakthroughs in AI have provided a set of powerful algorithms that are transforming biomedical research and clinical decision-making. DL models such as RNNs and GANs are increasingly applied in diagnostic and predictive medicine.[103,104] Esteban et al. (2017) found that long short-term memory (LSTM) RNNs could predict in-hospital mortality using ICU time-series data with AUCs between 0.84 and 0.88, outperforming standard models.[105] GANs have been employed in medical imaging to enrich training datasets; Frid-Adar et al. (2018) reported a 5%–10% boost in liver lesion classification accuracy when GAN-generated synthetic images were used.[106] Ronneberger et al. (2015) also contributed to image segmentation with the U-Net architecture, achieving Intersection-over-Union (IoU) scores exceeding 0.90 for cell and lesion identification tasks.[107,108]

Beyond image and text analysis, AI is used to optimize therapeutic strategies and accelerate drug development.[109] Peng et al. (2018) employed reinforcement learning to improve sepsis treatment planning, exhibiting a 20% simulated improvement in survival rates.[110] Graph neural networks (GNNs) have been used by Zeng et al. (2020) to study protein-protein interactions, discovering novel therapeutic targets with approximately 85% accuracy.[111] For increased clinical inference, Prosperi et al. (2020) underlined the necessity of causal inference techniques such as directed acyclic graphs (DAGs) to mitigate bias in observational health research.[112] Chen et al. (2021) found that incorporating CT scans, laboratory findings, and clinical notes into multi-modal AI models increased sepsis prediction AUC by 7% compared with single-source models.[113] Finally, digital twin technology is gaining ground in clinical simulations; Bruynseels et al. (2018) highlighted its potential in critical care, where virtual patient replicas were used to model and optimize ventilation strategies, paving the way for personalized treatment planning.[114]

3.12.2 Ethical AI and Explainable AI in Medicine

As AI becomes more ingrained in healthcare decision-making, ensuring its ethical and transparent deployment is vital. Traditional DL models often act as "black boxes," restricting interpretability and potentially compromising confidence.[115] To address this, interpretability tools like SHAP (Lundberg & Lee, 2017) and LIME (Ribeiro et al., 2016) have been developed – SHAP quantifies individual feature contributions in models such as cancer risk estimators,[116] while LIME explains predictions in domains like dermatology by locally approximating black-box models.[117] Ethical frameworks, such as those provided by Morley et al. (2020), stress foundational principles like beneficence, justice, and accountability, notably stressing that AI trained on biased datasets can systematically penalize vulnerable people.[118] This topic was highlighted by Obermeyer et al. (2019), who identified racial prejudice in a key health risk algorithm, later addressed using fairness-aware AI techniques.[119] These examples underline the significance of infusing ethical concepts directly into model creation and evaluation.

In parallel, securing patient privacy and ensuring model accountability are gaining priority. Differential privacy, as postulated by Dwork and Roth (2014), has been successfully deployed in sensitive disciplines like genomics, preserving individual identities while keeping data utility.[120] Algorithmic auditing, discussed by Veale and Binns (2017), plays a significant role post-deployment by detecting model drift and maintaining fairness over time.[121] To guarantee that AI systems augment rather than replace human judgment, Topol (2019) emphasizes human-centered design, arguing for intuitive interfaces that enhance clinical workflows.[122] Regulatory entities such as the U.S. FDA are also taking active roles, as demonstrated in their 2021 AI/ML Action Plan, which defines supervision mechanisms such as pre-market review, post-market surveillance, and algorithm update protocols. Together, these discoveries form a robust framework for deploying AI in healthcare in a way that is transparent, equitable, and aligned with human values.[123]

3.12.3 Integration of IoT, Blockchain, and AI for Healthcare Security

Combining AI with emerging technologies, like the Internet of Things (IoT) and blockchain, revolutionizes healthcare by enhancing real-time monitoring, data security, and patient autonomy. IoT devices such as smartwatches and biosensors continuously collect health indicators, while AI analyzes this data to predict adverse events and enable early interventions.[124,125] For instance, Zhou et al. (2019) reported that AI-enhanced wearable monitoring reduced heart failure readmissions by 25%.[126] As highlighted by Shi et al. (2016), Edge computing supports these roles by enabling low-latency processing at the data source – vital for critical care and remote situations.[127] Federated learning, emphasized by Sheller et al. (2020), enables institutions to collectively train AI models without disclosing raw patient data, ensuring privacy while achieving high predictive accuracy (Dice coefficient >0.87).[128] Moreover, homomorphic encryption (Acar et al., 2018) provides secure AI calculations on encrypted data, protecting sensitive genomic and clinical information.[129]

Blockchain further strengthens this ecosystem by offering secure, decentralized, and tamper-proof medical record management. Kuo et al. (2017) demonstrated how blockchain-based EHR systems, integrated with smart contracts, mitigate risks of data tampering and promote interoperability among providers.[130] Nguyen et al. (2020) showed that such contracts also automate permission management, reducing administrative overhead by almost 40%.[131] AI's function extends into cybersecurity, where models like those studied by Choraś et al. (2021) detect network anomalies with over 95% accuracy, improving attack response times by 60%.[132] Decentralized identification solutions like MedRec (Azaria et al., 2016) further empower patients, giving them a choice over data access while maintaining compliance and interoperability.[133] These technologies offer an integrated, secure, and intelligent foundation for next-generation healthcare systems.

3.13 CHALLENGES AND ETHICAL CONSIDERATIONS

3.13.1 Data Privacy and Security

The emergence of precision health has markedly amplified the quantity and sensitivity of personal data created, transferred, and evaluated. Ensuring data privacy and security has therefore become a foundational requirement in developing precision health platforms. These platforms typically operate in a decentralized, multi-stakeholder ecosystems, where data from clinical, genomic, and behavioral sources are collected and processed. This introduces several technical challenges, including secure data storage, confidential computation, and compliance with privacy regulations. To address these issues, a suite of privacy-preserving and secure computation techniques has emerged, each with distinct strengths and limitations depending on the context and computational demands.[134]

Homomorphic encryption (HE) is a technology that enables calculations on encrypted data without decryption, hence maintaining the confidentiality of sensitive information during the computational process. It can be classified into three primary categories: partially homomorphic encryption (PHE) facilitates a single operation type (either addition or multiplication); somewhat homomorphic encryption (SWHE) permits a restricted number of both operations; fully homomorphic encryption (FHE) enables unrestricted operations on ciphertexts. While FHE offers the strongest privacy guarantees, its implementation is still hindered by high computational overhead, especially in environments where data is heterogeneous and encrypted using different keys. Additionally, HE does not inherently provide verifiability of computation results and is vulnerable to collusion attacks in collaborative settings where the same public key is shared among multiple parties.[135,136]

Multiparty computation (MPC) offers an alternative method for privacy-preserving computation, allowing many untrusting individuals to collaboratively compute a function based on their inputs without disclosing them to one another. Compared with HE, MPC is computationally more efficient but suffers from high communication overhead and requires all parties to be online during computation. It also faces scalability limitations and potential leakage of private inputs through final results. MPC is often combined with techniques like differential privacy or secure enclaves to mitigate such risks to enhance privacy protection.[137–139]

Differential privacy (DP) introduces a mathematically rigorous mechanism for limiting the information inferred about any individual from aggregate data. This is generally accomplished by incorporating meticulously adjusted noise into the input or output of calculations. DP can be applied under local and global models, depending on whether the noise is added before or after computation. While DP provides strong privacy guarantees and is resistant to post-processing, its practical implementation faces challenges, such as determining optimal noise levels to maintain utility and computing global sensitivity values. Furthermore, privacy guarantees in DP degrade with repeated queries or larger group sizes, necessitating careful composition and budgeting of privacy loss.[140,141]

All three techniques – HE, MPC, and DP – have demonstrated significant potential in healthcare. HE has been used for secure genomic data analysis and predictive modeling in clinical settings. MPC has enabled privacy-preserving collaboration between hospitals and insurance companies without centralizing data. DP has been instrumental in safely releasing patient data for research and integrating privacy into distributed learning systems. Notably, hybrid approaches that combine these techniques are increasingly favored for their complementary strengths, offering more robust privacy and security frameworks for sensitive health data.

3.13.2 Algorithmic Bias and Fairness in Treatment

As AI increasingly informs clinical decision-making in precision medicine, algorithmic bias and fairness have emerged as critical ethical challenges. AI systems, especially those trained on large-scale health datasets, are only as unbiased as the data they learn from. Unfortunately, historical healthcare data often reflect systemic inequities – stemming from socioeconomic disparities, underrepresentation of minority groups, and unequal access to care – which may be inadvertently embedded into ML models.[142]

Bias in training data can arise due to imbalanced datasets, where certain demographic groups (e.g., racial minorities, women, and rural populations) are underrepresented. A prevalent commercial algorithm designed to inform healthcare decisions was discovered to allocate lower risk ratings to black patients relative to white patients with comparable health profiles, owing to its dependence on healthcare expenses as a proxy for need – an indirect and biased metric.[143] Such biases can result in differential treatment recommendations, delayed diagnoses, or exclusion from targeted therapies, thereby exacerbating health disparities.

In precision medicine, genomic and phenotypic data disparities further amplify concerns. Most genetic studies have disproportionately focused on individuals of European descent, limiting the generalizability of AI-driven insights to other populations.[144] This raises the risk that diagnostic tools or predictive models may perform suboptimally – or dangerously inaccurately – when applied to diverse patient populations.

From a technical standpoint, algorithmic fairness remains a complex and evolving field. Various definitions of fairness – such as equalized odds, demographic parity, and counterfactual fairness

– often conflict with each other and may not always align with clinical priorities.[145] Therefore, ensuring fair treatment requires methodological rigor and interdisciplinary oversight involving clinicians, data scientists, ethicists, and affected communities.

Mitigating algorithmic bias in healthcare requires a multifaceted strategy: diversifying training datasets, auditing models across subpopulations, incorporating fairness constraints during model development, and ensuring transparency through explainable AI methods.[146] Furthermore, regulators and institutional review boards must evolve to supervise the use of AI systems, ensuring adherence to ethical standards and preventing the reinforcement of existing imbalances. As AI continues to shape precision medicine, embedding fairness and accountability into algorithmic design is not just a technical necessity but a moral imperative.

3.13.3 Regulatory and Legal Considerations in AI Healthcare

3.13.3.1 Global Regulatory Initiatives

World Health Organization (WHO): In January 2024, the WHO published extensive guidelines on the ethics and governance of massive multi-modal AI models in healthcare. This document presents more than 40 recommendations for governments, technology firms, and healthcare providers to ensure the proper use of AI, emphasizing the need for transparency, accountability, and equity in AI-driven health solutions (WHO, 2024).[147]

3.13.3.2 U.S. Regulatory Landscape

Executive Actions: On October 30, 2023, President Joe Biden enacted a comprehensive executive order to regulate the research and application of safe and responsible AI across all industries, including healthcare. On March 28, 2024, the Office of Management and Budget (OMB) issued a final memorandum outlining agency mandates and directives for AI governance, innovation, and risk management (EY, 2024). AI in healthcare: regulatory and legislative outlook.[148]

FDA Guidelines: In December 2024, the U.S. FDA concluded recommendations to optimize the approval process for AI-driven medical devices. These guidelines allow manufacturers to update AI-enabled devices without resubmitting documentation, provided they adhere to predefined change control plans, facilitating timely improvements while maintaining safety standards (Goldman, 2024).[149]

State-Level Legislation: In 2024, state lawmakers introduced nearly 700 AI-related bills, several of which focused on healthcare applications. States like California, Utah, and Colorado have enacted laws addressing the application of AI in healthcare decision-making,[150] including utilization management and prior authorization processes (Kantrowitz et al., 2025).[151]

3.13.3.3 European Union's AI Act

The European Union's Artificial Intelligence Act, which took effect in August 2024, categorizes AI-based systems by risk levels and enforces rigorous rules for high-risk uses, particularly in healthcare. The Act requires transparency, accountability, and human oversight for AI-based systems to ensure ethical and safe deployment. However, concerns have been raised about the potential burden on small enterprises and the need for clear implementation guidelines (Esponoza, 2024).[152]

3.13.3.4 United Kingdom's Regulatory Approach

In October 2024, the U.K. government established the Regulatory Innovation Office (RIO) to accelerate the approval process for new technologies, including AI in healthcare. The RIO aims to streamline the introduction of advanced technologies, easing regulatory burdens to foster innovation and economic development (*Reuters*, 2024).[153]

3.13.4 Legal and Ethical Considerations

Beyond regulatory frameworks, several legal and ethical issues persist:

Data Privacy and Protection: Maintaining the confidentiality of patient data is of utmost importance. AI-based systems must adhere to data protection standards such as the GDPR in the European Union and the HIPAA in the United States.

Liability and Accountability: Establishing accountability in instances of AI-generated errors is intricate. Explicit rules are required to define the responsibilities of developers, healthcare professionals, and institutions.

Bias and Discrimination: AI-based systems trained on biased datasets can perpetuate health disparities. Regulatory bodies emphasize the need for fairness and equity in AI algorithms to prevent discrimination (Egan & Jieni, 2025).[154]

3.14 FUTURE PROSPECTS AND EMERGING TRENDS

3.14.1 Potential for AI in Global Health and Accessibility

AI possesses considerable potential in mitigating healthcare inequities, notably in environments with limited resources. AI-driven diagnostic tools are increasingly deployed in remote areas to mitigate the shortage of medical professionals. For instance, AI-assisted skin cancer detection devices have enhanced the diagnostic capabilities of non-specialist healthcare providers, improving early diagnosis rates in underserved regions. It was found that AI involvement greatly boosted the reliability of diagnosing non-medical professionals in identifying malignant skin lesions, elevating correct diagnoses from 47.6% to 87.5% without compromising specificity.[155]

Moreover, AI facilitates multilingual healthcare education by developing AI-assisted curricula that overcome language barriers, ensuring that essential medical knowledge is accessible to non-English-speaking healthcare providers. A linguistically tailored, multilingual curriculum employing generative language models has been created to support pediatric healthcare education, aligning with the WHO's Digital Health Guidelines.[156]

Even with the progress achieved, persistent challenges involve data security concerns, the demand for inclusive datasets, and the creation of robust legal and ethical frameworks to ensure ethical AI deployment. Tackling these issues is crucial to prevent the worsening of health disparities and foster fair healthcare delivery among diverse populations.[157]

3.14.2 Future Directions in AI-Powered Precision Medicine

Incorporating AI into precision medicine is accelerating, with fast-evolving advancements in genomics, drug discovery, and personalized treatment plans. AI algorithms can now analyze intricate genomic data to pinpoint potential treatment targets. For example, AI systems have been utilized to discover novel drug candidates for diseases like amyotrophic lateral sclerosis (ALS) and liver cancer,[158] significantly reducing the time required to develop drugs.[159] AI also contributes to more effective clinical trials by optimizing the enrollment process and predicting trial outcomes. DL models are being employed to stratify patients and forecast adverse events, thereby improving the effectiveness of clinical studies. The advancement of precision medicine depends on developing multi-modal AI systems that integrate various data types – genomic, imaging, and clinical to provide comprehensive and personalized healthcare solutions. These technologies have shown enhanced diagnostic precision and are anticipated to be crucial in promoting patient-centered treatment.[160]

3.14.3 Prospects of Human-AI Collaboration in Healthcare

The collaboration between humans and AI is reshaping healthcare delivery, focusing on enhancing human capabilities rather than replacing them. AI tools increasingly assist clinicians in diagnostics, treatment planning, and patient monitoring, enabling healthcare professionals to concentrate on complex decision-making and patient interaction. For instance, AI systems like the Articulate Medical Intelligence Explorer (AMIE) have been developed to conduct medical interviews, demonstrating performance comparable to human physicians in gathering patient information and suggesting diagnoses. In a study, AMIE outperformed primary care physicians in both diagnostic accuracy and communication skills during simulated patient interactions.[161]

Moreover, initiatives like the partnership between Philips and the Saudi Data & AI Authority focus on training local healthcare professionals and data scientists, cultivating a cooperative atmosphere where AI functions as an auxiliary resource for healthcare professionals.

Incorporating AI into healthcare requires meticulous examination of ethical ramifications, including maintaining patient trust, ensuring data privacy, and preventing over-reliance on AI systems. Continuous learning and skill development for medical professionals is essential to effectively work alongside AI technologies effectively and to harness their full potential in improving patient outcomes.[162]

3.15 CONCLUSION

Integrating AI into precision medicine and smart healthcare indicates a major advancement in the perception, execution, and personalization of medical treatment. Enhanced ML, DL, and big data capabilities have enabled AI to play a critical role in advancing diagnosis, therapy design, drug research, and the tracking of patient health. These technologies offer unprecedented opportunities for adapting medical care to suit individual genetic characteristics, lifestyle choices, and real-time health information.

Despite the remarkable progress, several challenges remain – from the protection of sensitive data and the mitigation of AI biases to navigating policy constraints and encouraging collaboration between different domains. Addressing these challenges will require robust ethical frameworks, transparent AI models, and strong partnerships among technologists, healthcare professionals, and policymakers.

Looking ahead, AI-powered precision medicine and smart healthcare systems are poised to redefine patient care, enabling proactive, predictive, and preventive healthcare. Continued research, innovation, and responsible implementation will be key to realizing AI's full potential in building a smarter, more equitable, and efficient healthcare ecosystem.

ABBREVIATIONS

AI	Artificial Intelligence
AMIE	Articulate Medical Intelligence Explorer
AMR	Antimicrobial Resistance
CBT	Cognitive Behavioral Therapy
CNN	Convolutional Neural Networks
COVID-19	Coronavirus Disease 2019
CT	Computed Tomography
ctDNA	Circulating Tumor Deoxyribonucleic Acid
DL	Deep Learning
DP	Differential Privacy
ECG	Electrocardiogram
FDA	United States Food and Drug Administration
FHE	Fully Homomorphic Encryption
GAN	Generative Adversarial Networks
GDPR	General Data Protection Regulation
EHR	Electronic Health Records
HIPAA	Health Insurance Portability and Accountability Act
IoMT	Internet of Medical Things
IoT	Internet of Things
IoU	Intersection-over-Union
LIME	Local Interpretable Model-agnostic Explanations
LSTM	Long Short-Term Memory
ML	Machine Learning
MPC	Multiparty Computation
MRI	Magnetic Resonance Imaging
NER	Named Entity Recognition
NLP	Natural Language Processing
PD-L1	Programmed Death-Ligand 1
PET	Positron Emission Tomography
PPG	Photoplethysmogram
RNN	Recurrent Neural Networks
SHAP	SHapley Additive exPlanations
SVM	Support Vector Machines
U.S. FDA	United States Food and Drug Administration
VHA	Virtual Health Assistants
WHO	World Health Organization

REFERENCES

[1] Carini C, Seyhan AA. Tribulations and future opportunities for artificial intelligence in precision medicine. *Journal of Translational Medicine* 2024;22(1):411. doi:10.1186/s12967-024-05067-0

[2] Chakraborty R, Afrose N, Bhowmick P, Bhowmick M. Biomedical sensors in wearable health technologies. In: Mahajan S, Pandit AK, eds. *Innovations in Biomedical Engineering*. Elsevier; 2025:159–184. doi:10.1016/B978-0-443-30146-9.00005-8

[3] Afrose N, Chakraborty R, Bhowmick P, Bhowmick M. Advancing drug delivery systems through biomedical engineering: Innovations and future directions. In: Mahajan S, Pandit AK, eds. *Innovations in Biomedical Engineering*. Elsevier; 2025:1–32. doi:10.1016/B978-0-443-30146-9.00001-0

[4] Giamarellos-Bourboulis EJ, Aschenbrenner AC, Bauer M, Bock C, Calandra T, Gat-Viks I, et al. The patho-physiology of sepsis and precision-medicine-based immunotherapy. *Nature Immunology* 2024;25(1):19–28. doi:10.1038/s41590-023-01660-5

[5] Yu X, Zhao H, Wang R, Chen Y, Ouyang X, Li W, et al. Cancer epigenetics: from laboratory studies and clinical trials to precision medicine. *Cell Death Discovery* 2024;10(1):28. doi:10.1038/s41420-024-01803-z

[6] Afrose N, Chakraborty R, Hazra A, Bhowmick P, Bhowmick M. AI-Driven Drug Discovery and Development. In: Khade SM, Mishra RG, eds. *Advances in Medical Diagnosis, Treatment, and Care.* IGI Global; 2024:259–277. doi:10.4018/979-8-3693-3629-8.ch013

[7] Satheeskumar R. AI-driven diagnostics and personalized treatment planning in oral oncology: Innovations and future directions. *Oral Oncology Reports* 2025;13:100704. doi:10.1016/j.oor.2024.100704

[8] Kothinti, RR. Deep learning in healthcare: Transforming disease diagnosis, personalized treatment, and clinical decision-making through AI-driven innovations. *World Journal of Advanced Research and Reviews* 2024;24(2):2841–2856. doi:10.30574/wjarr.2024.24.2.3435

[9] Tang Y, Zhang Y, Li J. A time series driven model for early sepsis prediction based on transformer module. *BMC Medical Research Methodology* 2024;24(1):23. doi:10.1186/s12874-023-02138-6

[10] Selmy HA, Mohamed HK, Medhat W. A predictive analytics framework for sensor data using time series and deep learning techniques. *Neural Computing and Applications* 2024;36(11):6119–6132. doi:10.1007/s00521-023-09398-9

[11] Chhabra A, Singh SK, Sharma A, Kumar S, Gupta BB, Arya V, et al. Sustainable and intelligent time-series models for epidemic disease forecasting and analysis. *Sustainable Technology and Entrepreneurship* 2024;3(2):100064. doi:10.1016/j.stae.2023.100064

[12] Adegoke, BO, Odugbose, T, Adeyemi, C. Data analytics for predicting disease outbreaks: A review of models and tools. *International Journal of Life Science Research Updates* 2024;2(2):001–009. doi:10.53430/ijlsru.2024.2.2.0023

[13] Asif S, Wenhui Y, ur-Rehman S, Ul-Ain Q, Amjad K, Yueyang Y, et al. Advancements and prospects of machine learning in medical diagnostics: Unveiling the future of diagnostic precision. *Archives of Computational Methods in Engineering* 2025;32(2):853–883. doi:10.1007/s11831-024-10148-w

[14] Nazir A, Hussain A, Singh M, Assad A. Deep learning in medicine: advancing healthcare with intelligent solutions and the future of holography imaging in early diagnosis. *Multimedia Tools and Applications.* July 5, 2024. doi:10.1007/s11042-024-19694-8

[15] Thakur GK, Thakur A, Kulkarni S, Khan N, Khan S. Deep learning approaches for medical image analysis and diagnosis. *Cureus.* May 2, 2024. doi:10.7759/cureus.59507

[16] Zheng H, Xu K, Zhou H, Wang Y, Su G. Medication recommendation system based on natural language processing for patient emotion analysis. *AJST* 2024;10(1):62–68. doi:10.54097/v160aa61

[17] Sim JA, Huang X, Horan MR, Baker JN, Huang IC. Using natural language processing to analyze unstructured patient-reported outcomes data derived from electronic health records for cancer populations: A systematic review. *Expert Review of Pharmacoeconomics & Outcomes Research* 2024;24(4):467–475. doi:10.1080/14737167.2024.2322664

[18] Chiang CC, Luo M, Dumkrieger G, Trivedi S, Chen YC, Chao CJ, et al. A large language model-based generative natural language processing framework finetuned on clinical notes accurately extracts headache frequency from electronic health records. October 3, 2023. doi:10.1101/2023.10.02.23296403

[19] Uddin MKS. A review of utilizing natural language processing and AI for advanced data visualization in real-time analytics. *GMJ* 2024;1(4):34–49. doi:10.62304/ijmisds.v1i04.185

[20] Zhang H, Shafiq MO. Survey of transformers and towards ensemble learning using transformers for natural language processing. *Journal of Big Data* 2024;11(1):25. doi:10.1186/s40537-023-00842-0

[21] Wubineh BZ, Deriba FG, Woldeyohannis MM. Exploring the opportunities and challenges of implementing artificial intelligence in healthcare: A systematic literature review. *Urologic Oncology: Seminars and Original Investigations* 2024;42(3):48–56. doi:10.1016/j.urolonc.2023.11.019

[22] Theodorakopoulos L, Theodoropoulou A, Stamatiou Y. A state-of-the-art review in big data management engineering: Real-life case studies, challenges, and future research directions. *Engineering* 2024;5(3):1266–1297. doi:10.3390/eng5030068

[23] Hussain M, Ajmal M, Subramanian G, Khan M, Anas S. Challenges of big data analytics for sustainable supply chains in healthcare – A resource-based view. *Benchmarking: An International Journal* 2024;31(9):2897–2918. doi:10.1108/BIJ-06-2022-0390

[24] Srivastava U, Kanchan S, Kesheri M, Gupta MK, Singh S. Types of omics data. In: Gupta MK, Katara P, Mondal S, Singh RL, eds. *Integrative Omics*. Elsevier; 2024:13–34. doi:10.1016/B978-0-443-16092-9.00002-3

[25] Lin YL, Chang PC, Hsu C, Hung MZ, Chien YH, Hwu WL, et al. Comparison of GATK and DeepVariant by trio sequencing. *Scientific Reports* 2022;12(1):1809. doi:10.1038/s41598-022-05833-4

[26] Kamya P, Ozerov IV, Pun FW, Tretina K, Fokina T, Chen S, et al. PandaOmics: An AI-driven platform for therapeutic target and biomarker discovery. *Journal of Chemical Information and Modeling* 2024;64(10):3961–3969. doi:10.1021/acs.jcim.3c01619

[27] Huang G, Liu L, Wang X, Wang L, Li H, Tu Z, et al. TranSmart: A practical interactive machine translation system. 2021. doi:10.48550/ARXIV.2105.13072

[28] Shebl A, Abriha D, Fahil AS, El-Dokouny HA, Elrasheed AA, Csámer Á. PRISMA hyperspectral data for lithological mapping in the Egyptian Eastern Desert: Evaluating the support vector machine, random forest, and XG boost machine learning algorithms. *Ore Geology Reviews* 2023;161:105652. doi:10.1016/j.oregeorev.2023.105652

[29] Hu W, Yii FSL, Chen R, Zhang X, Shang X, Kiburg K, et al. A systematic review and meta-analysis of applying deep learning in the prediction of the risk of cardiovascular diseases from retinal images. *Translational Vision Science & Technology* 2023;12(7):14. doi:10.1167/tvst.12.7.14

[30] Alum EU. AI-driven biomarker discovery: Enhancing precision in cancer diagnosis and prognosis. *Discover Oncology* 2025;16(1):313. doi:10.1007/s12672-025-02064-7

[31] Vadapalli S, Abdelhalim H, Zeeshan S, Ahmed Z. Artificial intelligence and machine learning approaches using gene expression and variant data for personalized medicine. *Briefings in Bioinformatics* 2022;23(5):bbac191. doi:10.1093/bib/bbac191

[32] Visan AI, Negut I. Integrating artificial intelligence for drug discovery in the context of revolutionizing drug delivery. *Lifestyles* 2024;14(2):233. doi:10.3390/life14020233

[33] Yang Z, Zeng X, Zhao Y, Chen R. AlphaFold2 and its applications in the fields of biology and medicine. *Signal Transduction and Targeted Therapy* 2023;8(1):115. doi:10.1038/s41392-023-01381-z

[34] Marques L, Costa B, Pereira M, Silva A, Santos J, Saldanha L, et al. Advancing precision medicine: A review of innovative in silico approaches for drug development, clinical pharmacology and personalized healthcare. *Pharmaceutics* 2024;16(3):332. doi:10.3390/pharmaceutics16030332

[35] Muhli H, Ala-Nissila T, Caro MA. Atom-wise formulation of the many-body dispersion problem for linear-scaling van der Waals corrections. *Physical Review B* 2025;111(5):054103. doi:10.1103/PhysRevB.111.054103

[36] Karpus J, Krüger A, Verba JT, Bahrami B, Deroy O. Algorithm exploitation: Humans are keen to exploit benevolent AI. *iScience* 2021;24(6):102679. doi:10.1016/j.isci.2021.102679

[37] Munk P, Brinch C, Møller FD, Petersen TN, Hendriksen RS, Seyfarth AM, et al. Genomic analysis of sewage from 101 countries reveals global landscape of antimicrobial resistance. *Nature Communications* 2022;13(1):7251. doi:10.1038/s41467-022-34312-7

[38] Wörheide MA, Krumsiek J, Kastenmüller G, Arnold M. Multi-omics integration in biomedical research – A metabolomics-centric review. *Analytica Chimica Acta* 2021;1141:144–162. doi:10.1016/j.aca.2020.10.038

[39] Yu Z, Wang K, Wan Z, Xie S, Lv Z. Popular deep learning algorithms for disease prediction: a review. *Cluster Computing* 2023;26(2):1231–1251. doi:10.1007/s10586-022-03707-y

[40] Tian M, He X, Jin C, He X, Wu S, Zhou R, et al. Transpathology: Molecular imaging-based pathology. *European Journal of Nuclear Medicine and Molecular Imaging* 2021;48(8):2338–2350. doi:10.1007/s00259-021-05234-1

[41] Sadybekov AV, Katritch V. Computational approaches streamlining drug discovery. *Nature* 2023; 616(7958):673–685. doi:10.1038/s41586-023-05905-z

[42] Mazza M, Caroppo E, De Berardis D, Marano G, Avallone C, Kotzalidis GD, et al. Psychosis in women: Time for personalized treatment. *JPM* 2021;11(12):1279. doi:10.3390/jpm11121279

[43] Junaid SB, Imam AA, Shuaibu AN, Basri S, Kumar G, Surakat YA, et al. Artificial intelligence, sensors and vital health signs: A Review. *Applied Sciences* 2022;12(22):11475. doi:10.3390/app122211475

[44] Li C, Han X, Zhang Q, Li M, Rao Z, Liao W, et al. State-of-health and remaining-useful-life estimations of lithium-ion battery based on temporal convolutional network-long short-term memory. *Journal of Energy Storage* 2023;74:109498. doi:10.1016/j.est.2023.109498

[45] Wang YC, Xu X, Hajra A, Apple S, Kharawala A, Duarte G, et al. Current advancement in diagnosing atrial fibrillation by utilizing wearable devices and artificial intelligence: A review study. *Diagnostics* 2022;12(3):689. doi:10.3390/diagnostics12030689

[46] Bhambri P, Khang A. Managing and monitoring patient's healthcare using AI and IoT technologies. In: Khang A, ed., *Advances in Medical Diagnosis, Treatment, and Care*. IGI Global;2024:1–23. doi:10.4018/979-8-3693-3679-3.ch001

[47] AI-powered virtual health assistants: Transforming patient care and engagement. *Global Insights in Artificial Intelligence and Computing* 2025;1(1):15–30. doi:10.70445/giaic.1.1.2025.15-30

[48] Yeh PL, Kuo WC, Tseng BL, Sung YH. Does the AI-driven chatbot work? Effectiveness of the Woebot app in reducing anxiety and depression in group counseling courses and student acceptance of technological aids. *Current Psychology* January 25, 2025. doi:10.1007/s12144-025-07359-0

[49] Reddy K, Gharde P, Tayade H, Patil M, Reddy LS, Surya D. Advancements in robotic surgery: A comprehensive overview of current utilizations and upcoming frontiers. *Cureus*. December 12, 2023. doi:10.7759/cureus.50415

[50] Cano S, González CS, Gil-Iranzo RM, Albiol-Pérez S. Affective communication for socially assistive robots (SARs) for children with autism spectrum disorder: A systematic review. *Sensors* 2021;21(15):5166. doi:10.3390/s21155166

[51] Han X, Zhou X, Tan B, Jiao L, Zhang R. AI-based next-generation sensors for enhanced rehabilitation monitoring and analysis. *Measurement* 2023;223:113758. doi:10.1016/j.measurement.2023.113758

[52] Singh V, Singh G, Sharma G. The future of the healthcare workforce in the age of automation. In: Kukreti M, Sehajpal S, Tiwari R, Sood K, eds. *Advances in human services and public health*. IGI Global; 2024:453–472. doi:10.4018/979-8-3373-0240-9.ch019

[53] Holland J, Kingston L, McCarthy C, Armstrong E, O'Dwyer P, Merz F, et al. Service robots in the healthcare sector. *Robotics* 2021;10(1):47. doi:10.3390/robotics10010047

[54] Kinoshita S, Abo M, Okamoto T, Miyamura K. Transitional and long-term care system in Japan and current challenges for stroke patient rehabilitation. *Frontiers in Neurology* 2022;12:711470. doi:10.3389/fneur.2021.711470

[55] Anwar SM, Majid M, Qayyum A, Awais M, Alnowami M, Khan MK. Medical image analysis using convolutional neural networks: A review. *Journal of Medical Systems* 2018;42(11):226. doi:10.1007/s10916-018-1088-1

[56] Iqbal SN, Qureshi A, Li J, Mahmood T. On the analyses of medical images using traditional machine learning techniques and convolutional neural networks. *Archives of Computational Methods in Engineering* 2023;30(5):3173–3233. doi:10.1007/s11831-023-09899-9

[57] Pezeshk A, Hamidian S, Petrick N, Sahiner B. 3-D Convolutional neural networks for automatic detection of pulmonary nodules in chest CT. *IEEE Journal of Biomedical and Health Informatics* 2019;23(5):2080–2090. doi:10.1109/JBHI.2018.2879449

[58] Khan US, Khan SUR. Boost diagnostic performance in retinal disease classification utilizing deep ensemble classifiers based on OCT. *Multimedia Tools and Applications*. July 29, 2024. doi:10.1007/s11042-024-19922-1

[59] Washington P, Park N, Srivastava P, Voss C, Kline A, Varma M, et al. Data-driven diagnostics and the potential of mobile artificial intelligence for digital therapeutic phenotyping in computational psychiatry. *Biological Psychiatry: Cognitive Neuroscience and Neuroimaging* 2020;5(8):759–769. doi:10.1016/j.bpsc.2019.11.015

[60] Mamidisetti S, Reddy M. Multimodal depression detection using audio, visual and textual cues: A survey. *NQ* 2022;20(4):325–336. doi:10.14704/nq.2022.20.4.NQ22127

[61] Ogwu MC, Izah SC. Technological advances for diagnosing tropical diseases. In: Ogwu MC, Izah SC. *Technological Innovations for Managing Tropical Diseases*. Health Information Science. Springer Nature Switzerland; 2025:27–56. doi:10.1007/978-3-031-82622-1_2

[62] Ogwu MC, Izah SC. Mobile health and telemedicine for tropical diseases. In: Ogwu MC, Izah SC. *Technological Innovations for Managing Tropical Diseases*. Health Information Science. Springer Nature Switzerland; 2025:183–211. doi:10.1007/978-3-031-82622-1_8

[63] Chang L, Liu J, Zhu J, Guo S, Wang Y, Zhou Z, et al. Advancing precision medicine: the transformative role of artificial intelligence in immunogenomics, radiomics, and pathomics for biomarker discovery and immunotherapy optimization. *Cancer Biology & Medicine* January 2, 2025:1–15. doi:10.20892/j.issn.2095-3941.2024.0376

[64] Lai B, Fu J, Zhang Q, Deng N, Jiang Q, Peng J. Artificial intelligence in cancer pathology: Challenge to meet increasing demands of precision medicine. *International Journal of Oncology* 2023;63(3):107. doi:10.3892/ijo.2023.5555

[65] McKinney SM, Sieniek M, Godbole V, Godwin J, Antropova N, Ashrafian H, et al. International evaluation of an AI system for breast cancer screening. *Nature* 2020;577(7788):89–94. doi:10.1038/s41586-019-1799-6

[66]   Esteva A, Kuprel B, Novoa RA, Ko J, Swetter SM, Blau HM, et al. Dermatologist-level classification of skin cancer with deep neural networks. *Nature* 2017;542(7639):115–118. doi:10.1038/nature21056

[67]   Cruz JA, Wishart DS. Applications of machine learning in cancer prediction and prognosis. *Cancer Informatics* 2006;2. doi:10.1177/117693510600200030

[68]   Jo T, Nho K, Saykin AJ. Deep learning in alzheimer's disease: Diagnostic classification and prognostic prediction using neuroimaging data. *Frontiers in Aging Neuroscience* 2019;11:220. doi:10.3389/fnagi.2019.00220

[69]   Fortman E, Hettinger AZ, Howe JL, Fong A, Pruitt Z, Miller K, et al. Varying rates of patient identity verification when using computerized provider order entry. *Journal of the American Medical Informatics Association* 2020;27(6):924–928. doi:10.1093/jamia/ocaa047

[70]   Hannun AY, Rajpurkar P, Haghpanahi M, Tison GH, Bourn C, Turakhia MP, et al. Cardiologist-level arrhythmia detection and classification in ambulatory electrocardiograms using a deep neural network. *Nature Medicine* 2019;25(1):65–69. doi:10.1038/s41591-018-0268-3

[71]   Wang L, Lin ZQ, Wong A. COVID-Net: A tailored deep convolutional neural network design for detection of COVID-19 cases from chest X-ray images. *Scientific Reports* 2020;10(1):19549. doi:10.1038/s41598-020-76550-z

[72]   Sheller MJ, Edwards B, Reina GA, Martin J, Pati S, Kotrotsou A, et al. Federated learning in medicine: facilitating multi-institutional collaborations without sharing patient data. *Scientific Reports* 2020;10(1):12598. doi:10.1038/s41598-020-69250-1

[73]   Mei X, Lee HC, Diao KY, Huang M, Lin B, Liu C, et al. Artificial intelligence-enabled rapid diagnosis of patients with COVID-19. *Nature Medicine* 2020;26(8):1224–1228. doi:10.1038/s41591-020-0931-3

[74]   Kavakiotis I, Tsave O, Salifoglou A, Maglaveras N, Vlahavas I, Chouvarda I. Machine learning and data mining methods in diabetes research. *Computational and Structural Biotechnology Journal* 2017;15:104–116. doi:10.1016/j.csbj.2016.12.005

[75]   Gulshan V, Peng L, Coram M, Stumpe MC, Wu D, Narayanaswamy A, et al. Development and validation of a deep learning algorithm for detection of diabetic retinopathy in retinal fundus photographs. *Journal of the American Medical Association* 2016;316(22):2402. doi:10.1001/jama.2016.17216

[76]   Ting DSW, Cheung CYL, Lim G, Tan GSW, Quang ND, Gan A, et al. Development and validation of a deep learning system for diabetic retinopathy and related eye diseases using retinal images from multiethnic populations with diabetes. *Journal of the American Medical Association* 2017;318(22):2211. doi:10.1001/jama.2017.18152

[77]   Topol EJ. High-performance medicine: The convergence of human and artificial intelligence. *Nature Medicine* 2019;25(1):44–56. doi:10.1038/s41591-018-0300-7

[78]   Libbrecht MW, Noble WS. Machine learning applications in genetics and genomics. *Nature Reviews. Genetics* 2015;16(6):321–332. doi:10.1038/nrg3920

[79]   Piro VC, Dadi TH, Seiler E, Reinert K, Renard BY. ganon: Precise metagenomics classification against large and up-to-date sets of reference sequences. *Bioinformatics* 2020;36(Supplement_1):i12–i20. doi:10.1093/bioinformatics/btaa458

[80]   DeFoor MT, Chung Y, Zadinsky JK, Dowling J, Sams RW. An interprofessional cohort analysis of student interest in medical ethics education: A survey-based quantitative study. *BMC Medical Ethics* 2020;21(1):26. doi:10.1186/s12910-020-00468-4

[81]   Esteva A, Kuprel B, Novoa RA, Ko J, Swetter SM, Blau HM, et al. Dermatologist-level classification of skin cancer with deep neural networks. *Nature* 2017;542(7639):115–118. doi:10.1038/nature21056

[82]   Khincha PP, Best AF, Fraumeni JF, Loud JT, Savage SA, Achatz MI. Reproductive factors associated with breast cancer risk in Li–Fraumeni syndrome. *European Journal of Cancer* 2019;116:199–206. doi:10.1016/j.ejca.2019.05.005

[83]   Sheller MJ, Edwards B, Reina GA, Martin J, Pati S, Kotrotsou A, et al. Federated learning in medicine: facilitating multi-institutional collaborations without sharing patient data. *Scientific Reports* 2020;10(1):12598. doi:10.1038/s41598-020-69250-1

[84]   Laguarta J, Hueto F, Subirana B. COVID-19 artificial intelligence diagnosis using only cough recordings. *IEEE Open Journal of Engineering in Medicine and Biology* 2020;1:275–281. doi:10.1109/OJEMB.2020.3026928

[85]   Rajkomar A, Oren E, Chen K, Dai AM, Hajaj N, Hardt M, et al. Scalable and accurate deep learning with electronic health records. *Npj Digital Medicine* 2018;1(1):18. doi:10.1038/s41746-018-0029-1

[86]   Adans-Dester C, Hankov N, O'Brien A, Vergara-Diaz G, Black-Schaffer R, Zafonte R, et al. Enabling precision rehabilitation interventions using wearable sensors and machine learning to track motor recovery. *Npj Digital Medicine* 2020;3:121. doi:10.1038/s41746-020-00328-w

[87] Kumar A. AI Driven precision oncology: Predictive biomarker discovery and personalized treatment optimization using genomic data. March 17, 2025. doi:10.5281/ZENODO.15037946

[88] Tran KA, Kondrashova O, Bradley A, Williams ED, Pearson JV, Waddell N. Deep learning in cancer diagnosis, prognosis and treatment selection. *Genome Medicine* 2021;13(1):152. doi:10.1186/s13073-021-00968-x

[89] Zhang B, Shi H, Wang H. Machine learning and AI in cancer prognosis, prediction, and treatment selection: A critical approach. *JMDH* 2023;16:1779–1791. doi:10.2147/JMDH.S410301

[90] Parikh RB, Manz C, Chivers C, Regli SH, Braun J, Draugelis ME, et al. Machine learning approaches to predict 6-month mortality among patients with cancer. *JAMA Network Open* 2019;2(10):e1915997. doi:10.1001/jamanetworkopen.2019.15997

[91] Kumar A. AI driven precision oncology: Predictive biomarker discovery and personalized treatment optimization using genomic data. March 17, 2025. doi:10.5281/ZENODO.15037946

[92] Gerratana L, Davis AA, Foffano L, Reduzzi C, Rossi T, Medford A, et al. Integrating machine learning-predicted circulating tumor cells (CTCs) and circulating tumor DNA (ctDNA) in metastatic breast cancer: A proof of principle study on endocrine resistance profiling. *Cancer Letters* 2025;609:217325. doi:10.1016/j.canlet.2024.217325

[93] Trebeschi S, Bodalal Z, Boellaard TN, Tareco Bucho TM, Drago SG, Kurilova I, et al. Prognostic value of deep learning-mediated treatment monitoring in lung cancer patients receiving immunotherapy. *Frontiers in Oncology* 2021;11:609054. doi:10.3389/fonc.2021.609054

[94] Dong X, Zheng W. Emerging technologies for drug repurposing: Harnessing the potential of text and graph embedding approaches. *Artificial Intelligence Chemistry* 2024;2(1):100060. doi:10.1016/j.aichem.2024.100060

[95] Cifci D, Veldhuizen GP, Foersch S, Kather JN. AI in computational pathology of cancer: Improving diagnostic workflows and clinical outcomes? *Annual Review of Cancer Biology* 2023;7(1):57–71. doi:10.1146/annurev-cancerbio-061521-092038

[96] Barrett M, Boyne J, Brandts J, Brunner-La Rocca HP, De Maesschalck L, De Wit K, et al. Artificial intelligence supported patient self-care in chronic heart failure: a paradigm shift from reactive to predictive, preventive and personalised care. *The EPMA Journal* 2019;10(4):445–464. doi:10.1007/s13167-019-00188-9

[97] Wang YC, Xu X, Hajra A, Apple S, Kharawala A, Duarte G, et al. Current advancement in diagnosing atrial fibrillation by utilizing wearable devices and artificial intelligence: A review study. *Diagnostics* 2022;12(3):689. doi:10.3390/diagnostics12030689

[98] Karan D. AI-powered patient education: Transforming type 2 diabetes management. *Indus Journal of Medical and Health Sciences* 2023;1(01):86–108. https://induspublishers.com/IJMHS/article/view/57. Accessed May 7, 2025

[99] Kalusivalingam AK, Sharma A, Patel N, Singh V. Enhancing patient care through IoT-enabled remote monitoring and AI-driven virtual health assistants: Implementing machine learning algorithms and natural language processing. *International Journal of AI and ML* 2021;2(3). https://www.cognitivecomputingjournal.com/index.php/IJAIML-V1/article/view/75. Accessed May 7, 2025

[100] Khalid U, Naeem M, Stasolla F, Syed M, Abbas M, Coronato A. Impact of AI-powered solutions in rehabilitation process: Recent improvements and future trends. *IJGM* 2024;17:943–969. doi:10.2147/IJGM.S453903

[101] Shilo S, Godneva A, Rachmiel M, Korem T, Kolobkov D, Karady T, et al. Prediction of personal glycemic responses to food for individuals with type 1 diabetes through integration of clinical and microbial data. *Diabetes Care* 2022;45(3):502–511. doi:10.2337/dc21-1048

[102] Pattabhi A. Risk intelligence: AI-powered financial risk management for a new era. *International Journal of Artificial Intelligence, Data Science, and Machine Learning* 2025;6(2):20–34. doi:10.63282/3050-9262.IJAIDSML-V6I2P103

[103] Celard P, Iglesias EL, Sorribes-Fdez JM, Romero R, Vieira AS, Borrajo L. A survey on deep learning applied to medical images: From simple artificial neural networks to generative models. *Neural Computing and Applications* 2023;35(3):2291–2323. doi:10.1007/s00521-022-07953-4

[104] Islam S, Aziz MdT, Nabil HR, Jim JR, Mridha MF, Kabir MdM, et al. Generative adversarial networks (GANs) in medical imaging: Advancements, applications, and challenges. *IEEE Access* 2024;12:35728–35753. doi:10.1109/ACCESS.2024.3370848

[105] Alshwaheen TI, Hau YW, Ass'Ad N, Abualsamen MM. A novel and reliable framework of patient deterioration prediction in intensive care unit based on long short-term memory-recurrent neural network. *IEEE Access* 2021;9:3894–3918. doi:10.1109/ACCESS.2020.3047186

[106] Chen Y, Lin H, Zhang W, Chen W, Zhou Z, Heidari AA, et al. ICycle-GAN: Improved cycle generative adversarial networks for liver medical image generation. *Biomedical Signal Processing and Control* 2024;92:106100. doi:10.1016/j.bspc.2024.106100

[107] Li D. Improving polygon image segmentation by enhancing U-Net architecture. *Journal of Physics: Conference Series* 2024;2711(1):012010. doi:10.1088/1742-6596/2711/1/012010

[108] Christopoulou F, Tran TT, Sahu SK, Miwa M, Ananiadou S. Adverse drug events and medication relation extraction in electronic health records with ensemble deep learning methods. *Journal of the American Medical Informatics Association* 2020;27(1):39–46. doi:10.1093/jamia/ocz101

[109] Blanco-González A, Cabezón A, Seco-González A, Conde-Torres D, Antelo-Riveiro P, Piñeiro Á, et al. The role of AI in drug discovery: Challenges, opportunities, and strategies. *Pharmaceuticals* 2023;16(6):891. doi:10.3390/ph16060891

[110] Li L, Albert-Smet I, Faisal AA. Optimizing medical treatment for sepsis in intensive care: from reinforcement learning to pre-trial evaluation. 2020. doi:10.48550/ARXIV.2003.06474

[111] Zhang Z, Chen L, Zhong F, Wang D, Jiang J, Zhang S, et al. Graph neural network approaches for drug-target interactions. *Current Opinion in Structural Biology* 2022;73:102327. doi:10.1016/j.sbi.2021.102327

[112] Prosperi M, Guo Y, Bian J. Bagged random causal networks for interventional queries on observational biomedical datasets. *Journal of Biomedical Informatics* 2021;115:103689. doi:10.1016/j.jbi.2021.103689

[113] Tong L, Shi W, Isgut M, Zhong Y, Lais P, Gloster L, et al. Integrating multi-omics data with EHR for precision medicine using advanced artificial intelligence. *IEEE Reviews in Biomedical Engineering* 2024;17:80–97. doi:10.1109/RBME.2023.3324264

[114] Zhong X, Babaie Sarijaloo F, Prakash A, Park J, Huang C, Barwise A, et al. A multidisciplinary approach to the development of digital twin models of critical care delivery in intensive care units. *International Journal of Production Research* 2022;60(13):4197–4213. doi:10.1080/00207543.2021.2022235

[115] Rudin C. Stop explaining black box machine learning models for high stakes decisions and use interpretable models instead. *Nature Machine Intelligence* 2019;1(5):206–215. doi:10.1038/s42256-019-0048-x

[116] Lundberg SM, Lee SI. A unified approach to interpreting model predictions. In: von Luxburg U, Bengio S, Fergus R, Garnett R, Guyon I, Wallach H, Vishwanathan SVN, eds. *Advances in Neural Information Processing Systems*. Vol 30. Curran Associates 2017. https://dl.acm.org/doi/10.5555/3295222.3295230. Accessed September 1, 2025

[117] Ribeiro MT, Singh S, Guestrin C. "Why Should I Trust You?": Explaining the predictions of any classifier. In: *Proceedings of the 22nd ACM SIGKDD International Conference on Knowledge Discovery and Data Mining*. ACM; 2016:1135–1144. doi:10.1145/2939672.2939778

[118] Morley J, Machado CCV, Burr C, Cowls J, Joshi I, Taddeo M, et al. The ethics of AI in health care: A mapping review. *Social Science & Medicine* 2020;260:113172. doi:10.1016/j.socscimed.2020.113172

[119] Obermeyer Z, Powers B, Vogeli C, Mullainathan S. Dissecting racial bias in an algorithm used to manage the health of populations. *Science* 2019;366(6464):447–453. doi:10.1126/science.aax2342

[120] Dwork C, Roth A. The algorithmic foundations of differential privacy. *FNT in Theoretical Computer Science* 2013;9(3–4):211–407. doi:10.1561/0400000042

[121] Veale M, Binns R. Fairer machine learning in the real world: Mitigating discrimination without collecting sensitive data. *Big Data & Society* 2017;4(2):205395171774353. doi:10.1177/2053951717743530

[122] Topol EJ. High-performance medicine: the convergence of human and artificial intelligence. *Nature Medicine* 2019;25(1):44–56. doi:10.1038/s41591-018-0300-7

[123] Goktas P, Grzybowski A. Shaping the future of healthcare: Ethical clinical challenges and pathways to trustworthy AI. *JCM* 2025;14(5):1605. doi:10.3390/jcm14051605

[124] Dave Sh, Dave A, Radhakrishnan S, Das J, Dave Su. Biosensors for healthcare: an artificial intelligence approach. In: Das J, Dave S, Radhakrishnan S, Mohanty P, eds. *Biosensors for Emerging and Re-emerging Infectious Diseases*. Elsevier; 2022:365–383. doi:10.1016/B978-0-323-88464-8.00008-7

[125] Hosain MN, Kwak YS, Lee J, Choi H, Park J, Kim J. IoT-enabled biosensors for real-time monitoring and early detection of chronic diseases. *Physical Activity and Nutrition* 2024;28(4):060–069. doi:10.20463/pan.2024.0033

[126] Nasri N, Orts-Escolano S, Gomez-Donoso F, Cazorla M. Inferring static hand poses from a low-cost non-intrusive sEMG sensor. *Sensors* 2019;19(2):371. doi:10.3390/s19020371

[127] Shahriari B, Swersky K, Wang Z, Adams RP, De Freitas N. Taking the human out of the loop: A review of Bayesian optimization. *Proceedings of the IEEE* 2016;104(1):148–175. doi:10.1109/JPROC.2015.2494218

[128] Sheller MJ, Edwards B, Reina GA, Martin J, Pati S, Kotrotsou A, et al. Federated learning in medicine: Facilitating multi-institutional collaborations without sharing patient data. *Scientific Reports* 2020;10(1):12598. doi:10.1038/s41598-020-69250-1

[129] Acar A, Aksu H, Uluagac AS, Conti M. A survey on homomorphic encryption schemes: Theory and implementation. *ACM Computing Surveys* 2018;51(4):1–35. doi:10.1145/3214303

[130] Kuo TT, Kim HE, Ohno-Machado L. Blockchain distributed ledger technologies for biomedical and health care applications. *Journal of the American Medical Informatics Association* 2017;24(6):1211–1220. doi:10.1093/jamia/ocx068

[131] Chien W, De Jesus J, Taylor B, Dods V, Alekseyev L, Shoda D, et al. The last mile: DSCSA solution through blockchain technology: Drug tracking, tracing, and verification at the last mile of the pharmaceutical supply chain with BRUINchain. *BHTY*. March 12, 2020. doi:10.30953/bhty.v3.134

[132] Nemes G, Chiffoleau Y, Zollet S, Collison M, Benedek Z, Colantuono F, et al. The impact of COVID-19 on alternative and local food systems and the potential for the sustainability transition: Insights from 13 countries. *Sustainable Production and Consumption* 2021;28:591–599. doi:10.1016/j.spc.2021.06.022

[133] Azaria A, Ekblaw A, Vieira T, Lippman A. MedRec: Using blockchain for medical data access and permission management. In: *2016 2nd International Conference on Open and Big Data (OBD)*. IEEE; 2016:25–30. doi:10.1109/OBD.2016.11

[134] Odeh A, Abdelfattah E, Salameh W. Privacy-preserving data sharing in telehealth services. *Applied Sciences* 2024;14(23):10808. doi:10.3390/app142310808

[135] Acar A, Aksu H, Uluagac AS, Conti M. A survey on homomorphic encryption schemes: Theory and implementation. *ACM Computing Surveys* 2019;51(4):1–35. doi:10.1145/3214303

[136] Gong Y, Chang X, Mišić J, Mišić VB, Wang J, Zhu H. Practical solutions in fully homomorphic encryption: a survey analyzing existing acceleration methods. *Cybersecurity* 2024;7(1):5. doi:10.1186/s42400-023-00187-4

[137] Gamiz I, Regueiro C, Lage O, Jacob E, Astorga J. Challenges and future research directions in secure multiparty computation for resource-constrained devices and large-scale computations. *International Journal of Information Security* 2025;24(1):27. doi:10.1007/s10207-024-00939-4

[138] Damgård I, Ishai Y. Scalable secure multiparty computation. In: Dwork C, ed. *Advances in cryptology - CRYPTO 2006*. Vol 4117. Lecture Notes in Computer Science. Springer Berlin Heidelberg; 2006:501–520. doi:10.1007/11818175_30

[139] Evans D, Kolesnikov V, Rosulek M. A pragmatic introduction to secure multi-party computation. *FNT in Privacy and Security* 2018;2(2–3):70–246. doi:10.1561/3300000019

[140] Dwork C, Roth A. The algorithmic foundations of differential privacy. *FNT in Theoretical Computer Science* 2014;9(3–4):211–407. doi:10.1561/0400000042

[141] Abowd JM. The U.S. census bureau adopts differential privacy. In: *Proceedings of the 24th ACM SIGKDD International Conference on Knowledge Discovery & Data Mining*. ACM; 2018:2867. doi:10.1145/3219819.3226070

[142] Obermeyer Z, Powers B, Vogeli C, Mullainathan S. Dissecting racial bias in an algorithm used to manage the health of populations. *Science* 2019;366(6464):447–453. doi:10.1126/science.aax2342

[143] Obermeyer Z, Powers B, Vogeli C, Mullainathan S. Dissecting racial bias in an algorithm used to manage the health of populations. *Science* 2019;366(6464):447–453. doi:10.1126/science.aax2342

[144] Popejoy AB, Fullerton SM. Genomics is failing on diversity. *Nature* 2016;538(7624):161–164. doi:10.1038/538161a

[145] Mehrabi N, Morstatter F, Saxena N, Lerman K, Galstyan A. A survey on bias and fairness in machine learning. *ACM Computing Surveys* 2022;54(6):1–35. doi:10.1145/3457607

[146] Rajkomar A, Hardt M, Howell MD, Corrado G, Chin MH. Ensuring fairness in machine learning to advance health equity. *Annals of Internal Medicine* 2018;169(12):866–872. doi:10.7326/M18-1990

[147] WHO releases AI ethics and governance guidance for large multi-modal models. https://www.who.int/news/item/18-01-2024-who-releases-ai-ethics-and-governance-guidance-for-large-multi-modal-models. Accessed May 6, 2025

[148] Safe, secure, and trustworthy development and use of artificial intelligence. *Federal Register*. November 1, 2023. https://www.federalregister.gov/documents/2023/11/01/2023-24283/safe-secure-and-trustworthy-development-and-use-of-artificial-intelligence. Accessed May 15, 2025

[149] Goldman M. FDA streamlines approval of AI-powered devices. *Axios*. December 4, 2024. https://www.axios.com/2024/12/04/fda-ai-device-guidance. Accessed May 6, 2025

[150] Ali KA, Mohin S, Mondal P, Goswami S, Ghosh S, Choudhuri S. Influence of artificial intelligence in modern pharmaceutical formulation and drug development. *Future Journal of Pharmaceutical Sciences* 2024;10(1):53. doi:10.1186/s43094-024-00625-1

[151] Kantrowitz, R, Meintjes, R, McCallum, K. Considering the future of AI regulation on health sector. Kirkland & Ellis LLP. https://www.kirkland.com/publications/article/2025/03/considering-the-future-of-ai-regulation-on-health-sector. Accessed May 6, 2025

[152] Espinoza J. Europe's rushed attempt to set the rules for AI. *Financial Times*. July 16, 2024

[153] Britain sets up regulatory innovation office to boost growth. *Reuters*. https://www.reuters.com/world/uk/britain-sets-up-regulatory-innovation-office-boost-growth-2024-10-07/ Accessed May 6, 2025

[154] Egan, D, Jieni, J. AI in healthcare: Legal and ethical considerations in this new frontier. International Bar Association. February 11, 2025. https://www.ibanet.org/ai-healthcare-legal-ethical. Accessed May 6, 2025

[155] Wei ML, Tada M, So A, Torres R. Artificial intelligence and skin cancer. *Frontiers in Medicine* 2024;11:1331895. doi:10.3389/fmed.2024.1331895

[156] Benboujja F, Hartnick E, Zablah E, Hersh C, Callans K, Villamor P, et al. Overcoming language barriers in pediatric care: A multilingual, AI-driven curriculum for global healthcare education. *Frontiers in Public Health* 2024;12:1337395. doi:10.3389/fpubh.2024.1337395

[157] Dankwa-Mullan I. Health equity and ethical considerations in using artificial intelligence in public health and medicine. *Preventing Chronic Disease* 2024;21:240245. doi:10.5888/pcd21.240245

[158] Ali KA, Choudhuri S, Roy SK, Reza KH, Mandal D, Chakraborty R, et al. AI-generated nanoparticles in immunological research: Opportunities and challenges. In: Chen JT, ed. *Advances in Computational Intelligence and Robotics*. IGI Global; 2025:295–322. doi:10.4018/979-8-3693-9725-1.ch011

[159] Fu C, Chen Q. The future of pharmaceuticals: Artificial intelligence in drug discovery and development. *Journal of Pharmaceutical Analysis*. February 2025:101248. doi:10.1016/j.jpha.2025.101248

[160] Chen YM, Hsiao TH, Lin CH, Fann YC. Unlocking precision medicine: Clinical applications of integrating health records, genetics, and immunology through artificial intelligence. *Journal of Biomedical Science* 2025;32(1):16. doi:10.1186/s12929-024-01110-w

[161] Tu T, Schaekermann M, Palepu A, Saab K, Freyberg J, Tanno R, et al. Towards conversational diagnostic artificial intelligence. *Nature* 2025. doi:10.1038/s41586-025-08866-7

[162] Harishbhai Tilala M, Kumar Chenchala P, Choppadandi A, Kaur J, Naguri S, Saoji R, et al. Ethical considerations in the use of artificial intelligence and machine learning in health care: A comprehensive review. *Cureus* 2024. doi:10.7759/cureus.62443

4 Precision Medicine and Omics Approaches for Managing Infectious Diseases

Technologies, Applications, and Current Achievements

Cláudia S. Oliveira and Irina Negut

4.1 INTRODUCTION

Infectious diseases are illnesses caused by pathogens, which can spread from an infected person, an infected animal, or a contaminated object to a susceptible host [1]. Environmental, social, and economic factors play critical roles in the spread and management of infectious diseases. In fact, urbanization, global travel, climate change, and weak health infrastructures have all played a critical role in the emergence and spread of infectious diseases, imposing a significant rise in their global incidence. Besides, antimicrobial resistance (AMR) further exacerbates the global threat by diminishing the efficacy of standard treatments, leading to prolonged illnesses, higher healthcare costs, and increased mortality rate [2].

For centuries, infectious disease-causing pathogens, including viruses, bacteria, fungi, protozoan parasites, and worms, have remained a persistent threat to human health, and they will continue to challenge global well-being in the future. For instance, acute respiratory infections, diarrheal diseases, immunodeficiency syndrome, malaria, tuberculosis, and various neglected tropical diseases remain the leading causes of infectious mortality worldwide, with a severe impact in Africa [3]. Consequently, global efforts are necessary to combat infectious diseases through a combination of accessible preventive measures, such as vaccination, sanitation, and education, along with advancements in diagnostics, therapeutics, and prevention strategies.

In this sense, the integration of emerging technologies, including precision medicine and omics-based approaches, holds significant promise for developing more targeted and effective interventions.

Precision medicine has revolutionized healthcare, offering tailored diagnostic, therapeutic, and preventive strategies based on individual parameters, lifestyle, and environment [4]. This is especially crucial in infectious disease management and treatment, where a deep understanding of the complex interactions between host, pathogen, and environmental factors can significantly improve treatment efficacy, disease outcomes, and strategies for control and prevention. Furthermore, the rapid identification of disease-causing pathogens and their antimicrobial resistance profiles are essential parameters for selecting the most effective treatment, ensuring optimal patient management, control, and spread.

The integration of omics technologies, such as genomics, transcriptomics, proteomics, metabolomics, microbiomics, and beyond, has provided key advancements in precision medicine. Regarding infectious diseases, these tools have enabled the comprehensive analysis of host and pathogen molecular profiles, uncovering biomarkers for disease susceptibility, drug resistance, and therapeutic response. By integrating these insights, precision medicine enables the design of personalized treatments that address the complexity and heterogeneity of infectious diseases while reducing adverse effects and antimicrobial resistance. These advancements have the potential to significantly improve infectious management and global health outcomes, although disparities in access to innovations worldwide remain challenging.

This chapter provides a comprehensive overview of precision medicine and omics-based strategies in managing infectious diseases, highlighting their applications in early diagnosis, targeted therapies, biomarker discovery, vaccine and drug target development, and personalized treatment approaches – while also addressing current challenges and future directions.

4.2 FOUNDATIONS OF PRECISION MEDICINE IN INFECTIOUS DISEASES

The evolution of precision medicine in infectious diseases has been marked by groundbreaking milestones, beginning with early understandings of pathogen behavior and advancing to genomic insights that revolutionize patient care. This approach tailors interventions to the individual, addressing pathogen variability, host response, and environmental influences.

Precision medicine owes its origins to the systematic study of diseases by early physicians. The germ theory of disease, established in the 19th century by Louis Pasteur and Robert Koch, provided the first scientific framework for understanding infectious diseases. Koch's postulates formalized the link between pathogens and diseases, laying the groundwork for modern diagnostics [5].

DOI: 10.1201/9781003615699-4

The discovery of the structure of deoxyribonucleic acid (DNA) in 1953 by Watson and Crick heralded a new era in medicine, providing tools to explore genetic influences on infection susceptibility. By the late 20th century, advancements in polymerase chain reaction (PCR) and DNA sequencing enabled precise identification of pathogens, significantly improving diagnostics [6].

Also, the completion of the Human Genome Project in 2003 was a pivotal moment in precision medicine. It offered insights into human genetic variability and its role in disease susceptibility. This was critical for understanding host-pathogen interactions, particularly in infections where the immune response varies significantly among individuals [7].

In the early 21st century, pathogen genomics emerged as a vital tool in infectious disease management. Whole genome sequencing allowed rapid identification of drug-resistant strains and facilitated the development of targeted therapies, especially in combating antimicrobial resistance [8]. In addition, launched in 2015 by President Barack Obama, the Precision Medicine Initiative highlighted the importance of integrating genomics, data analytics, and patient-centric approaches in healthcare. This initiative catalyzed research into infectious diseases, promoting advancements in diagnostics, therapeutics, and personalized care [9].

The journey of precision medicine in infectious diseases is ongoing. Future milestones include the integration of multi-omics data to enhance pathogen and host profiling, the development of vaccines tailored to individual immune responses, and overcoming barriers to access and equity. As genomic technologies become more accessible, their application in low-resource settings will be critical to addressing global health disparities [10].

4.2.1 Integration of Medicine, Data Science, and Biotechnology

The integration of medicine, data science, and biotechnology forms the backbone of precision medicine in infectious diseases. By leveraging cutting-edge tools and approaches, this interdisciplinary collaboration aims to address the complexities of infectious pathogens and host responses. This synergy is transforming prevention, diagnosis, and treatment strategies through personalized approaches tailored to individual patients.

In precision medicine, clinicians move beyond traditional diagnostics to integrate molecular insights, such as genetic predispositions and immune profiles, into patient care. The application of molecular pathological epidemiology combines epidemiological methods with molecular diagnostics to uncover how environmental and genetic factors contribute to disease susceptibility and progression [11].

The role of data science in precision medicine cannot be overlooked. Advances in artificial intelligence (AI) and machine learning (ML) enable the processing of vast datasets, including genomics, proteomics, and electronic health records. These technologies identify actionable patterns and biomarkers, allowing for precise predictions of disease progression and therapeutic responses. For example, AI-driven models analyze genomic and proteomic data to refine diagnostics and guide therapeutic interventions [12]. AI tools enhance decision-making by predicting treatment responses and identifying optimal intervention windows. These models are crucial in managing infections – namely, sepsis – where rapid diagnosis and tailored treatment can significantly improve outcomes [13].

Biotechnology innovations, including high-throughput sequencing and Clustered Regularly Interspaced Short Palindromic Repeats (CRISPR) gene-editing tools, enable precise identification of pathogens and the development of targeted treatments. Bioinformatics platforms integrate this data to optimize vaccine design and identify molecular targets for antimicrobial therapies [14]. Moreover, advances in biotechnology have revolutionized vaccine development, with M Ribonucleic Acid (mRNA) vaccines being a prime example. These technologies allow rapid adaptation to emerging pathogens, as seen during the COVID-19 pandemic [15].

Successful integration of these domains requires not only computational advancements but also coordinated policy, ethical governance, and interdisciplinary collaboration. As outlined by Ginsburg and Haga, a robust infrastructure combining clinical, genomic, and digital health systems is necessary to realize the full potential of precision medicine in infectious diseases [16].

4.2.2 Principles of Precision Medicine

Precision medicine aims to tailor medical decisions, treatments, and diagnostics to individual variability in genes, environment, and lifestyle. In infectious diseases, this paradigm enables clinicians to move beyond empirical treatment and instead apply data-driven, personalized strategies to optimize patient outcomes.

4.2.2.1 Genomic Profiling for Pathogen and Host

Genomic sequencing technologies are indispensable in precision medicine for characterizing both pathogen and host factors that influence infection risk, disease severity, and therapeutic response. By identifying pathogen-specific genetic markers and host susceptibility loci, these technologies empower clinicians to move beyond generalized treatment protocols and apply patient-specific strategies.

For pathogens, whole genome sequencing (WGS) has revolutionized infectious disease management by enabling real-time tracking of antimicrobial resistance (AMR) genes and mapping pathogen evolution and transmission. For example, genomic epidemiology has been used to monitor the global spread of resistance genes such as *blaCTX-M*, *mecA*, and *blaKPC* in pathogens like *Escherichia coli* and *Klebsiella pneumoniae*, thereby guiding public health interventions and antibiotic stewardship programs [17]. Sequencing *Mycobacterium tuberculosis* isolates can detect rifampicin and isoniazid resistance, improving therapeutic outcomes [18]. Similarly, host genetic polymorphisms, such as those affecting immune receptors like TLR4, are associated with varying susceptibilities to infections like sepsis or dengue [19]. For example, genome sequencing can guide the selection of antibiotics by predicting resistance patterns [8].

Similarly, genomic surveillance platforms like CZ ID now enable the simultaneous detection of pathogens and AMR genes using cloud-based workflows and metagenomic data. These tools are used in both clinical diagnostics and environmental surveillance to trace hospital-acquired infections and monitor resistance spread in wastewater and sepsis cases [20].

On the host side, human genetic variation plays a critical role in determining susceptibility and immune response. Studies have demonstrated how single nucleotide polymorphisms (SNPs) in immune genes, such as those encoding TLR4, Human Leukocyte Antigen alleles, and cytokine receptors, affect individual responses to infections like COVID-19, tuberculosis, and malaria [21]. These host-pathogen interactions are increasingly being mapped using multi-omics approaches and genome-wide association studies (GWAS).

Dual transcriptomics and host-pathogen genomic integration allow simultaneous assessment of how both genomes respond during infection. This systems-level understanding informs vaccine design, therapeutic targeting, and identification of high-risk patients[22].

Genomic profiling is also central to One Health strategies, linking human, animal, and environmental health through pathogen genome tracking. Studies have emphasized the need for international data sharing, standardized genomic methods, and integrated bioinformatics to manage zoonotic threats and AMR globally [23].

4.2.2.2 Integration of Multi-omics Data

Precision medicine relies heavily on systems biology, particularly the integration of multiple layers of molecular data (genomics, transcriptomics, proteomics, metabolomics, epigenomics, and microbiomics) to obtain a holistic understanding of host-pathogen interactions. This approach, often termed multi-omics, improves diagnostic accuracy, enables biomarker discovery, and facilitates the design of highly personalized interventions in infectious diseases.

One of the key strengths of multi-omics is its ability to resolve complex biological networks that govern disease progression and immune response. For example, multi-omics profiling has been used to understand immune signaling in viral infections and identify patients at risk of severe outcomes by linking immune transcriptomics with plasma proteomics and metabolomics data [24].

A comprehensive review by Abdelaziz et al. describes the full pipeline for integrating omics data in precision medicine, from dimensionality reduction and ML algorithms to biomarker identification and survival prediction. The authors emphasize the need for harmonized databases and robust analytical models to translate multi-omics into actionable clinical insights [25].

Network-based systems biology approaches have proven particularly valuable in uncovering molecular drivers of infectious diseases. These methods reduce complexity by modeling interactions across omics layers, improving the precision of drug targeting and therapeutic development [26]. For example, the integration of pharmacogenomics with transcriptomic and proteomic data can reveal personalized responses to antimicrobial agents. This has been demonstrated in tuberculosis and hepatitis B virus infections, where host genomic variation in drug-metabolizing enzymes impacts therapy efficacy and toxicity [27].

Advanced AI-driven frameworks such as cross-modal neural networks and attention-based deep learning (DL) models have been developed to harmonize disparate omics layers while retaining critical biological signals. These models improve prediction accuracy and uncover hidden relationships across data types [28].

Importantly, multi-omics integration is also transforming diagnostics. For instance, microbiomics combined with host metabolomics has improved the early diagnosis of *Clostridium difficile* infection and inflammatory bowel disease flares [29, 30]. Similarly, proteogenomics is being used to stratify sepsis patients and identify those who may benefit from immunomodulatory therapies [31].

Despite its promise, challenges remain, including data heterogeneity, lack of standardization, and ethical concerns regarding the integration of genomics with personal health records. However, the increasing availability of interoperable bioinformatics tools and cloud-based platforms is gradually overcoming these barriers [32].

4.2.2.3 Biomarker-Driven Therapy

Biomarkers play a pivotal role in precision medicine, serving as tools for stratifying patients based on disease susceptibility, progression, and treatment responsiveness. In the case of infectious diseases, biomarkers offer a bridge between molecular insights and actionable clinical decisions, enabling earlier diagnoses, optimized therapies, and reduced treatment-related toxicity.

Despite their potential, biomarker implementation faces practical challenges such as variability in tissue acquisition, sample degradation, and analytical inconsistencies. As outlined by Pritzker, inconsistencies in fixation, sampling, and informatics pipelines can reduce diagnostic precision, particularly when applied to fragile or scarce clinical samples [33].

To overcome these limitations, liquid biopsies have emerged as non-invasive alternatives, detecting circulating nucleic acids or microbial signatures directly from blood. This method improves patient compliance and enables dynamic monitoring of treatment efficacy. It is especially useful in tracking viral loads (e.g., human immunodeficiency virus (HIV), hepatitis B/C) and resistance mutations without repeated tissue biopsies [34].

In addition, single-cell transcriptomics offers high-resolution views of cellular heterogeneity during infection. It has been used to uncover immune signatures specific to severe versus mild COVID-19 outcomes and to predict responsiveness to immunomodulatory therapies [35].

Biomarker panels integrating proteomics and metabolomics are also advancing infectious disease care. For example, early sepsis biomarkers such as procalcitonin and interleukin-6, combined with genomic profiling, allow for faster triaging and targeted antibiotic therapy. These multiplex platforms reduce diagnostic delays, especially in critical care settings [36].

Another key development is risk biomarker modeling, wherein population-derived data and ML are used to predict infectious disease risk and stratify patients for surveillance or preemptive interventions. This is especially relevant in public health initiatives for chronic infections and emerging pathogens [37].

The biomarker pipeline is expanding through initiatives such as the Immunoscore, which integrates immune cell density into prognostic assessments, initially in oncology but now explored in infectious diseases such as tuberculosis and hepatitis for immune-based stratification [33].

The host's genetic and immunological profile significantly influences disease outcomes. Advances in immunotherapy, inspired by oncology, show promise for targeting host pathways exploited by pathogens, offering novel strategies to combat antimicrobial resistance [7].

High-throughput sequencing technologies, including precision metagenomics, enable fast and accurate identification of pathogens in clinical samples. These methods provide critical data on antimicrobial resistance and facilitate personalized treatment plans [38].

4.2.2.4 Artificial Intelligence and Predictive Modeling

The fusion of AI and precision medicine is playing a transformative role in the management of infectious diseases. With the help of ML algorithms on high-dimensional datasets, from genomic sequencing to clinical and environmental records, AI offers rapid, adaptive, and personalized solutions to longstanding challenges in diagnostics, epidemiology, and treatment optimization.

One of the most comprehensive frameworks in this field was proposed by Fatima et al., who demonstrated how AI can unify genomics, proteomics, microbiomics, and clinical data to support precise diagnosis, risk prediction, and individualized therapy in infectious diseases. Their model emphasized a multidisciplinary strategy that allows clinicians to anticipate disease course and deliver more personalized care [12].

Rayan highlighted the potential of AI tools like IBM Watson in precision medicine, particularly in infectious disease decision-making. He notes how the selection of appropriate AI models

– depending on the task, whether diagnosis or therapy optimization – is key to effective integration into healthcare systems [39].

AI has also been used to guide drug repurposing and vaccine design in infectious disease outbreaks. For instance, predictive modeling helped identify potential antiviral compounds during the early stages of SARS-CoV-2 by simulating host-pathogen interactions and screening drug-target networks [40].

In diagnostic applications, AI has improved the detection of pathogens using metagenomic data, especially in cases where conventional diagnostics failed. These tools integrate sequencing reads with clinical features to provide more accurate microbial identification, particularly for polymicrobial or rare infections [41].

Lastly, explainable AI (XAI) frameworks are enabling clinicians to interpret the rationale behind model predictions, fostering greater trust in automated decision systems – an essential factor for clinical uptake. This is particularly useful in predicting antibiotic resistance and understanding patient-specific responses to treatments [42].

4.2.2.5 Precision Metagenomics and Infectious Disease Diagnostics

Precision metagenomics, powered by high-throughput next-generation sequencing (NGS), has significantly enhanced infectious disease diagnostics by enabling unbiased detection of pathogens and AMR markers directly from clinical samples. Unlike traditional culture-dependent methods, metagenomics allows for rapid, comprehensive profiling of microbial communities, including rare, fastidious, or novel pathogens, in a single assay.

A foundational study by Afshinnekoo et al. introduced the concept of "precision metagenomics," detailing protocols for DNA/RNA extraction, library preparation, sequencing, and bioinformatics workflows that provide not only pathogen detection but also resistance gene identification and metabolic profiling within 24–72 hours depending on the analysis depth [38].

In a clinical setting, metagenomics has proven effective in guiding antibiotic therapy. For instance, the Explify® Urinary Pathogen ID Panel, a precision metagenomics platform, enabled physicians to detect over 170 pathogens and 2,000 AMR markers from urine samples. It not only showed high agreement with traditional culture but also uncovered additional pathogens and resistance genes missed by standard testing, altering treatment decisions in 28% of cases [43].

The diagnostic superiority of metagenomics was also demonstrated in a comparative analysis, where it identified 62 distinct uropathogens from patient samples, vastly outperforming PCR (19 organisms) and traditional culture (13 organisms). This not only improved diagnostic accuracy in polymicrobial and culture-negative infections but also broadened insight into resistance profiles [44].

Beyond UTIs, metagenomic sequencing has shown immense utility in respiratory, bloodstream, and central nervous system infections, where time-critical decisions are vital. In these cases, real-time metagenomics delivers actionable results faster than conventional diagnostics, improving survival and antimicrobial stewardship.

Moreover, tools like MetaGeneMiner now allow targeted extraction of AMR-related sequences from large metagenomic datasets. In intensive care unit patients, this tool accurately identified resistance genes for *Acinetobacter baumannii* and *Herpes simplex* virus -1, proving its utility in critical care settings [45].

A broader systematic review confirmed the high diagnostic performance of metagenomics, with a pooled sensitivity of 88% and specificity of 86% across diverse infectious conditions. Despite challenges in cost, contamination control, and human DNA background, the review predicts metagenomic sequencing will become a cornerstone of clinical diagnostics in the near future [46].

4.2.2.6 Pharmacogenomics and Therapeutic Individualization

Pharmacogenomics, the study of how genetic variation affects individual responses to medications, has become a cornerstone of precision medicine in infectious diseases. It allows clinicians to tailor treatment regimens based on patient-specific genetic markers, optimizing drug efficacy while minimizing adverse effects.

In the management of tuberculosis, pharmacogenomic research has identified polymorphisms in genes such as NAT2 and SLCO1B1, which influence isoniazid metabolism and drug-induced hepatotoxicity. Studies at the Center for Personalized Precision Medicine of Tuberculosis (cPMTb) in South Korea have shown that individualized drug dosing based on these polymorphisms can improve treatment outcomes and reduce toxicity [47].

Similarly, a precision medicine model proposed for drug-resistant tuberculosis emphasizes using host genetic profiles and pathogen genomics to guide drug selection, dose optimization, and treatment duration, redefining tuberculosis management in resource-limited settings [48].

In HIV therapy, pharmacogenomics plays a vital role in managing antiretroviral treatment (ART). For instance, Cytochrome (CYP) 2B6 polymorphisms significantly influence plasma levels of efavirenz, a commonly used ART drug. Patients with slow-metabolizing alleles may experience neurotoxicity unless doses are adjusted accordingly [49]. Precision dosing based on these genotypes improves adherence, reduces toxicity, and increases viral suppression rates [50].

Pharmacogenomics is also valuable in co-infection scenarios. In patients with HIV-tuberculosis co-infection, the risk of drug-drug interactions is high. Pharmacogenetic testing facilitates safer co-administration of ART and anti-tuberculosis drugs, particularly in settings with limited treatment alternatives [51].

Ongoing global initiatives are expanding the application of pharmacogenomics in infectious diseases. For example, in Africa, researchers are working to validate PGx panels in tuberculosis and HIV populations where drug toxicity and resistance are prevalent. This includes characterizing NAT2, CYP, and ABC transporter polymorphisms in ethnically diverse populations to reduce ADRs and improve efficacy [52].

As the costs of genotyping decline and infrastructure improve, pharmacogenomics can become a standard tool in individualized infectious disease therapy, reducing treatment failure, preventing resistance, and improving patient safety.

4.3 OMICS TECHNOLOGIES IN PRECISION MEDICINE FOR INFECTIOUS DISEASE MANAGEMENT

Over the past decade, omics technologies have been extremely useful in the management of infectious diseases, driven by advancements in genomics, transcriptomics, proteomics, metabolomics, and microbiomics [53]. These approaches integrate high-throughput biochemical assays, advanced computational tools, and extensive biological databases that enable complex and multidimensional analyses. By leveraging diverse biological samples, including blood, saliva, urine, cerebrospinal fluid, feces, tissue biopsies, and cultured cells (Table 4.1), omics technologies have amplified the understanding of infectious diseases. They facilitate precise diagnostics, biomarker identification, targeted therapy development, and personalized treatment strategies, ultimately advancing precision medicine while improving patient outcomes.

Genomic technologies, such as NGS and WGS have emerged as powerful tools for the sequencing of both pathogen and host genomes [54]. Pathogen genome sequencing facilitates the identification of new microorganisms, endemic pathogens, virulence factors, drug resistance genes, and transmission patterns, playing a crucial role in outbreak surveillance and vaccine development. Meanwhile, host genome analysis provides valuable insights into genetic susceptibility, immune responses, and vaccine-derived immunity, with implications for clinical decision-making, and personalized treatment approaches.

In contrast, transcriptomic analyses evaluate the presence, abundance, and dynamics of RNA molecules within a specific physiological or experimental context, providing insights into gene expression patterns and regulatory mechanisms [55]. This tool plays a pivotal role in understanding gene expression changes during infection, offering critical insights into host-pathogen interactions and immune responses. Single-cell RNA sequencing (scRNA-seq) enables the analysis of transcriptional changes at a single-cell level, refining our understanding of cellular responses to infection with unprecedented resolution [56]. Meanwhile, dual RNA sequencing (dual RNA-Seq) allows the simultaneous evaluation of both host and pathogen transcriptomes from the same biological sample, providing a comprehensive view of their dynamic interactions. To achieve this, various biological samples derived from uninfected hosts, infected hosts, and pathogens have been evaluated and screened, providing valuable insights about host responses and infection mechanisms (Figure 4.1).

For instance, by analyzing mRNA expression profiles, researchers have identified key regulatory pathways activated during infection, uncovering potential therapeutic targets and biomarkers for disease progression [57]. This approach has been particularly valuable in studying viral infections, such as SARS-CoV-2, where transcriptome analyses have shed light on viral replication mechanisms and host immune evasion strategies. As transcriptome profiling continues to evolve, it holds immense potential for advancing precision medicine and developing targeted interventions against infectious diseases.

Proteomics plays a crucial role in deciphering the structure, function, and interactions of proteins, providing deep insights into an organism's biology [58]. In the field of infectious diseases, mass

TABLE 4.1 Overview of Key Omics Technologies in Precision Medicine for Infectious Disease Management

Omics Technology	Concept	Biological Samples	Advantages	Disadvantages
Genomics	Study of the complete genome, including DNA sequencing and genetic variations	Blood, saliva, tissue biopsies, cultured cells	Identifies genetic mutations. Useful for disease risk assessment	Does not capture dynamic gene expression or environmental influences
Transcriptomics	Analysis of RNA expression to study gene activity	Bulk tissue, single cells, biofluids (e.g., plasma, cerebrospinal fluid)	Reveals gene expression patterns and regulatory mechanisms	Sensitive to sample quality, requires normalization for accurate interpretation
Proteomics	Large-scale study of proteins, their structures, and functions	Blood, urine, cerebrospinal fluid, tissues	Ability to reflect biological functions; Identification of disease biomarkers	Protein stability issues require high-resolution mass spectrometry
Metabolomics	Comprehensive analysis of small molecules involved in cellular metabolism	Plasma, urine, cerebrospinal fluid, feces, tissues	Provides insight into metabolic pathways, links genotype to phenotype	Highly dynamic, requires strict sample handling and complex data analysis
Microbiomics	Study the ecological community of microorganisms that symbiotically or pathologically live on plants or animals, including bacteria, archaea, protozoa, fungi, and viruses	Feces, skin swabs, oral samples, environmental samples	Identifies host-microbe interactions, essential for microbiome-based therapies	Requires deep sequencing, high inter-individual variability

spectrometry-based proteomics has been instrumental in detecting pathogen-derived proteins and host inflammatory markers, enabling early and accurate diagnosis. Beyond diagnostics, proteomics has significantly contributed to vaccine development by identifying immunogenic proteins capable of triggering protective immune responses [59]. The immune response induced by vaccines operates like a complex "social network," involving the coordinated action of both innate and adaptive immune cells. By integrating high-throughput cellular and molecular omics technologies, modern vaccine research has enabled a comprehensive analysis of immune responses, deepening our understanding of vaccine-induced immunity and paving the way for more effective immunization strategies.

Translating to cellular metabolism, the metabolic profile offers a dynamic snapshot of the intricate interplay between the genome, environment, and biochemical processes. Metabolites are small molecules (50 to 1,500 Daltons) generated through cellular processes, acting as essential regulators of cell activity and functioning as key bioindicators of cellular physiology [60]. By analyzing metabolites, such as lipids, amino acids, and carbohydrates, *metabolomics* has identified specific metabolic signatures associated with infections. These metabolic fingerprints serve as potential biomarkers for disease diagnosis, prognosis, and treatment monitoring. For instance, advances in high-throughput techniques, such as nuclear magnetic resonance and mass spectrometry, have enabled the identification of metabolic pathways exploited by pathogens, offering new avenues for antimicrobial drug discovery.

Finally, *microbiomics* explores the complex interactions between the human microbiota and infectious agents. The gut, skin, and respiratory microbiomes play crucial roles in modulating immune responses and influencing susceptibility to infections. However, imbalances in microbial communities, known as dysbiosis, have been linked to increased vulnerability to bacterial, viral, and fungal infections. Understanding microbiome composition and function provides valuable insights into host-pathogen interactions, paving the way for microbiome-targeted therapies, such as probiotics, prebiotics, and fecal microbiota transplantation, and microbial-based drugs [61]. Furthermore,

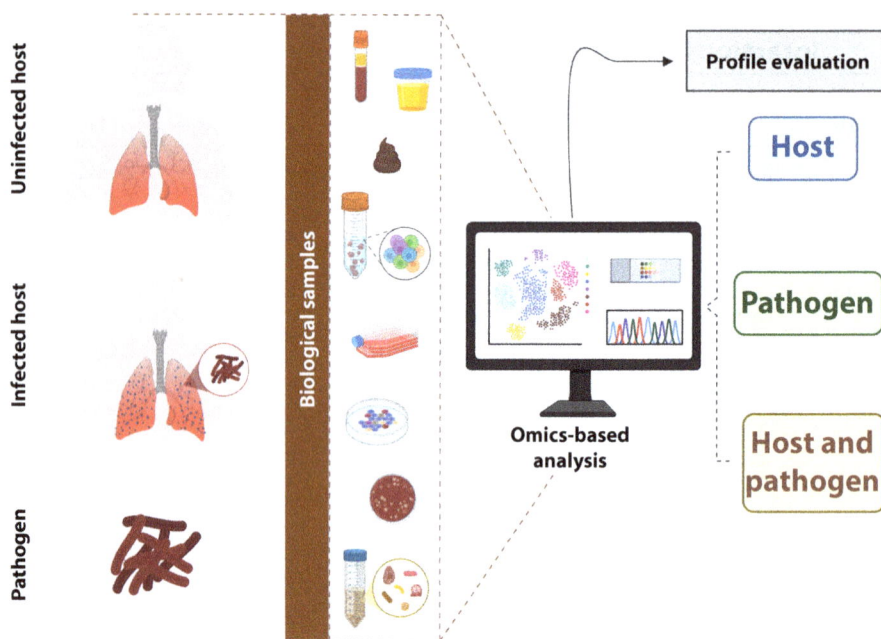

Figure 4.1 Representation of sample collections for omics analysis. Various biological samples, including blood, urine, feces, biopsy, cultured cells, cultured pathogens, and so on, can be collected from uninfected and infected hosts, as well as pathogens, for omics-based analysis. Depending on the approach, individual profiles of the host or pathogen can be generated, both profiles can be analyzed simultaneously. (Created in BioRender. Oliveira, C. (2025) https://BioRender.com/mm3rorx.)

microbiomics contributes to the development of novel antimicrobial strategies by identifying beneficial microbes with protective properties against pathogens.

4.4 APPLICATIONS OF OMICS TECHNOLOGIES IN INFECTIOUS DISEASES

As previously described, the use of omics technologies has revolutionized infectious disease management by providing a comprehensive understanding of pathogen biology, antimicrobial resistance mechanisms, host responses, and human-pathogen interactions. These advancements have significantly contributed to precise diagnosis, disease surveillance, optimized/personalized therapeutic interventions, vaccine development, and strengthened prevention strategies (Figure 4.2).

4.4.1 Diagnosis

Traditional diagnostic methods, such as microbial cultures and serological tests, often require extended processing times and have poor sensitivity? In contrast, omics-driven molecular diagnostics significantly improve the speed, sensitivity, and accuracy of pathogen detection and identification. For this, the integration of high-throughput sequencing, bioinformatics, and AI-driven predictive analytics has enabled the development of next-generation diagnostic tools. For this, the main techniques used are:

i. **CRISPR-based detection**: Provides rapid identification of specific nucleic acid sequences in pathogens within minutes;

ii. **NGS**: Allows real-time sequencing for pathogen identification without extensive sample preparation;

iii. **Multiplex PCR and microarrays**: Simultaneous detection of multiple pathogens and resistance genes in a single test; and

iv. **Proteomic biomarkers and mass spectrometry (MS)**: Identification of pathogens based on unique protein signatures, reducing reliance on genomic data alone.

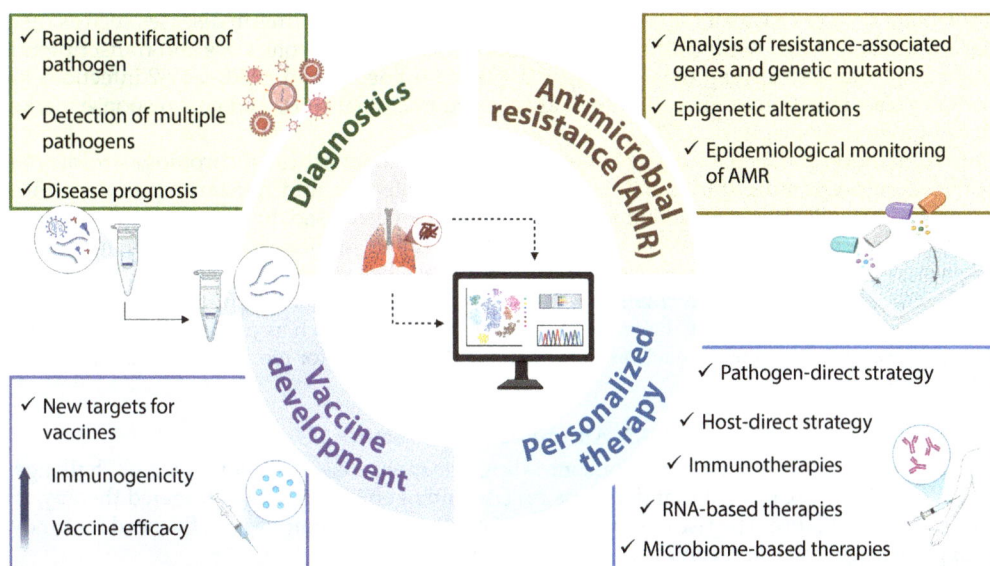

Figure 4.2 Key applications of omics technologies in infectious diseases management, highlighting their role in diagnostics, antimicrobial resistance strategies, personalized therapies, and vaccine development. (Created in BioRender. Oliveira, C. (2025) https://biorender.com/h17h1q5.)

For example, the COVID-19 pandemic highlighted the critical role of genomics and proteomics in pathogen characterization, particularly in identifying viral mutations and post-translational modifications [62]. These insights were pivotal in developing highly sensitive and specific SARS-CoV-2 diagnostic tools.

Currently, reverse transcription polymerase chain reaction (RT-PCR) is the gold standard for SARS-CoV-2 detection. However, its efficacy varies across biological samples, with certain matrices, such as urine, presenting limitations for detection. As an alternative, MS-based proteomic approaches have emerged as powerful tools due to their rapid and high-sensitivity capabilities. Targeted proteomic analyses have facilitated the identification of viral peptides as biomarkers for SARS-CoV-2 detection. A notable example is the development of a protein-based microarray for high-throughput characterization of proteome-wide antibody responses, which has been instrumental in identifying diagnostic and therapeutic targets for COVID-19 [63]. This approach has demonstrated high sensitivity in detecting SARS-CoV-2 nucleocapsid proteins across various biological samples, including nasopharyngeal swabs, respiratory specimens, gargle solutions, and serum.

Similarly, to complement conventional diagnostic tools, proteomic analysis of *Plasmodium spp.*, including both parasite pellet and secreted proteins in plasma, has been utilized for malaria diagnostics [64]. Additionally, the host protein profile serves as a prognostic tool, as a panel of differentially regulated host proteins in malaria patients, compared to healthy volunteers, provides insights into disease progression and severity [64]. In this case, multiplexed quantitative proteomics identifies differentially altered host proteins, revealing key pathways affected in severe and non-severe malaria conditions. These include disruptions in lipid metabolism, complement activation, platelet degranulation, and homeostasis, providing valuable insights into malaria pathophysiology. Hence, a comprehensive proteomic analysis encompassing both parasite and host proteins offers a powerful approach for precise diagnostics, disease prognosis, and the identification of optimal therapeutic strategies.

Metagenomic sequencing has allowed for unbiased pathogen detection directly from clinical samples, bypassing the need for culture and enabling identification in cases of unknown or complex infections [65]. In sepsis, omics-based biomarkers have shown promise for early and precise diagnosis, improving patient stratification and therapy selection [66].

Transcriptomics has been employed to distinguish bacterial from viral infections using host gene expression profiles, leading to better diagnostic accuracy and antibiotic stewardship [67].

Proteomic analyses have identified diagnostic serum protein signatures in diseases such as tuberculosis and malaria [68] while multi-omics approaches integrating proteomics and transcriptomics have been pivotal in mapping the host response during influenza and SARS-CoV-2 infections [69]. In urinary tract infections, omics profiling has revealed microbial shifts and host response markers that can guide early diagnostics [70].

Additionally, omics has played a key role in personalized diagnostics for chronic gastrointestinal infections, outperforming traditional tests in pathogen detection and resistance profiling [65]. In wastewater surveillance, metagenomic approaches have been used to detect and map human pathogens and resistance genes in environmental samples, aiding public health monitoring. Moreover, companion diagnostics using omics circuits are being developed for real-time, patient-specific disease monitoring and treatment decisions in infectious and inflammatory diseases [71]. In veterinary infectious disease diagnostics, omics technologies such as proteomics are also being increasingly applied for early detection of zoonotic pathogens and for monitoring disease spread in livestock [72].

4.4.2 Antimicrobial Resistance

While accurate diagnosis is essential for controlling infectious diseases, another major challenge is AMR, a growing global threat that demands precise pathogen identification, targeted therapy, and continuous surveillance. The rise of multidrug-resistant pathogens emphasizes the need for innovative approaches to track and combat resistance mechanisms.

Omics technologies play a crucial role in uncovering the physiological mechanisms underlying AMR. By enabling high-resolution analysis of resistance genes, genetic mutations, and epigenetic modifications, these approaches provide fundamental insights into the evolution and adaptation of resistant microorganisms [73]. This understanding is essential for developing targeted strategies and monitoring and controlling AMR.

These advances are made possible through cutting-edge tools, including the following:

i. **WGS for resistance profiling**: Identifies known and novel resistance genes that facilitate the development of personalized antimicrobial therapies.

ii. **Transcriptomic analyses**: Highlight overall alterations in gene expression in response to environmental events, providing a snapshot of the genes that act under a given condition.

iii. **Metagenomic surveillance**: detects resistance elements in environmental and clinical samples, supporting epidemiological tracking of AMR trends and monitoring.

iv. **Epigenomics in pathogen adaptation**: Evaluates DNA methylation and histone modifications that influence pathogen virulence and drug resistance mechanisms.

v. **MS-based proteomics**: Identifies resistance-associated protein signatures, allowing functional validation of genomic findings and rapid characterization of resistant strains.

vi. **Metabolomics for drug resistance screening**: Analyzes metabolic alterations in resistant strains to identify biomarkers that predict treatment efficacy and potential therapeutic targets.

For instance, recent transcriptomic evaluation has investigated key AMR in *Escherichia coli*, focusing on compensatory mutations that arise in response to the loss of major antibiotic efflux pumps, *AcrEF* and *AcrAB* [74]. Cho et al. identified mutations in four key regulatory genes (*rpoB*, *baeS*, *hns*, and *crp*) which may activate alternative pathways to enhance antibiotic resistance. RNA-seq analysis revealed a downregulation of DNA and protein biosynthesis pathways, while stress response pathways were upregulated, suggesting an adaptive mechanism to counteract cellular stress. Notably, the study highlighted that these compensatory mutations can interact synergistically, restoring antibiotic resistance to levels comparable to those of the efflux pump-competent parental strain [74].

4.4.3 Personalized Therapeutics

Usually, traditional antimicrobial treatments adopt a one-size-fits-all approach, which can lead to ineffective treatments, prolonged infections, and the emergence of AMR. In contrast, omics technologies have revolutionized the development of precision antimicrobial therapies and host-targeted treatments by enabling a deeper understanding of pathogen-host interactions, resistance mechanisms, and individual patient responses. The use of omics technologies, namely genomics, transcriptomics, proteomics, and metabolomics, enhance therapeutic efficacy while reducing adverse effects and the risk of AMR.

For instance, WGS of *Mycobacterium tuberculosis* has been employed to detect resistance to both first- and second-line anti-tuberculosis drugs [75]. This approach provides valuable insights for the clinical management of drug-resistant tuberculosis cases by enabling the rapid identification of resistance-associated mutations. SNPs in genes like CYP450 are critical in predicting drug metabolism rates, allowing clinicians to adjust dosages and avoid toxicity [76]. Similarly, in tuberculosis treatment, genomic profiling helps to identify mutations in the rpoB gene of *Mycobacterium tuberculosis* that confer resistance to rifampicin, guiding clinicians in selecting more effective therapies [77]. Additionally, WGS facilitates the selection of more effective drug combinations, optimizing treatment strategies and reducing the risk of further transmission of resistant strains.

Another example is malaria, one of the three major global health threats alongside tuberculosis and HIV. The emergence of drug resistance poses a significant challenge to the treatment, making the search for efficient therapeutic strategies a high priority in malaria management and control. To address this, omics technologies have provided a comprehensive understanding of the biological mechanisms underlying malaria infection. For instance, the integration of multi-omics data has identified key molecular pathways, drug resistance markers, and potential therapeutic targets, ultimately guiding the development of more effective and personalized treatment strategies for patients with such conditions [78]. The emergence of chloroquine-resistant strains has been linked to genetic alterations, particularly in the *Plasmodium falciparum* chloroquine resistance transporter (PfCRT) gene, the ABC transporter *P. falciparum* multidrug resistance (PfMDR1) gene, and modifications in the chloroquine-transporter protein CG2. These genetic changes contribute to reduced drug efficacy, necessitating alternative therapeutic strategies to chloroquine. Consequently, a combination of therapies incorporating chloroquine with other drug agents has been proposed to mitigate resistance to *P. falciparum* and enhance treatment effectiveness [79].

Beyond pathogen-directed strategies, omics technologies may also drive host-directed therapies (HDTs), which aim to modulate the host immune response or specific cellular pathways to enhance infection control. In this line, the most important approaches include the following:

i. **Host genomics for precision immunotherapy**: GWAS and single-cell sequencing help identify host genetic variations that influence infection susceptibility and treatment response. These tolls have contributed to the development of personalized immunotherapies.

ii. **Transcriptomics-guided immune modulation**: Understanding host gene expression patterns during infection allows for targeted interventions, such as cytokine therapy or immune checkpoint modulation, to enhance immune responses against pathogens. For instance, in malaria, transcriptomic analysis of patient samples has helped categorize responses to different antimalarial treatments and predict the efficacy of vaccine candidates like malaria vaccine (RTS, S) [80]. In chronic viral infections such as HIV and hepatitis, transcriptomics has been used to assess immune escape mechanisms, enabling the development of therapies that are tailored to the patient's viral load and immune status [81]. Such insights aid in personalized immune modulation, helping clinicians adjust immune therapy to enhance patient outcomes.

iii. **RNA therapeutics and epigenetic modifications**: Advances in RNA sequencing and epigenomics enable the development of novel RNA-based drugs and epigenetic modulators that enhance host defense mechanisms.

iv. **Metabolomics and microbiome-based therapeutics**: Profiling host metabolic and microbiome changes during infection can guide probiotic or metabolic interventions to restore immune homeostasis and prevent secondary infections. Personalized metabolic profiling is also increasingly being used to monitor antibiotic resistance in bacterial infections. For instance, in *Pseudomonas aeruginosa* infections, metabolomics has been used to track the changes in metabolic pathways that accompany antibiotic resistance, leading to the development of tailored interventions [82].

In general, HDTs enhance defense mechanisms by strengthening the immune response against pathogens. They achieve this by targeting virulence factors and modulating disease-related inflammation, ultimately leading to improved clinical outcomes, including reduced organ damage, mortality and morbidity [83].

Although HDTs offer significant benefits in combating various infectious diseases, including HIV, SARS-CoV-2, and *M. tuberculosis*, their clinical application presents certain challenges. Since host factors play essential roles in numerous physiological processes, targeting them may lead to unintended side effects. Therefore, a key challenge in HDT implementation is the need for personalized therapy, which requires comprehensive health profile screening to tailor treatments to individual patients while minimizing risks. Furthermore, another crucial consideration is that HDTs should not be used as standalone therapies but rather as adjuncts to pathogen-directed strategies. Given the dynamic nature of immune responses, the effectiveness of HDTs depends on the disease stage, necessitating a strategic combination with antimicrobial treatments to achieve optimal clinical outcomes.

4.4.4 Vaccine Development and Drug Target Discovery

Conventional vaccine development often involves labor-intensive processes such as pathogen attenuation, inactivation, or subunit antigen identification. However, in the last decade, omics technologies have impulsed vaccine research, enabling rapid identification of potential vaccine targets, enhancing immunogenicity, and improving vaccine efficacy. This section explores then the main omics technologies in vaccine development and highlights some successful applications.

4.4.4.1 Genomics and Reverse Vaccinology

Among the various advances in infectious disease research, genomics has significantly enhanced our understanding of pathogen genomes. Reverse vaccinology leverages genomic data to predict surface-exposed or virulence-associated antigens without the need for culturing pathogens, significantly accelerating vaccine development.

WGS, in particular, allows the comprehensive analysis of a pathogen's genetic composition and the prediction of antigenic components capable of eliciting robust immune responses. A striking example is the rapid sequencing of the SARS-CoV-2 genome in early 2020, which accelerated the development of mRNA vaccines, such as Pfizer-BioNTech® and Moderna® [84]. These vaccines deliver a nanoparticle-encapsulated mRNA platform to instruct host cells to synthesize the viral spike protein and trigger a targeted immune response involving both B and T lymphocytes. It has also been used for *Streptococcus agalactiae* to identify conserved proteins for neonatal vaccines [85].

Another example of WGS application is its use in analyzing *Neisseria meningitidis* isolates to identify and select candidate vaccine antigens. This genomic approach led to the development of the MenB vaccine (Bexsero®), a multivalent formulation containing three recombinant antigens: Neisserial Heparin Binding Antigen, Factor H Binding Protein, and Neisseria Adhesin A [86]. These antigens were strategically chosen based on their ability to enhance immune recognition and protection against meningococcal infections.

In parasitic infections, reverse vaccinology has identified novel vaccine targets in *Plasmodium falciparum* for malaria [87]. In leishmaniasis, reverse vaccinology has led to the identification of conserved proteins across *Leishmania major* and *L. infantum*. These antigens have shown promising binding affinity to human major histocompatibility complex (MHC) class I and II alleles and are now being tested as synthetic polyepitope vaccines [88]. A more recent study used genome-wide prediction in *Leishmania major* to identify 25 proteins containing signal peptides and glycosylphosphatidylinositol anchors, with robust T-cell epitope predictions using tools like NetCTL and OptiTope [89]. Also, T-cell epitopes in *Leishmania spp.* have been identified [90]. In cystic fibrosis–related lung infections, chronic colonization by *Pseudomonas aeruginosa*, *Burkholderia cepacia*, and *Haemophilus influenzae* contributes to significant morbidity. Reverse vaccinology has been used to identify surface-expressed, non-human-homologous proteins in these pathogens as targets for multi-pathogen vaccine design. These strategies aim to reduce biofilm formation and improve respiratory health in affected individuals [91].

Moreover, reverse vaccinology has shown promise in less-explored areas, such as *Rhipicephalus microplus* (cattle tick) control, where over 700 potential vaccine candidates were identified using genomics and immunoinformatics [92], and in *Bacillus anthracis* (anthrax), where in silico epitope mining tools have facilitated the design of potent recombinant vaccine candidates [93].

4.4.4.2 Proteomics and Immunoproteomics

Proteomics and immunoproteomics have become indispensable tools in vaccine discovery, enabling the identification of pathogen proteins that are actively expressed during infection and are capable of eliciting immune responses. By combining mass spectrometry-based protein profiling with serological screening, these approaches help uncover immunodominant antigens critical for protective immunity. For instance, in *Opisthorchis viverrini* infections, proteomic and immunoinformatic methods were used to predict antigens targeting both B and T cells [94]. In *Neisseria meningitidis* serogroup B, immunoproteomics revealed surface-expressed proteins that were validated in murine immunization models [95].

Similarly, *Bacillus anthracis* was studied using proteomic analysis to identify vaccine candidates through epitope mapping and host-pathogen interaction data [96]. In the case of *Streptococcus spp.*, proteomic strategies enabled the discovery of conserved antigens relevant to multiple strains, contributing to cross-protective vaccine development [97].

In viral diseases, *Neisseria gonorrhoeae* has been extensively profiled using immunoproteomics and structure-function antigen analysis to reveal envelope proteins suitable for subunit vaccine design [98]. Additionally, structural and surface proteomics have facilitated antigen discovery for pathogens like *Bordetella pertussis* and *Propionibacterium acnes*, revealing novel immunogenic proteins suitable for recombinant vaccine development [99, 100]. In zoonotic pathogens such as ticks (*Rhipicephalus spp.*), proteomics has been employed to identify both tick- and pathogen-derived vaccine candidates targeting livestock diseases [101].

Furthermore, the application of proteomics has proven valuable in identifying surface-exposed and secreted proteins as potential vaccine candidates, such as for tuberculosis. Given that *M. tuberculosis* relies on multiple secreted proteins to mediate its virulence, targeting these proteins has facilitated the development of attenuated immunogenic vaccines. This approach has emerged as a promising alternative to the Bacillus Calmette-Guérin vaccine, which provides protection against childhood tuberculosis but is ineffective in preventing pulmonary tuberculosis. Notably, proteomics-driven research on *M. tuberculosis* has led to the identification of immunogenic proteins, contributing to the development of novel vaccine candidates such as *Mycobacterium tuberculosis* vaccine, a live-attenuated vaccine currently undergoing clinical trials [102].

4.4.4.3 Transcriptomics for Dynamic Target Identification

While genomics has been pivotal in identifying pathogen genomes and vaccine targets, transcriptomics has further deepened our understanding of key virulence mechanisms and immune evasion strategies. This has facilitated the development of more targeted and effective vaccines.

4.4.4.3.1 Bacterial Infections

Transcriptomic technologies have significantly enhanced our ability to identify druggable and vaccine-relevant targets by profiling gene expression patterns under infection-relevant conditions. This is particularly useful in elucidating the biology of persistent pathogens and uncovering hidden therapeutic opportunities.

This approach has been extensively applied to *Mycobacterium tuberculosis*, where transcriptomics has helped identify drug-responsive genes and regulatory pathways associated with latency and antibiotic resistance [103]. Similarly, a systems-based omics review highlighted how transcriptomics and proteomics have revealed host-pathogen interactions and stress adaptations during infection, particularly Mtb's manipulation of host immunity and nutrient scavenging within the intracellular niche. These findings support efforts to design anti-tuberculosis drugs targeting metabolic cross-talk and dormancy pathways [104]. Metabolic flux analysis has also shown how Mtb reprograms carbon and nitrogen utilization under host-imposed stress, adapting central pathways like glycolysis, the tricarboxylic acid cycle, and glutamine metabolism to persist during treatment – highlighting these pathways as potential drug targets for both active and latent tuberculosis [105].

Similarly, spatial transcriptomics, an emerging branch of the field, has enhanced resolution in identifying infection sites and localized immune responses, proving valuable in drug discovery and host-pathogen mapping [106]. Transcriptomic analysis has been pivotal in uncovering genes involved in dormancy, latency, and resistance to first-line drugs. Studies have identified high-in-degree proteins within the transcriptional network that function as regulatory hubs, representing attractive drug targets due to their essentiality in bacterial survival and stress adaptation [107]. Additional work using subtractive proteomics and conserved gene filtering across multiple *M. tuberculosis* strains has shortlisted membrane proteins and enzymes involved in redox and lipid metabolism as potential broad-spectrum targets [108].

In the case of *Streptococcus pyogenes*, metabolic pathway analysis combined with transcriptomic data revealed essential enzymes, such as 6-phosphofructokinase, which serve as candidate drug targets [109]. In *Staphylococcus aureus*, whole-transcriptome profiling revealed 94 essential non-homologous genes, many associated with cell wall biosynthesis and folate metabolism. These genes were further prioritized as vaccine and drug targets using subcellular localization and pathway enrichment methods [110].

In *Mycoplasma hyopneumoniae*, the transcriptome-driven metabolic analysis revealed unique pathways absent in humans, leading to the identification of 42 druggable proteins and 21 vaccine-candidate proteins [110].

Comparative transcriptomics in *Mycoplasma genitalium* also helped identify 67 non-homologous essential proteins, enriching the pool of potential therapeutic targets for sexually transmitted infections [111]. In silico transcriptomic analysis across five strains of *M. genitalium* identified 19 vaccine and 7 drug target candidates. This approach focused on conserved, essential, non-host homologous proteins, validated via docking with known drug libraries [112].

Further, transcriptomic tools have been applied to *Clostridium botulinum* to identify essential proteins, some with homology across multiple pathogens, thus serving as broad-spectrum antimicrobial targets [113]. Lastly, genome-scale transcriptomic analysis of *Vibrio cholerae* enabled the selection of metabolic enzymes and membrane proteins absent in humans for downstream drug docking studies [114].

A landmark study by Mujawar et al. demonstrated the power of a systems biology and interactome-driven framework for identifying both molecular vaccine candidates and novel drug targets in multidrug-resistant *Acinetobacter baumannii*. Using an integrative pipeline combining transcriptomics, proteomics, and protein-protein interaction networks, the researchers analyzed 3,766 proteins from the *A. baumannii* ATCC19606 strain. The study began by mapping known vaccine antigens and virulence factors from related strains onto the target genome and then explored their positions in the global protein interactome. This systems-level approach allowed the identification of both well-characterized and previously uncharacterized proteins that are highly central or influential in the network. These proteins were evaluated based on essentiality, localization, connectivity, and role in virulence or antibiotic resistance pathways. The outcome was a refined list of potentially druggable and immunogenic proteins, providing a robust theoretical foundation for future therapeutic development. The study emphasized the importance of incorporating network analysis and omics data synergy to rationalize target selection in complex, resistant pathogens like *A. baumannii* [115].

Microarray-based co-expression analysis has emerged as a powerful strategy for discovering novel drug targets in infectious diseases. In a study by Bora et al., a genome-wide computational pipeline was applied to analyze expression profiles derived from microarray datasets. From an initial set of 4,508 expressed genes, the study employed differential expression analysis and functional filtering to exclude human homologous genes, ultimately identifying 13 novel gene targets that are pathogen-specific and potentially druggable. These targets were associated with essential virulence functions, making them strong candidates for further experimental validation in antimicrobial development [116].

This approach presents the utility of high-throughput transcriptomics and bioinformatics filtering to narrow down therapeutic targets in silico, saving time, cost, and resources in the early phases of drug discovery. Furthermore, integrating gene expression patterns with functional annotations and network connectivity can reveal co-expressed gene modules involved in critical virulence pathways, improving our understanding of pathogen biology and identifying new points for pharmacological intervention [117].

A study by Birhanu et al. applied a subtractive and comparative transcriptomics approach to identify essential, non-homologous genes in *Actinobacillus pleuropneumoniae*, a key respiratory pathogen in swine. From an initial dataset of genes involved in 20 unique metabolic pathways, 122 essential proteins were identified. These included 95 cytoplasmic and 11 transmembrane proteins, prioritized for their potential as drug targets and vaccine candidates, respectively. Notably, three of these proteins shared homology with targets in *Mycoplasma hyopneumoniae*, suggesting their value as broad-spectrum targets across respiratory pathogens. The study used multiple in silico databases, including Kyoto Encyclopedia of Genes and Genomes, Differentially Expressed Genes (DEG), DrugBank, and Swiss-Prot, to conduct metabolic pathway enrichment, subtractive analysis against the host proteome, and functional annotation. This robust strategy exemplifies how computational systems biology can streamline vaccine and drug discovery, particularly for pathogens with limited existing therapeutics [118].

A study by Telkar et al. employed a genome-scale, synteny-based comparative approach to systematically identify novel drug targets in *Streptococcus gordonii*, an opportunistic pathogen involved in dental plaque formation and infective endocarditis. From an initial pool of 534 non-homologous genes, they integrated metabolic pathway annotation, subcellular localization, and host similarity filtering to narrow down to 16 high-confidence, non-human homologous targets. These proteins were prioritized using structural features and their unique roles in essential bacterial pathways, providing a streamlined computational framework for future antibacterial drug discovery targeting commensal pathogens with opportunistic behavior [119].

4.4.4.3.2 *Viral Infections*

Transcriptomics is becoming a cornerstone in virology, particularly for mapping host-pathogen interactions and identifying immune response signatures with unprecedented resolution. This high-throughput approach allows researchers to capture genome-wide RNA expression dynamics during infection, offering insights into both viral mechanisms of pathogenicity and the host's immune modulation in response.

For instance, in influenza, transcriptomics has revealed unique immunosignatures that distinguish it from other respiratory infections. These signatures can even predict the severity of illness and differentiate responses between live-attenuated and inactivated vaccines, thus informing more effective vaccine design [120].

In tuberculosis, studies have used transcriptome data to identify interferon-inducible gene signatures that correlate with disease activity and treatment outcomes. These findings enhance diagnosis and help stratify patients based on progression risk [121].

In malaria, transcriptomic profiling has elucidated severity-associated signatures in rodent models, uncovering mechanisms such as anemia and lung inflammation, which precede clinical symptoms [122].

Transcriptomic analyses have also provided valuable insights into how different influenza strains interact with the host immune system. By examining gene expression profiles, researchers have identified immune signatures that predict vaccine responsiveness, contributing to the development of more effective seasonal flu vaccines [123]. A novel approach using transcriptome signature reversion identified compounds like nifurtimox and chrysin as antiviral agents against influenza A. These drugs were selected based on their ability to reverse the host gene expression profile associated with infection [124].

In murine cytomegalovirus infection, a dual RNA-Seq analysis of both viral and host transcriptomes revealed unexpected immune pathways and novel viral transcripts, including one that functions both as a noncoding RNA and an mRNA. This study identified numerous transcriptionally active host genes involved in inflammation and immunity, offering new antiviral and vaccine targets [125].

In the case of Peste des petits ruminants virus vaccine studies, transcriptomic profiling of infected blood cells revealed 985 DEG associated with immune regulation, apoptosis, and transcription factors. These findings highlighted key regulators of host-virus interaction and immune modulation that could be targeted for improving vaccine efficacy [126].

A pivotal study by Bouwman et al. demonstrated the use of mRNA-based assays targeting the Janus Kinase (JAK) - Signal Transducer and Activator of Transcription (STAT) signaling pathway to quantitatively measure cellular immune responses in viral infections and after vaccination. This transcriptomics approach was applied to blood samples from patients with infections such as influenza, respiratory syncytial virus (RSV), dengue, yellow fever, and rotavirus, as well as from vaccinated individuals. The JAK-STAT1/2 pathway activity was found to be selectively increased in viral infections, not bacterial ones, offering a precise method to assess immune activation specificity. Interestingly, higher JAK-STAT1/2 activity was observed in influenza compared to RSV, reflecting known immunogenicity differences. Conversely, elevated JAK-STAT3 activity was associated with more severe RSV cases. In vaccine comparisons, the live-attenuated yellow fever vaccine-induced stronger JAK-STAT1/2 activation than the inactivated influenza vaccine, indicating a higher immunogenic profile. The assay also established baseline JAK-STAT pathway activity in healthy individuals, enabling interpretation without the need for reference controls. These results suggest that JAK-STAT pathway assays can serve as robust biomarkers for predicting vaccine responsiveness, monitoring disease severity, and supporting in vitro vaccine development. The method holds significant promise for precision immunology and antiviral therapy design [127].

During the COVID-19 pandemic, transcriptomics revealed key immune signatures associated with disease severity and vaccine responsiveness. Single-cell RNA sequencing further allowed high-resolution profiling of innate and adaptive immune cells, uncovering dysregulated molecular pathways linked to SARS-CoV-2 infection [128]. Transcriptomic modeling has also been used to simulate the immune dynamics of SARS-CoV-2, correlating IL-6 levels and T-cell memory to vaccine protection efficacy. This computational approach supports vaccine design strategies that enhance long-term immune memory.

In the field of African swine fever virus (ASFV) research, transcriptomic and functional studies have shed crucial light on the virus's sophisticated immune evasion strategies, particularly its ability to suppress host interferon signaling and manipulate inflammatory responses. These insights are instrumental in guiding the development of next-generation ASFV vaccines. A notable review by He et al. comprehensively summarizes how ASFV encodes multiple proteins that modulate host innate immune responses, especially by targeting interferon (IFN) pathways and inflammatory mediators. The study highlights several ASFV genes involved in suppressing IFN production, inhibiting stimulator of interferon genes (STING) signaling, and manipulating cytokine expression, all of which play a role in delaying or evading host immune detection. These immune antagonistic functions are pivotal in determining viral pathogenicity and are being actively explored as molecular targets for vaccine design. By uncovering how ASFV evades immune surveillance – such as by preventing STAT1 nuclear translocation, degrading IRF3, or disrupting STING oligomerization, transcriptomic analyses offer a functional roadmap for designing attenuated strains or subunit vaccines that lack these immune-modulating proteins. This body of research not only advances ASFV vaccine development but also provides a broader model for understanding immune evasion in large DNA viruses [129].

4.4.4.3.3 Parasitic Diseases

Transcriptomics has become a powerful tool in parasitology, complementing genomic and proteomic methods to uncover both stage-specific and conserved gene expression signatures. These insights facilitate the identification of novel drug targets and vaccine candidates by revealing DEG across life stages, tissues, and host interactions. In parasitic diseases like visceral leishmaniasis, transcriptomics and proteomics have together accelerated antigen discovery and informed the design of novel leishmanicidal agents [130].

In a groundbreaking study, Gramberg et al. presented the first spatial transcriptome of the liver fluke *Fasciola hepatica*, offering a high-resolution molecular atlas of gene expression across eight distinct tissues. This spatially resolved analysis has significantly advanced the understanding of parasite biology by identifying both tissue-specific gene expression patterns and functionally important genes relevant for therapeutic intervention. Among the key findings were vaccine-candidate genes, such as Ly6 family proteins, which showed specific expression in tegumental tissues, surfaces most exposed to host immunity. In parallel, the study also pinpointed drug resistance-associated genes, including glutathione S-transferases and ABC transporters, which were enriched in tissues linked to detoxification and metabolism. One of the most promising therapeutic targets, a tegumental PKCβ kinase, was validated through small molecule inhibition, resulting in parasite death, showcasing the translational potential of spatial transcriptomics in drug discovery. This tissue-specific transcriptomic resource sets a new benchmark for parasitology, enabling the prioritization of vaccine antigens and druggable proteins with unprecedented anatomical resolution. It provides a valuable platform for the rational design of next-generation antiparasitic therapies, especially in the context of increasing resistance to current drugs and the absence of effective commercial vaccines for *Fasciola hepatica* [131].

In *Haemonchus contortus*, a blood-feeding nematode, genome-scale transcriptomics identified parasite-specific genes involved in blood digestion and neurological signaling, enabling the discovery of vaccine antigens and novel drug targets such as ion channels and kinases [132].

Dictyocaulus viviparus, a bovine lungworm, was studied using stage- and gender-specific transcriptomics. The analysis identified key metabolic enzymes uniquely expressed in infectious larval stages, which may represent novel drug targets. Several proteins had lethal RNAi phenotypes in *C. elegans*, supporting their functional importance [133].

In *Schistosoma mansoni*, a DNA microarray covering over 7,000 transcripts revealed sex-specific gene expression patterns in adult worms. This analysis identified nearly 200 gender-biased transcripts, aiding in the discovery of sex-linked vaccine and therapeutic targets related to reproduction and immune evasion [134].

In *Plasmodium falciparum*, transcriptomic data from blood-stage parasites were used to prioritize genes for drug screening and vaccine design. The studies revealed stage-specific transcriptional control that governs immune evasion and drug susceptibility [135].

4.4.4.4 Multi-omics Integration for Target Discovery

Although individual omics technologies have valuable insights, their integration through multi-omics approaches offers a more comprehensive understanding of vaccine responses. By combining genomic, transcriptomic, proteomic, and metabolomic data, researchers can define predictive models of vaccine efficacy, identify key immune signatures, optimize immunization strategies for diverse populations, and develop personalized vaccines. This approach enhances the precision of vaccine enabling the development of more effective vaccination strategies.

For example, integrated omics was used to uncover metabolic vulnerabilities and immune-modulating targets in *Plasmodium spp.*, supporting antimalarial drug development [136].

For *Mycobacterium tuberculosis*, omics-driven models have mapped drug resistance mechanisms and host-pathogen metabolic interactions, aiding novel therapeutic strategies [137]. In COVID-19, multi-platform omics (transcriptomic, proteomic, and metabolomic) identified molecular signatures like ANXA1 and CLEC3B as potential therapeutic targets and severity biomarkers [138]. Similar frameworks were applied to *Pseudomonas aeruginosa* and *Haemophilus influenzae* for cystic fibrosis-associated infections, revealing targets involved in chronic biofilm formation [91].

Moreover, multi-omics-guided systems biology was used to identify immunogenic proteins for vaccine development against *Rickettsia* species, showing high conservation across strains and favorable immune signatures [139]. The zebrafish model has also been combined with omics data to validate anti-infective drug targets discovered through integrative analyses [140]. The integration of omics with network pharmacology has enabled the discovery of synergistic drug targets and biomarkers across multiple infectious agents and resistance mechanisms [141].

A notable application of this approach is the yellow fever 17D vaccine, one of the most effective vaccines in history. In this case, multi-omics analyses have been employed to investigate early immune mechanisms following vaccination, uncovering key immune pathways activated by the vaccine [142]. This knowledge has been valuable in guiding the design of novel vaccines against other infectious diseases.

4.4.4.5 Computational Tools and Bioinformatics

Computational tools and bioinformatics have become central to modern vaccine development, supporting epitope prediction, antigen selection, immune simulation, and structural modeling to streamline and enhance the vaccine discovery pipeline. Platforms such as iVAX provide integrated immunoinformatics workflows for T-cell epitope prediction, immunogenicity scoring, and vaccine optimization, and have been successfully applied to vaccines for Q fever, swine influenza, and malaria [143]. The Immune Epitope Database, along with tools like NetCTL and VaxiJen, are routinely used for B- and T-cell epitope prediction, helping researchers assess antigenicity, allergenicity, and population coverage for viral targets like Zika and HIV [144, 145].

In the case of dengue virus, immunoinformatics-guided vaccine design led to the identification of CD4+ T-cell epitopes that showed strong docking interaction with immune receptors like TLR5 [146]. Computational pipelines have also been developed for complex pathogens such as human papillomavirus and *Mycobacterium tuberculosis*, combining genome and proteome data with structural modeling and immune simulations [147, 148]. Reviews have shown how tools like MODELLER, PatchDock, FireDock, and Chimera assist in refining vaccine constructs and simulating their interaction with MHC molecules, critical for immune activation [149].

Other innovative platforms include Vaxign and EpiJen, used in designing peptide vaccines for influenza, SARS-CoV-2, and Lassa virus [150], while computational vaccinology has enabled rapid antigen design during pandemics, significantly reducing the time from target identification to pre-clinical testing [151]. Predictive modeling has also played a role in HIV vaccine research, where epitope maps inform the design of vaccines resilient to viral mutation [152].

Beyond existing platforms like iVAX and VaxiJen, numerous computational pipelines now support rational, genome-informed vaccine design. For example, Vaxign is a web-based reverse vaccinology platform that integrates protein subcellular localization, transmembrane domains, adhesion probability, and binding prediction, and has been applied to *Brucella* and other bacterial pathogens [153]. The JanusMatrix tool, showcased at ICoVax 2013, was designed to analyze T-cell epitope cross-conservation between pathogens and the human genome, aiding in minimizing autoimmune risk in vaccine design [154].

Computational frameworks like pBone/pView, developed for *Epstein-Barr* virus, guide epitope selection through system biology-informed target prioritization, integrating viral proteomes and host interaction data [155]. In a broader context, structural vaccinology merges 3D structural data with in silico modeling to refine antigen design and predict epitope accessibility – this approach has revolutionized subunit vaccine development for pathogens like RSV and Zika [156].

Additionally, recent innovations incorporate AI and DL for epitope clustering, peptide-MHC interaction scoring, and immune response prediction, dramatically accelerating candidate screening for difficult targets like HIV, hepatitis C virus, and emerging zoonoses [157]. These tools are often organized into multi-layer workflows, combining antigenicity prediction (e.g., ANTIGENpro), allergenicity checkers (e.g., AllerTOP), population coverage (Immune Epitope Database tools), and protein-protein docking (e.g., PatchDock, FireDock) [151].

Recent reviews also emphasize systems vaccinology, a field that integrates omics, ML, and big data analytics, to predict vaccine efficacy and adverse reactions across diverse populations. This systems approach has been key to designing adaptive vaccines for pandemics and for personalizing immunization in cancer and infectious diseases [158].

4.5 CHALLENGES AND LIMITATIONS

Omics platforms (e.g., genomics, transcriptomics, proteomics) generate massive and high-dimensional datasets. Integrating these with clinical data such as EHRs, imaging, and lifestyle factors poses a significant analytical challenge. Discrepancies in data quality, missing values, and inconsistent sample annotation can lead to irreproducible outcomes across studies [151].

Many research centers and health systems use different ontologies, data formats, and metadata standards. This lack of interoperability makes sharing and combining datasets difficult, slowing down collaborative efforts and delaying clinical implementation. The use of machine-readable, harmonized ontologies is strongly recommended to facilitate standardized data integration [159].

Translating multi-omics data into actionable clinical insights remains difficult. While many biomarkers and targets are identified, their relevance in diverse patient populations is often unclear. Many findings lack sufficient validation across cohorts, delaying their adoption in clinical workflows.

High-throughput sequencing, cloud storage, and advanced analytics require robust infrastructure and funding. These tools are often unaffordable in low- and middle-income countries – regions that disproportionately bear the burden of infectious diseases [160].

Precision medicine often requires genetic testing and the use of predictive algorithms, both of which face regulatory uncertainty. Inconsistent international regulations and data classification standards complicate the approval and implementation of diagnostics and personalized therapies. Regulatory reform is essential for enabling cross-border precision health efforts [161].

The granular nature of omics data raises serious privacy and consent concerns. There is growing anxiety around the potential misuse of genetic data by employers, insurers, or state entities. A 2020 study emphasized that data protection laws and transparency in algorithmic use must evolve alongside scientific advances to safeguard individuals and build trust [162].

Most genomic and transcriptomic studies are based on populations of European descent. This underrepresentation creates a knowledge gap and risks exacerbating health disparities. Precision interventions developed using such biased data may be ineffective, or even harmful, for underrepresented groups [163].

Although many omics-based discoveries have been made, few have made it into clinical guidelines or practice. Reasons include a lack of clinician training, unclear reimbursement frameworks, and insufficient prospective trials demonstrating real-world benefits [164].

4.6 CONCLUSIONS

The integration of precision medicine and omics technologies into infectious disease management is reshaping the paradigm of diagnosis, treatment, and prevention. Unlike traditional approaches, precision medicine emphasizes patient-specific variability, utilizing multi-omics data – genomics, transcriptomics, proteomics, metabolomics, and microbiomics – to gain granular insights into host-pathogen interactions, disease progression, and treatment responses.

Omics platforms have enabled early pathogen identification, the discovery of novel therapeutic targets, and the development of next-generation vaccines through methods such as reverse vaccinology and immunoinformatics. They have also facilitated the stratification of patients based on immune responses, risk profiles, and drug metabolism pathways, contributing to more effective and safer therapeutic regimens.

The convergence of advanced computational tools, AI, and systems biology has further expanded the scope and depth of data analysis, enabling real-time decision-making, outbreak forecasting, and drug repurposing. Personalized therapeutic strategies, biomarker-driven interventions, and precision diagnostics now form the core of modern infectious disease management.

Despite these advancements, numerous challenges remain – including data integration, infrastructure disparities, regulatory barriers, ethical considerations, and clinical translation gaps. The effective implementation of precision medicine will require continued investment in bioinformatics infrastructure, standardization efforts, and international collaborations that bridge the gap between innovation and practice.

In conclusion, precision medicine and omics technologies hold immense potential to revolutionize how infectious diseases are understood and managed. As these approaches continue to mature, they promise not only improved patient outcomes but also broader public health benefits through targeted interventions, optimized resource allocation, and more resilient healthcare systems.

ACKNOWLEDGMENTS

I.N. acknowledges a grant of the Ministry of Research, Innovation and Digitalization, CCCDI – UEFISCDI, project number PN-IV-P2-2.1-TE-2023-0993, within PNCDI IV. This research was also founded by the Romanian Ministry of Research, Innovation and Digitalization under the Romanian National Nucleu Program LAPLAS VII – contract No. 30N/2023. C.S.O. acknowledges the financial support from the integrated project be@t – textile Bioeconomy (TC-C12-i01, Sustainable Bioeconomy No. 02/C12 – i01.01/20229, promoted by the Recovery and Resilience plan (PRR), Next Generation EU, for the 2021–2026 period). C.S.F.O was also supported by FCT – Fundação para a Ciência e a Tecnologia, through the Project UIDB/50016/2020. B.B. acknowledges the Core Program with the National Research Development and Innovation Plan 2022–2027, carried out with the support of MCID, project no. PN 23 05 and by the Ministry of Research and Innovation through Program I – Development of the National R&D System, Subprogram 1.2 – Institutional Performance – Projects for Excellence Financing in RDI, contract no. 18PFE/30.12.2021.

ABBREVIATIONS

AI	Artificial Intelligence
AMR	Antimicrobial Resistance
ASFV	African Swine Fever Virus
COVID-19	Coronavirus Disease 2019
CRISPR	Clustered Regularly Interspaced Short Palindromic Repeats
CYP	Cytochrome
DEG	Differentially Expressed Genes
DL	Deep Learning
DNA	Deoxyribonucleic Acid
GWAS	Genome-Wide Association Studies
HDT	Host-Directed Therapy
HIV	Human Immunodeficiency Virus
IFN	Interferon
JAK	Janus Kinase
MHC	Major Histocompatibility Complex
ML	Machine Learning
MS	Mass Spectrometry
NGS	Next-Generation Sequencing
PCR	Polymerase Chain Reaction
RNA	Ribonucleic Acid
RSV	Respiratory Syncytial Virus
STAT	Signal Transducer and Activator of Transcription
STING	Stimulator of Interferon Genes
WGS	Whole Genome Sequencing

REFERENCES

1. van Seventer, J.M.; Hochberg, N.S. Principles of Infectious Diseases: Transmission, Diagnosis, Prevention, and Control. *Int. Encycl. Public Health* 2017, 22–39, doi:10.1016/B978-0-12-803678-5.00516-6

2. Salam, M.A.; Al-Amin, M.Y.; Salam, M.T.; Pawar, J.S.; Akhter, N.; Rabaan, A.A.; Alqumber, M.A.A. Antimicrobial Resistance: A Growing Serious Threat for Global Public Health. *Healthcare* 2023, *11*, doi:10.3390/healthcare11131946

3. Baker, R.E.; Mahmud, A.S.; Miller, I.F.; Rajeev, M.; Rasambainarivo, F.; Rice, B.L.; Takahashi, S.; Tatem, A.J.; Wagner, C.E.; Wang, L.-F.; et al. Infectious Disease in an Era of Global Change. *Nat. Rev. Microbiol.* 2022, *20*, 193–205, doi:10.1038/s41579-021-00639-z

4. Molla, G.; Bitew, M. Revolutionizing Personalized Medicine: Synergy With Multi-Omics Data Generation, Main Hurdles, and Future Perspectives. *Biomedicines* 2024, *12*, 2750, doi:10.3390/biomedicines12122750

5. Sakai, T.; Morimoto, Y. The History of Infectious Diseases and Medicine. *Pathogens* 2022, *11*, 1147, doi:10.3390/pathogens11101147

6. Schuhladen, K.; Stich, L.; Schmidt, J.; Steinkasserer, A.; Boccaccini, A.R.; Zinser, E. Cu, Zn Doped Borate Bioactive Glasses: Antibacterial Efficacy and Dose-Dependent in Vitro Modulation of Murine Dendritic Cells. *Biomater. Sci.* 2020, *8*, 2143–2155, doi:10.1039/C9BM01691K

7. Mahon, R.N.; Hafner, R. Applying Precision Medicine and Immunotherapy Advances From Oncology to Host-Directed Therapies for Infectious Diseases. *Front. Immunol.* 2017, *8*, doi:10.3389/fimmu.2017.00688

8. Ladner, J.T.; Grubaugh, N.D.; Pybus, O.G.; Andersen, K.G. Precision Epidemiology for Infectious Disease Control. *Nat. Med.* 2019, *25*, 206–211, doi:10.1038/s41591-019-0345-2

9. Twa, M.D. Precision Medicine and the Future of Health Care. *Optom. Vis. Sci.* 2017, *94*, 635, doi:10.1097/OPX.0000000000001089

10. Liu, X.; Luo, X.; Jiang, C.; Zhao, H. Difficulties and Challenges in the Development of Precision Medicine. *Clin. Genet.* 2019, *95*, 569–574, doi:10.1111/cge.13511

11. Nishi, A.; Milner, Danny A. Jr; Giovannucci, Edward L.; Nishihara, Reiko; Tan Andy, S; Kawachi, Ichiro; Ogino, S. Integration of Molecular Pathology, Epidemiology and Social Science for Global Precision Medicine. *Expert Rev. Mol. Diagn.* 2016, *16*, 11–23, doi:10.1586/14737159.2016.1115346

12. Fatima, G.; Allami, R.H.; Yousif, M.G. Integrative AI-Driven Strategies for Advancing Precision Medicine in Infectious Diseases and Beyond: A Novel Multidisciplinary Approach 2023, *arXiv*:2307.15228, doi: 10.48550/arXiv.2307.15228

13. Ahmed, Z. Practicing Precision Medicine with Intelligently Integrative Clinical and Multi-Omics Data Analysis. *Hum. Genomics* 2020, *14*, 35, doi:10.1186/s40246-020-00287-z

14. Olorunsogo, T.O.; Balogun, O.D.; Ayo-Farai, O.; Ogundairo, O.; Maduka, C.P.; Okongwu, C.C.; Onwumere, C. Bioinformatics and Personalized Medicine in the U.S.: A Comprehensive Review: Scrutinizing the Advancements in Genomics and Their Potential to Revolutionize Healthcare Delivery. *World J. Adv. Res. Rev.* 2024, *21*, 335–351, doi:10.30574/wjarr.2024.21.1.0016

15. Equils, O.; Bakaj, Angela; Wilson-Mifsud, Brittany; Chatterjee, A. Restoring Trust: The Need for Precision Medicine in Infectious Diseases, Public Health and Vaccines. *Hum. Vaccines Immunother.* 2023, *19*, 2234787, doi:10.1080/21645515.2023.2234787

16. Ginsburg, G.S.; Haga, S.B. 2 - Foundations and Application of Precision Medicine. In *Emery and Rimoin's Principles and Practice of Medical Genetics and Genomics* (7th ed.); Pyeritz, R.E., Korf, B.R., Grody, W.W., Eds.; Academic Press, 2019; pp. 21–45.

17. Seelall, J. Genomic Epidemiology of Antimicrobial Resistance: Tracking the Global Spread of Resistant Pathogens. *Int. J. Sci. Res. Manag. IJSRM* 2024, *12*, 1214–1227, doi:10.18535/ijsrm/v12i10.mp04

18. Streeter, O.E.; Beron, P.J.; Iyer, P.N. Precision Medicine: Genomic Profiles to Individualize Therapy. *Otolaryngol. Clin. North Am.* 2017, *50*, 765–773, doi:10.1016/j.otc.2017.03.012

19. Kessler, C. Genomics and Precision Medicine. *AACN Adv. Crit. Care* 2018, *29*, 26–27, doi:10.4037/aacnacc2018823

20. Lu, D.; Kalantar, K.L.; Chu, V.T.; Glascock, A.L.; Guerrero, E.S.; Bernick, N.; Butcher, X.; Ewing, K.; Fahsbender, E.; Holmes, O.; et al. Simultaneous Detection of Pathogens and Antimicrobial Resistance Genes with the Open Source, Cloud-Based, CZ ID Pipeline 2024, 2024.04.12.589250.

21. Kwok, A.J.; Mentzer, A.; Knight, J.C. Host Genetics and Infectious Disease: New Tools, Insights and Translational Opportunities. *Nat. Rev. Genet.* 2021, *22*, 137–153, doi:10.1038/s41576-020-00297-6

22. Shannon, M.F. Chapter 57 – Genomic Approaches to the Host Response to Pathogens. In *Essentials of Genomic and Personalized Medicine*; Ginsburg, G.S., Willard, H.F., Eds.; Academic Press, 2010; pp. 733–743.

23. *Research Inventions* journals.

24. Babu, M.; Snyder, M. Multi-Omics Profiling for Health. *Mol. Cell. Proteomics* 2023, *22*, 100561, doi:10.1016/j.mcpro.2023.100561

25. Abdelaziz, E.H.; Ismail, R.; Mabrouk, M.S.; Amin, E. Multi-Omics Data Integration and Analysis Pipeline for Precision Medicine: Systematic Review. *Comput. Biol. Chem.* 2024, *113*, 108254, doi:10.1016/j.compbiolchem.2024.108254

26. Turanli, B.; Karagoz, K.; Gulfidan, G.; Sinha, R.; Mardinoglu, A.; Arga, K.Y. A Network-Based Cancer Drug Discovery: From Integrated Multi-Omics Approaches to Precision Medicine. *Curr. Pharm. Des* 24, 3778–3790, doi:10.2174/1381612824666181106095959

27. Zafari, N.; Bathaei, P.; Velayati, M.; Khojasteh-Leylakoohi, F.; Khazaei, M.; Fiuji, H.; Nassiri, M.; Hassanian, S.M.; Ferns, G.A.; Nazari, E.; et al. Integrated Analysis of Multi-Omics Data for the Discovery of Biomarkers and Therapeutic Targets for Colorectal Cancer. *Comput. Biol. Med.* 2023, *155*, 106639, doi:10.1016/j.compbiomed.2023.106639

28. Zaghlool, S.B.; Attallah, O. A Review of Deep Learning Methods for Multi-Omics Integration in Precision Medicine. In *Proceedings of the 2022 IEEE International Conference on Bioinformatics and Biomedicine (BIBM)*; December 2022; pp. 2208–2215.

29. Antharam, V.C.; McEwen, D.C.; Garrett, T.J.; Dossey, A.T.; Li, E.C.; Kozlov, A.N.; Mesbah, Z.; Wang, G.P. An Integrated Metabolomic and Microbiome Analysis Identified Specific Gut Microbiota Associated With Fecal Cholesterol and Coprostanol in Clostridium Difficile Infection. *PLOS ONE* 2016, *11*, e0148824, doi:10.1371/journal.pone.0148824

30. Robinson, J.I.; Weir, W.H.; Crowley, J.R.; Hink, T.; Reske, K.A.; Kwon, J.H.; Burnham, C.-A.D.; Dubberke, E.R.; Mucha, P.J.; Henderson, J.P. Metabolomic Networks Connect Host-Microbiome Processes to Human *Clostridioides difficile* Infections. *J. Clin. Invest.* 2019, *129*, 3792–3806, doi:10.1172/JCI126905

31. Liu, A.C.; Patel, K.; Vunikili, R.D.; Johnson, K.W.; Abdu, F.; Belman, S.K.; Glicksberg, B.S.; Tandale, P.; Fontanez, R.; Mathew, O.K.; et al. Sepsis in the Era of Data-Driven Medicine: Personalizing Risks, Diagnoses, Treatments and Prognoses. *Brief. Bioinform.* 2020, *21*, 1182–1195, doi:10.1093/bib/bbz059

32. Nam, Y.; Kim, J.; Jung, S.-H.; Woerner, J.; Suh, E.H.; Lee, D.; Shivakumar, M.; Lee, M.E.; Kim, D. Harnessing Artificial Intelligence in Multimodal Omics Data Integration: Paving the Path for the Next Frontier in Precision Medicine. *Annu. Rev. Biomed. Data Sci.* 2024, *7*, 225–250, doi:10.1146/annurev-biodatasci-102523-103801

33. Pritzker, K. Biomarker Imprecision in Precision Medicine. *Expert Rev. Mol. Diagn.* 2018, *18*, 685–687, doi:10.1080/14737159.2018.1493379

34. Wilson, J.L.; Altman, R.B. Biomarkers: Delivering on the Expectation of Molecularly Driven, Quantitative Health. *Exp. Biol. Med.* 2018, *243*, 313–322, doi:10.1177/1535370217744775

35. Roda, A.; Michelini, E.; Caliceti, C.; Guardigli, M.; Mirasoli, M.; Simoni, P. Advanced Bioanalytics for Precision Medicine. *Anal. Bioanal. Chem.* 2018, *410*, 669–677, doi:10.1007/s00216-017-0660-8

36. Ielapi, N.; Andreucci, M.; Licastro, N.; Faga, T.; Grande, R.; Buffone, G.; Mellace, S.; Sapienza, P.; Serra, R. Precision Medicine and Precision Nursing: The Era of Biomarkers and Precision Health. *Int. J. Gen. Med.* 2020, *13*, 1705–1711, doi:10.2147/IJGM.S285262

37. Ollier, W.; Muir, K.R.; Lophatananon, A.; Verma, A.; Yuille, M. Risk Biomarkers Enable Precision in Public Health. *Pers. Med.* 2018, *15*, 329–342, doi:10.2217/pme-2017-0068

38. Afshinnekoo, E.; Chou, C.; Alexander, N.; Ahsanuddin, S.; Schuetz, A.N.; Mason, C.E. Precision Metagenomics: Rapid Metagenomic Analyses for Infectious Disease Diagnostics and Public Health Surveillance. *J. Biomol. Tech. JBT* 2017, *28*, 40–45, doi:10.7171/jbt.17-2801-007

39. Rayan, R. Precision Medicine in the Context of Artificial Intelligence, *SSRN Electronic Journal*. 2019, 9.

40. Xuanyu, Liu Artificial Intelligence-Based Drug Development and Precision Medicine Exploration. *Front. Med. Sci. Res.* 2024, *6*, doi:10.25236/FMSR.2024.061006

41. Whirl-Carrillo, M.; Klein, C.; Klein, T.E. Chapter Two – Scientific Evidence and Sources of Knowledge for Pharmacogenomics. In *Clinical Decision Support for Pharmacogenomic Precision Medicine*; Devine, B., Boyce, R.D., Wiisanen, K., Eds.; Academic Press, 2022; pp. 19–51.

42. Giacobbe, D.R.; Zhang, Yudong; de la Fuente, J. Explainable Artificial Intelligence and Machine Learning: Novel Approaches to Face Infectious Diseases Challenges. *Ann. Med.* 2023, *55*, 2286336, doi:10.1080/07853890.2023.2286336

43. Hanson, A.P.; Stinnett, R.C.; Bhasin, A.; Conrad, H.; Nguyen, D.C.; Ramchandar, N.; Boswell, M.; Stauffer, S.; Farnaes, L.; Briggs, B.; et al. 526. Antimicrobial Selection Based on Precision Metagenomics Compared with Standard Urine Culture/Susceptibility: A Reliability and Inter-Rater Agreement Analysis for Application in Patients with Urinary Tract Infections. *Open Forum Infect. Dis.* 2022, *9*, ofac492.581, doi:10.1093/ofid/ofac492.581

44. Carpenter, R.E.; Almas, S.; Tamrakar, V.K.; Sharma, R. Dataset for Comparative Analysis of Precision Metagenomics and Traditional Methods in Urinary Tract Infection Diagnostics. *Data Brief* 2025, *59*, 111339, doi:10.1016/j.dib.2025.111339

45. Liu, C.; Tang, Z.; Li, L.; Kang, Y.; Teng, Y.; Yu, Y. Enhancing Antimicrobial Resistance Detection With MetaGeneMiner: Targeted Gene Extraction from Metagenomes. *Chin. Med. J. (Engl.)* 2024, *137*, 2092, doi:10.1097/CM9.0000000000003182

46. Govender, K.N.; Street, T.L.; Sanderson, N.D.; Eyre, D.W. Metagenomic Sequencing as a Clinical Diagnostic Tool for Infectious Diseases: A Systematic Review and Meta-Analysis 2020, 10.1101/2020.03.30.20043901

47. Cho, Y.-S.; Shin, J.-G. Research Work at the Center for Personalized Precision Medicine of Tuberculosis (cPMTb). *Impact* 2021, *2021*, 25–27, doi:10.21820/23987073.2021.8.25

48. Mahomed, S.; Padayatchi, N.; Singh, J.; Naidoo, K. Precision Medicine in Resistant Tuberculosis: Treat the Correct Patient, at the Correct Time, With the Correct Drug. *J. Infect.* 2019, *78*, 261–268, doi:10.1016/j.jinf.2019.03.006

49. Getahun, K.A.; Angaw, D.A.; Asres, M.S.; Kahaliw, W.; Petros, Z.; Abay, S.M.; Yimer, G.; Berhane, N. The Role of Pharmacogenomics Studies for Precision Medicine Among Ethiopian Patients and Their Clinical Implications: A Scoping Review. *Pharmacogenomics Pers. Med.* 2024, *17*, 347–361, doi:10.2147/PGPM.S454328

50. Mu, Y.; Kodidela, S.; Wang, Y.; Kumar, S.; Cory, T.J. The Dawn of Precision Medicine in HIV: State of the Art of Pharmacotherapy. *Expert Opin. Pharmacother.* 2018, *19*, 1581–1595, doi:10.1080/14656566.2018.1515916

51. Zuur, M.A.; Akkerman, O.W.; Touw, D.J.; van der Werf, T.S.; Cobelens, F.; Burger, D.M.; Grobusch, M.P.; Alffenaar, J.-W.C. Dried Blood Spots Can Help Decrease the Burden on Patients Dually Infected With Multidrug-Resistant Tuberculosis and HIV. *Eur. Respir. J.* 2016, *48*, 932–934, doi:10.1183/13993003.00599-2016

52. Oelofse, C.; Ndong Sima, C.A.A.; Möller, M.; Uren, C. Pharmacogenetics as Part of Recommended Precision Medicine for Tuberculosis Treatment in African Populations: Could It Be a Reality? *Clin. Transl. Sci.* 2023, *16*, 1101–1112, doi:10.1111/cts.13520

53. Magro, D.; Venezia, M.; Rita Balistreri, C. The Omics Technologies and Liquid Biopsies: Advantages, Limitations, Applications. *Med. Omics* 2024, *11*, 100039, doi:10.1016/j.meomic.2024.100039

54. Pronyk, P.M.; de Alwis, R.; Rockett, R.; Basile, K.; Boucher, Y.F.; Pang, V.; Sessions, O.; Getchell, M.; Golubchik, T.; Lam, C.; et al. Advancing Pathogen Genomics in Resource-Limited Settings. *Cell Genomics* 2023, *3*, doi:10.1016/j.xgen.2023.100443

55. de Anda-Jáuregui, G.; Hernández-Lemus, E. Computational Oncology in the Multi-Omics Era: State of the Art. *Front. Oncol.* 2020, *10*, doi:10.3389/fonc.2020.00423

56. Nathan, S. Transcriptome Profiling to Understand Host-Bacteria Interactions: Past, Present and Future. *ScienceAsia* 2020, *46*, 503, doi:10.2306/scienceasia1513-1874.2020.083

57. Khalid, Z.; Huan, M.; Sohail Raza, M.; Abbas, M.; Naz, Z.; Kombe Kombe, A.J.; Zeng, W.; He, H.; Jin, T. Identification of Novel Therapeutic Candidates Against SARS-CoV-2 Infections: An Application of RNA Sequencing Toward mRNA Based Nanotherapeutics. *Front. Microbiol.* 2022, *13*, doi:10.3389/fmicb.2022.901848

58. Al-Amrani, S.; Al-Jabri, Z.; Al-Zaabi, A.; Alshekaili, J.; Al-Khabori, M. Proteomics: Concepts and Applications in Human Medicine. *World J. Biol. Chem.* 2021, *12*, 57–69, doi:10.4331/wjbc.v12.i5.57

59. Wang, Y.; Wang, X.; Luu, L.D.W.; Chen, S.; Jin, F.; Wang, S.; Huang, X.; Wang, L.; Zhou, X.; Chen, X.; et al. Proteomic and Metabolomic Signatures Associated With the Immune Response in Healthy Individuals Immunized With an Inactivated SARS-CoV-2 Vaccine. *Front. Immunol.* 2022, *13*, doi:10.3389/fimmu.2022.848961

60. Diray-Arce, J.; Conti, M.G.; Petrova, B.; Kanarek, N.; Angelidou, A.; Levy, O. Integrative Metabolomics to Identify Molecular Signatures of Responses to Vaccines and Infections. *Metabolites* 2020, *10*, 492, doi:10.3390/metabo10120492

61. Yaqub, M.O.; Jain, A.; Joseph, C.E.; Edison, L.K. Microbiome-Driven Therapeutics: From Gut Health to Precision Medicine. *Gastrointest. Disord.* 2025, *7*, 7, doi:10.3390/gidisord7010007

62. Yang, J.; Yan, Y.; Zhong, W. Application of Omics Technology to Combat the COVID-19 Pandemic. *MedComm* 2021, *2*, 381–401, doi:10.1002/mco2.90

63. Jiang, H.; Li, Y.; Zhang, H.; Wang, W.; Yang, X.; Qi, H.; Li, H.; Men, D.; Zhou, J.; Tao, S. SARS-CoV-2 Proteome Microarray for Global Profiling of COVID-19 Specific IgG and IgM Responses. *Nat. Commun.* 2020, *11*, 3581, doi:10.1038/s41467-020-17488-8

64. Aggarwal, S.; Peng, W.K.; Srivastava, S. Multi-Omics Advancements towards Plasmodium Vivax Malaria Diagnosis. *Diagnostics* 2021, *11*, 2222, doi:10.3390/diagnostics11122222

65. Schneeberger, P.H.H. Development and Application of Omics and Bioinformatics Approaches for a Deeper Understanding of Infectious Diseases Systems. Thesis, University of Basel, 2015.

66. Průcha, M.; Zazula, R.; Russwurm, S. Sepsis Diagnostics in the Era of "Omics" Technologies. *Prague Med. Rep.* 2018, *119*, 9–29, doi:10.14712/23362936.2018.2

67. Ko, E.R.; Yang, W.E.; McClain, M.T.; Woods, C.W.; Ginsburg, G.S.; Tsalik, E.L. What Was Old Is New Again: Using the Host Response to Diagnose Infectious Disease. *Expert Rev. Mol. Diagn.* 2015, *15*, 1143–1158, doi:10.1586/14737159.2015.1059278

68. Moshkovskii, S.A. Omics Biomarkers and Early Diagnostics. *Biomeditsinskaya Khimiya* 2017, *63*, 369–372, https://pubmed.ncbi.nlm.nih.gov/29080866

69. Foster, S.; Luciani, F. Omics in Immunology. *Immunol. Cell Biol.* 2021, *99*, 133–134, doi:10.1111/imcb.12435

70. Panunzio, A.; Tafuri, A.; Princiotta, A.; Gentile, I.; Mazzucato, G.; Trabacchin, N.; Antonelli, A.; Cerruto, M.A. Omics in Urology: An Overview on Concepts, Current Status and Future Perspectives. *Urol. J.* 2021, *88*, 270–279, doi:10.1177/03915603211022960

71. Chen, J.; Subramaniam, S.; Wishart, D.; Wong, S. Guest Editorial—Special Issue on "-Omics" Based Companion Diagnostics for Personalized Medicine. *IEEE Trans. Biomed. Circuits Syst.* 2014, *8*, 1–3, doi:10.1109/TBCAS.2014.2308993

72. Elrashedy, A.; Mousa, W.; Nayel, M.; Salama, A.; Zaghawa, A.; Elsify, A.; Hasan, M.E. Advances in Bioinformatics and Multi-Omics Integration: Transforming Viral Infectious Disease Research in Veterinary Medicine. *Virol. J.* 2025, *22*, 22, doi:10.1186/s12985-025-02640-x

73. Rajesh, A.M.; Pawar, S.S.; Doriya, K.; Dandela, R. Combating Antibiotic Resistance: Mechanisms, Challenges, and Innovative Approaches in Antibacterial Drug Development. *Explor. Drug Sci.* 2025, *3*, 100887, doi:10.37349/eds.2025.100887

74. Cho, H.; Misra, R. Mutational Activation of Antibiotic-Resistant Mechanisms in the Absence of Major Drug Efflux Systems of Escherichia Coli. *J. Bacteriol.* 2021, *203*, doi:10.1128/jb.00109-21

75. Papaventsis, D.; Casali, N.; Kontsevaya, I.; Drobniewski, F.; Cirillo, D.M.; Nikolayevskyy, V. Whole Genome Sequencing of *Mycobacterium tuberculosis* for Detection of Drug Resistance: A Systematic Review. *Clin. Microbiol. Infect.* 2017, *23*, 61–68, doi:10.1016/j.cmi.2016.09.008

76. Becker, D.; Bharatam, P.V.; Gohlke, H. Molecular Mechanisms Underlying Single Nucleotide Polymorphism-Induced Reactivity Decrease in CYP2D6. *J. Chem. Inf. Model.* 2024, *64*, 6026–6040, doi:10.1021/acs.jcim.4c00276

77. Elton, L.; Aydin, A.; Stoker, N.; Rofael, S.; Wildner, L.M.; Abdul, J.B.P.A.A.; Tembo, J.; Hamid, M.A.; Chastel, M.M.C.; Canseco, J.O.; et al. A Pragmatic Pipeline for Drug Resistance and Lineage Identification in *Mycobacterium tuberculosis* Using Whole Genome Sequencing. *PLOS Glob. Public Health* 2025, *5*, e0004099, doi:10.1371/journal.pgph.0004099

78. Tasleem Raza, S.; Fatima, K.; Srivsatava, S.; Ouhtit, A. Chapter 5 - Multiomics Approaches in the Development of Therapy against Malaria. In *Falciparum Malaria*; Qidwai, T., Ed.; Academic Press, 2024; pp. 77–86.

79. Abumsimir, B.; Al-Qaisi, T.S. The Next Generation of Malaria Treatments: The Great Expectations. *Future Sci. OA* 2023, *9*, FSO834, doi:10.2144/fsoa-2023-0018

80. Laurenson, A.J.; Laurens, M.B. A New Landscape for Malaria Vaccine Development. *PLOS Pathog.* 2024, *20*, e1012309, doi:10.1371/journal.ppat.1012309

81. Brunner, N.; Bruggmann, P. Trends of the Global Hepatitis C Disease Burden: Strategies to Achieve Elimination. *J. Prev. Med. Pub. Health* 2021, *54*, 251–258, doi:10.3961/jpmph.21.151

82. Khatri, S.; Sazinas, P.; Strube, M.L.; Ding, L.; Dubey, S.; Shivay, Y.S.; Sharma, S.; Jelsbak, L. Pseudomonas is a Key Player in Conferring Disease Suppressiveness in Organic Farming. *Plant Soil* 2024, *503*, 85–104, doi:10.1007/s11104-023-05927-6

83. Wallis, R.S.; O'Garra, A.; Sher, A.; Wack, A. Host-Directed Immunotherapy of Viral and Bacterial Infections: Past, Present and Future. *Nat. Rev. Immunol.* 2023, *23*, 121–133, doi:10.1038/s41577-022-00734-z

84. Demongeot, J.; Fougère, C. mRNA COVID-19 Vaccines – Facts and Hypotheses on Fragmentation and Encapsulation. *Vaccines* 2023, *11*, 40, doi:10.3390/vaccines11010040

85. Seib, K.L.; Zhao, X.; Rappuoli, R. Developing Vaccines in the Era of Genomics: A Decade of Reverse Vaccinology. *Clin. Microbiol. Infect.* 2012, *18*, 109–116, doi:10.1111/j.1469-0691.2012.03939.x

86. Masignani, V.; Pizza, M.; Moxon, E.R. The Development of a Vaccine Against Meningococcus B Using Reverse Vaccinology. *Front. Immunol.* 2019, *10*, doi:10.3389/fimmu.2019.00751

87. Pritam, M.; Singh, G.; Swaroop, S.; Singh, A.K.; Singh, S.P. Exploitation of Reverse Vaccinology and Immunoinformatics as Promising Platform for Genome-Wide Screening of New Effective Vaccine Candidates against Plasmodium Falciparum. *BMC Bioinformatics* 2019, *19*, 468, doi:10.1186/s12859-018-2482-x

88. John, L.; John, G.J.; Kholia, T. A Reverse Vaccinology Approach for the Identification of Potential Vaccine Candidates from Leishmania Spp. *Appl. Biochem. Biotechnol.* 2012, *167*, 1340–1350, doi:10.1007/s12010-012-9649-0

89. Singh, S.P.; Roopendra, K.; Mishra, B.N. Genome-Wide Prediction of Vaccine Candidates for Leishmania Major: An Integrated Approach. *J. Trop. Med.* 2015, *2015*, 709216, doi:10.1155/2015/709216

90. Lew-Tabor, A.E.; Rodriguez Valle, M. A Review of Reverse Vaccinology Approaches for the Development of Vaccines against Ticks and Tick Borne Diseases. *Ticks Tick-Borne Dis.* 2016, *7*, 573–585, doi:10.1016/j.ttbdis.2015.12.012

91. Cocorullo, M.; Chiarelli, L.R.; Stelitano, G. Improving Protection to Prevent Bacterial Infections: Preliminary Applications of Reverse Vaccinology Against the Main Cystic Fibrosis Pathogens. *Vaccines* 2023, *11*, 1221, doi:10.3390/vaccines11071221

92. Maritz-Olivier, C.; van Zyl, W.; Stutzer, C. A Systematic, Functional Genomics, and Reverse Vaccinology Approach to the Identification of Vaccine Candidates in the Cattle Tick, *Rhipicephalus Microplus*. *Ticks Tick-Borne Dis.* 2012, *3*, 179–187, doi:10.1016/j.ttbdis.2012.01.003

93. Kashikar, P.; Dipke, C. Insilico Design and Development of Vaccine by Reverse Vaccinology Approach for Anthrax.

94. Kafle, A.; Ojha, S.C. Advancing Vaccine Development against Opisthorchis Viverrini: A Synergistic Integration of Omics Technologies and Advanced Computational Tools. *Front. Pharmacol.* 2024, *15*, doi:10.3389/fphar.2024.1410453

95. Williams, E.; Seib, K.L.; Fairley, C.K.; Pollock, G.L.; Hocking, J.S.; McCarthy, J.S.; Williamson, D.A. Neisseria Gonorrhoeae Vaccines: A Contemporary Overview. *Clin. Microbiol. Rev.* 2024, *37*, e00094-23, doi:10.1128/cmr.00094-23

96. Jagusztyn-Krynicka, E.K.; Dadlez, M.; Grabowska, A.; Roszczenko, P. Proteomic Technology in the Design of New Effective Antibacterial Vaccines. *Expert Rev. Proteomics* 2009, *6*, 315–330, doi:10.1586/epr.09.47

97. Chao, P.; Zhang, X.; Zhang, L.; Yang, A.; Wang, Y.; Chen, X. Proteomics-Based Vaccine Targets Annotation and Design of Multi-Epitope Vaccine against Antibiotic-Resistant Streptococcus Gallolyticus. *Sci. Rep.* 2024, *14*, 4836, doi:10.1038/s41598-024-55372-3

98. Baarda, B.I.; Martinez, F.G.; Sikora, A.E. Proteomics, Bioinformatics and Structure-Function Antigen Mining for Gonorrhea Vaccines. *Front. Immunol.* 2018, *9*, doi:10.3389/fimmu.2018.02793

99. Dwivedi, P.; Alam, S.I.; Tomar, R.S. Secretome, Surfome and Immunome: Emerging Approaches for the Discovery of New Vaccine Candidates against Bacterial Infections. *World J. Microbiol. Biotechnol.* 2016, *32*, 155, doi:10.1007/s11274-016-2107-3

100. Huang, C.-P.; Liu, Y.-T.; Nakatsuji, T.; Shi, Y.; Gallo, R.R.; Lin, S.-B.; Huang, C.-M. Proteomics Integrated with *Escherichia coli* Vector-Based Vaccines and Antigen Microarrays Reveals the Immunogenicity of a Surface Sialidase-like Protein of *Propionibacterium acnes*. *PROTEOMICS – Clin. Appl.* 2008, *2*, 1234–1245, doi:10.1002/prca.200780103

101. Villar, M.; Marina, Anabel; de la Fuente, J. Applying Proteomics to Tick Vaccine Development: Where Are We? *Expert Rev. Proteomics* 2017, *14*, 211–221, doi:10.1080/14789450.2017.1284590

102. Veerapandian, R.; Gadad, S.S.; Jagannath, C.; Dhandayuthapani, S. Live Attenuated Vaccines against Tuberculosis: Targeting the Disruption of Genes Encoding the Secretory Proteins of Mycobacteria. *Vaccines* 2024, *12*, 530.

103. Goff, A.; Cantillon, D.; Muraro Wildner, L.; Waddell, S.J. Multi-Omics Technologies Applied to Tuberculosis Drug Discovery. *Appl. Sci.* 2020, *10*, 4629, doi:10.3390/app10134629

104. Borah, K.; Xu, Y.; McFadden, J. Dissecting Host-Pathogen Interactions in TB Using Systems-Based Omic Approaches. *Front. Immunol.* 2021, *12*, doi:10.3389/fimmu.2021.762315

105. Xu, Y.; Borah, K. *Mycobacterium tuberculosis* Carbon and Nitrogen Metabolic Fluxes. *Biosci. Rep.* 2022, *42*, BSR20211215, doi:10.1042/BSR20211215

106. Cao, J.; Li, C.; Cui, Z.; Deng, S.; Lei, T.; Liu, W.; Yang, H.; Chen, P. Spatial Transcriptomics: A Powerful Tool in Disease Understanding and Drug Discovery. *Theranostics* 2024, *14*, 2946–2968, doi:10.7150/thno.95908

107. Cui, T.; Zeng, J.; He, Z.-G. Anti-Tuberculosis Drug Target Discovery by Targeting the Higher in-Degree Proteins (HidPs) of the Pathogen's Transcriptional Network. *J. Tuberc.* 2018, *1*, 1001.

108. Nalamolu, R.M.; Pasala, C.; Katari, S.K.; Amineni, U. Discovery of Common Putative Drug Targets and Vaccine Candidates for *Mycobacterium tuberculosis* Sp. *J. Drug Deliv. Ther.* 2019, *9*, 67–71, doi:10.22270/jddt.v9i2-s.2603

109. Singh, N.; Kanojia, H.; Singh, S.; Verma, D.; Gautam, B.; Wadhwa, G. *Structure Prediction of Drug Target Identified by Metabolic Pathway Analysis of Streptococcus Pyogenes*. New Delhi Publishers, 2013, *1*, 79–85.

110. Ghosh, S.; Prava, J.; Samal, H.B.; Suar, M.; Mahapatra, R.K. Comparative Genomics Study for the Identification of Drug and Vaccine Targets in *Staphylococcus Aureus*: MurA Ligase Enzyme as a Proposed Candidate. *J. Microbiol. Methods* 2014, *101*, 1–8, doi:10.1016/j.mimet.2014.03.009

111. Butt, A.M.; Tahir, S.; Nasrullah, I.; Idrees, M.; Lu, J.; Tong, Y. *Mycoplasma genitalium*: A Comparative Genomics Study of Metabolic Pathways for the Identification of Drug and Vaccine Targets. *Infect. Genet. Evol.* 2012, *12*, 53–62, doi:10.1016/j.meegid.2011.10.017

112. Nogueira, W.G.; Jaiswal, A.K.; Tiwari, S.; Ramos, R.T.J.; Ghosh, P.; Barh, D.; Azevedo, V.; Soares, S.C. Computational Identification of Putative Common Genomic Drug and Vaccine Targets in *Mycoplasma genitalium*. *Genomics* 2021, *113*, 2730–2743, doi:10.1016/j.ygeno.2021.06.011

113. Sudha, R.; Katiyar, A.; Katiyar, P.; Singh, H.; Prasad, P. Identification of Potential Drug Targets and Vaccine Candidates in Clostridium Botulinum Using Subtractive Genomics Approach. *Bioinformation* 2019, *15*, 18–25, doi:10.6026/97320630015018

114. Chawley, P.; Samal, H.B.; Prava, J.; Suar, M.; Mahapatra, R.K. Comparative Genomics Study for Identification of Drug and Vaccine Targets in *Vibrio Cholerae*: MurA Ligase as a Case Study. *Genomics* 2014, *103*, 83–93, doi:10.1016/j.ygeno.2013.12.002

115. Mujawar, S.; Mishra, R.; Pawar, S.; Gatherer, D.; Lahiri, C. Delineating the Plausible Molecular Vaccine Candidates and Drug Targets of Multidrug-Resistant Acinetobacter Baumannii. *Front. Cell. Infect. Microbiol.* 2019, *9*, doi:10.3389/fcimb.2019.00203

116. Serban, N.; Wasserman, L.; Peters, D.; Spirtes, P.; O'Doherty, R.; Handley, D.; Scheines, R.; Glymour, C. *Analysis of Microarray Data for Treated Fat Cells.* 2013. https://philpapers.org/rec/SERAOM

117. Gao, X.; Arodz, T. Detecting Differentially Co-Expressed Genes for Drug Target Analysis. *Procedia Comput. Sci.* 2013, *18*, 1392–1401, doi:10.1016/j.procs.2013.05.306

118. Birhanu, B.T.; Lee, S.-J.; Park, N.-H.; Song, J.-B.; Park, S.-C. In Silico Analysis of Putative Drug and Vaccine Targets of the Metabolic Pathways of Actinobacillus Pleuropneumoniae Using a Subtractive/Comparative Genomics Approach. *J. Vet. Sci.* 2018, *19*, 188–199, doi:10.4142/jvs.2018.19.2.188

119. Telkar, S.; Hs, S.K.; D, N.; Mahmood, R. Strategic Genome-Scale Prioritization of Unique Drug Targets: A Case Study of Streptococcus Gordonii. *Bioinformation* 2013, *9*, 983–987, doi:10.6026/97320630009983

120. Rao, S.; Ghosh, D.; Asturias, E.J.; Weinberg, A. What Can We Learn about Influenza Infection and Vaccination from Transcriptomics? *Hum. Vaccines Immunother.* 2019, *15*, 2615–2623, doi:10.1080/21645515.2019.1608744

121. Singhania, A.; Wilkinson, R.J.; Rodrigue, M.; Haldar, P.; O'Garra, A. The Value of Transcriptomics in Advancing Knowledge of the Immune Response and Diagnosis in Tuberculosis. *Nat. Immunol.* 2018, *19*, 1159–1168, doi:10.1038/s41590-018-0225-9

122. Lin, J.; Sodenkamp, J.; Cunningham, D.; Deroost, K.; Tshitenge, T.C.; McLaughlin, S.; Lamb, T.J.; Spencer-Dene, B.; Hosking, C.; Ramesar, J.; et al. Signatures of Malaria-Associated Pathology Revealed by High-Resolution Whole-Blood Transcriptomics in a Rodent Model of Malaria. *Sci. Rep.* 2017, *7*, 41722, doi:10.1038/srep41722

123. Aydillo, T.; Gonzalez-Reiche, A.S.; Stadlbauer, D.; Amper, M.A.; Nair, V.D.; Mariottini, C.; Sealfon, S.C.; van Bakel, H.; Palese, P.; Krammer, F.; et al. Transcriptome Signatures Preceding the Induction of Anti-Stalk Antibodies Elicited after Universal Influenza Vaccination. *Npj Vaccines* 2022, *7*, 1–14, doi:10.1038/s41541-022-00583-w

124. Xin, Y.; Chen, S.; Tang, K.; Wu, Y.; Guo, Y. Identification of Nifurtimox and Chrysin as Anti-Influenza Virus Agents by Clinical Transcriptome Signature Reversion. *Int. J. Mol. Sci.* 2022, *23*, 2372.

125. Maekawa, S.; Wang, P.-C.; Chen, S.-C. Comparative Study of Immune Reaction Against Bacterial Infection From Transcriptome Analysis. *Front. Immunol.* 2019, *10*, doi:10.3389/fimmu.2019.00153

126. Manjunath, S.; Kumar, G.R.; Mishra, B.P.; Mishra, B.; Sahoo, A.P.; Joshi, C.G.; Tiwari, A.K.; Rajak, K.K.; Janga, S.C. Genomic Analysis of Host – Peste Des Petits Ruminants Vaccine Viral Transcriptome Uncovers Transcription Factors Modulating Immune Regulatory Pathways. *Vet. Res.* 2015, *46*, 15, doi:10.1186/s13567-015-0153-8

127. Bouwman, W.; Verhaegh, W.; Holtzer, L.; van de Stolpe, A. Measurement of Cellular Immune Response to Viral Infection and Vaccination. *Front. Immunol.* 2020, *11*, doi:10.3389/fimmu.2020.575074

128. Ghosh, S.; Chatterjee, A.; Maitra, A. An Insight into COVID-19 Host Immunity at Single-Cell Resolution. *Int. Rev. Immunol* 1–16, doi:10.1080/08830185.2024.2443420

129. He, W.-R.; Yuan, J.; Ma, Y.-H.; Zhao, C.-Y.; Yang, Z.-Y.; Zhang, Y.; Han, S.; Wan, B.; Zhang, G.-P. Modulation of Host Antiviral Innate Immunity by African Swine Fever Virus: A Review. *Animals* 2022, *12*, 2935.

130. Kumari, S.; Kumar, A.; Samant, M.; Singh, N.; Dube, A. Discovery of Novel Vaccine Candidates and Drug Targets Against Visceral Leishmaniasis Using Proteomics and Transcriptomics. *Curr. Drug Targets 9*, 938–947, doi:10.2174/138945008786786091

131. Gramberg, S.; Puckelwaldt, O.; Schmitt, T.; Lu, Z.; Haeberlein, S. Spatial Transcriptomics of a Parasitic Flatworm Provides a Molecular Map of Vaccine Candidates, Drug Targets and Drug Resistance Genes 2023, doi:10.1101/2023.12.11.571084

132. Laing, R.; Kikuchi, T.; Martinelli, A.; Tsai, I.J.; Beech, R.N.; Redman, E.; Holroyd, N.; Bartley, D.J.; Beasley, H.; Britton, C.; et al. The Genome and Transcriptome of Haemonchus Contortus, a Key Model Parasite for Drug and Vaccine Discovery. *Genome Biol.* 2013, *14*, R88, doi:10.1186/gb-2013-14-8-r88

133. Cantacessi, C.; Gasser, R.B.; Strube, C.; Schnieder, T.; Jex, A.R.; Hall, R.S.; Campbell, B.E.; Young, N.D.; Ranganathan, S.; Sternberg, P.W.; et al. Deep Insights into *Dictyocaulus Viviparus* Transcriptomes Provides Unique Prospects for New Drug Targets and Disease Intervention. *Biotechnol. Adv.* 2011, *29*, 261–271, doi:10.1016/j.biotechadv.2010.11.005

134. Fitzpatrick, J.M.; Johnston, D.A.; Williams, G.W.; Williams, D.J.; Freeman, T.C.; Dunne, D.W.; Hoffmann, K.F. An Oligonucleotide Microarray for Transcriptome Analysis of *Schistosoma Mansoni* and Its Application/Use to Investigate Gender-Associated Gene Expression. *Mol. Biochem. Parasitol.* 2005, *141*, 1–13, doi:10.1016/j.molbiopara.2005.01.007

135. Young, J.A.; Winzeler, E.A. Using Expression Information to Discover New Drug and Vaccine Targets in the Malaria Parasite Plasmodium Falciparum. *Pharmacogenomics* 2005, *6*, 17–26, doi:10.1517/14622416.6.1.17

136. Verma, D.; Kapoor, S. Chapter 11 - Omics Approach for Personalized and Diagnostics Medicine. In *Integrative Omics*; Gupta, M.K., Katara, P., Mondal, S., Singh, R.L., Eds.; Academic Press, 2024; pp. 175–185.

137. Du, P.; Fan, R.; Zhang, N.; Wu, C.; Zhang, Y. Advances in Integrated Multi-Omics Analysis for Drug-Target Identification. *Biomolecules* 2024, *14*, 692, doi:10.3390/biom14060692

138. Li, Y.; Hou, G.; Zhou, H.; Wang, Y.; Tun, H.M.; Zhu, A.; Zhao, J.; Xiao, F.; Lin, S.; Liu, D.; et al. Multi-Platform Omics Analysis Reveals Molecular Signature for COVID-19 Pathogenesis, Prognosis and Drug Target Discovery. *Signal Transduct. Target. Ther.* 2021, *6*, 1–11, doi:10.1038/s41392-021-00508-4

139. Ivanisevic, T.; Sewduth, R.N. Multi-Omics Integration for the Design of Novel Therapies and the Identification of Novel Biomarkers. *Proteomes* 2023, *11*, 34, doi:10.3390/proteomes11040034

140. Nishimura, Yuhei Target Discovery Using Omics and Zebrafish. 日本薬理学会年会要旨集 2019, *92*, JKL-04, doi:10.1254/jpssuppl.92.0_JKL-04

141. Yang, Y.; Adelstein, S.J.; Kassis, A.I. Integrated Bioinformatics Analysis for Cancer Target Identification. In *Bioinformatics for Omics Data: Methods and Protocols*; Mayer, B., Ed.; Humana Press, 2011; pp. 527–545.

142. Querec, T.D.; Akondy, R.S.; Lee, E.K.; Cao, W.; Nakaya, H.I.; Teuwen, D.; Pirani, A.; Gernert, K.; Deng, J.; Marzolf, B.; et al. Systems Biology Approach Predicts Immunogenicity of the Yellow Fever Vaccine in Humans. *Nat. Immunol.* 2009, *10*, 116–125, doi:10.1038/ni.1688

143. De Groot, A.S.; Moise, L.; Terry, F.; Gutierrez, A.H.; Hindocha, P.; Richard, G.; Hoft, D.F.; Ross, T.M.; Noe, A.R.; Takahashi, Y.; et al. Better Epitope Discovery, Precision Immune Engineering, and Accelerated Vaccine Design Using Immunoinformatics Tools. *Front. Immunol.* 2020, *11*, doi:10.3389/fimmu.2020.00442

144. Janahi, E.M.; Dhasmana, A.; Srivastava, V.; Sarangi, A.N.; Raza, S.; Arif, J.M.; Bhatt, M.L.B.; Lohani, M.; Areeshi, M.Y.; Saxena, A.M.; et al. In Silico CD4+, CD8+ T-Cell and B-Cell Immunity Associated Immunogenic Epitope Prediction and HLA Distribution Analysis of Zika Virus. *EXCLI J.* 2017, *16*, 63–72, doi:10.17179/excli2016-719

145. Abdulla, F.; Adhikari, U.K.; Uddin, M.K. Exploring T & B-Cell Epitopes and Designing Multi-Epitope Subunit Vaccine Targeting Integration Step of HIV-1 Lifecycle Using Immunoinformatics Approach. *Microb. Pathog.* 2019, *137*, 103791, doi:10.1016/j.micpath.2019.103791

146. Bano, N.; Kumar, A. Immunoinformatics Study to Explore Dengue (DENV-1) Proteome to Design Multi-Epitope Vaccine Construct by Using CD4+ Epitopes. *J. Genet. Eng. Biotechnol.* 2023, *21*, 128, doi:10.1186/s43141-023-00592-9

147. Najafi, A.; Ataee, M.H.; Farzanehpour, M.; Esmaeili, H.; Ghaleh, G. A Comparative Analysis of Computational Strategies in Multi-Epitope Vaccine Design Against Human Papillomavirus and Cervical Cancer. *Cell J. Yakhteh* 2024, *26*, 403–426.

148. Kardani, K.; Bolhassani, A.; Namvar, A. An Overview of in Silico Vaccine Design Against Different Pathogens and Cancer. *Expert Rev. Vaccines* 2020, *19*, 699–726, doi:10.1080/14760584.2020.1794832

149. Arya, H.; Bhatt, T.K. Chapter 20 – Role of Bioinformatics in Subunit Vaccine Design. In *Molecular Docking for Computer-Aided Drug Design*; Coumar, M.S., Ed.; Academic Press, 2021; pp. 425–439.

150. Kaliamurthi, S.; Selvaraj, G.; Junaid, M.; Khan, A.; Gu, K.; Wei, D.-Q. Cancer Immunoinformatics: A Promising Era in the Development of Peptide Vaccines for Human Papillomavirus-Induced Cervical Cancer. *Curr. Pharm. Des.* 24, 3791–3817, doi:10.2174/1381612824666181106094133

151. Basmenj, E.R.; Pajhouh, S.R.; Ebrahimi Fallah, A.; Naijian, R.; Rahimi, E.; Atighy, H.; Ghiabi, S.; Ghiabi, S. Computational Epitope-Based Vaccine Design with Bioinformatics Approach; A Review. *Heliyon* 2025, *11*, e41714, doi:10.1016/j.heliyon.2025.e41714

152. Prediction of Epitopes of Viral Antigens Recognized by Cytotoxic T Lymphocytesas an Immunoinformatics Approach to Anti-HIV/Aids Vaccine Design. *Int. J. Vaccines Vaccin.* 2015, *1*, https://acquire.cqu.edu.au/articles/journal_contribution/Prediction_of_epitopes_of_viral_antigens_recognized_by_cytotoxic_T_Lymphocytes_as_an_immunoinformatics_approach_to_anti-HIV_AIDS_vaccine_design/13435388

153. Xiang, Z.; He, Y. Vaxign: A Web-Based Vaccine Target Design Program for Reverse Vaccinology. *Procedia Vaccinol.* 2009, *1*, 23–29, doi:10.1016/j.provac.2009.07.005

154. Groot, A.; Groot, P.D.; He, Y. ICoVax 2013: The 3rd ISV Pre-Conference Computational Vaccinology Workshop. *BMC Bioinformatics* 2014, *15*, 1–1, doi:10.1186/1471-2105-15-S4-I1

155. Söllner, J.; Heinzel, A.; Summer, G.; Fechete, R.; Stipkovits, L.; Szathmary, S.; Mayer, B. Concept and Application of a Computational Vaccinology Workflow. *Immunome Res.* 2010, *6*, S7, doi:10.1186/1745-7580-6-S2-S7

156. Liljeroos, L.; Malito, E.; Ferlenghi, I.; Bottomley, M.J. Structural and Computational Biology in the Design of Immunogenic Vaccine Antigens. *J. Immunol. Res.* 2015, *2015*, 156241, doi:10.1155/2015/156241

157. Srivastava, P.; Jain, C.K. Computer Aided Reverse Vaccinology: A Game-Changer Approach for Vaccine Development. *Comb. Chem. High Throughput Screen.* 26, 1813–1821, doi:10.2174/1386207325666220930124013

158. Zameer, S. System Vaccinology: Applications, Trends, and Perspectives. *i-Manager's Journal on Life Sciences* 2024, 3(2), doi: 10.26634/jls.3.2.21038

159. Wang, Z.; He, Y. Precision Omics Data Integration and Analysis with Interoperable Ontologies and Their Application for COVID-19 Research. *Brief. Funct. Genomics* 2021, *20*, 235–248, doi:10.1093/bfgp/elab029

160. Naithani, N.; Atal, A.T.; Tilak, T.V.S.V.G.K.; Vasudevan, B.; Misra, P.; Sinha, S. Precision Medicine: Uses and Challenges. *Med. J. Armed Forces India* 2021, *77*, 258–265, doi:10.1016/j.mjafi.2021.06.020

161. Pettitt, D.; Smith, J.; Meadow, N.; Arshad, Z.; Schuh, A.; DiGiusto, D.; Bountra, C.; Holländer, G.; Barker, R.; Brindley, D. Regulatory Barriers to the Advancement of Precision Medicine. *Expert Rev. Precis. Med. Drug Dev.* 2016, *1*, 319–329, doi:10.1080/23808993.2016.1176526

162. Beauvais, M.; Knoppers, B.M. When Information Is the Treatment? Precision Medicine in Healthcare. *Healthc. Manage. Forum* 2020, *33*, 120–125, doi:10.1177/0840470419859017

163. Bylstra, Y.; Davila, S.; Lim, W.K.; Wu, R.; Teo, J.X.; Kam, S.; Lysaght, T.; Rozen, S.; Teh, B.T.; Yeo, K.K.; et al. Implementation of Genomics in Medical Practice to Deliver Precision Medicine for an Asian Population. *Npj Genomic Med.* 2019, *4*, 1–7, doi:10.1038/s41525-019-0085-8

164. Chen, R.; Snyder, M. Promise of Personalized Omics to Precision Medicine. *WIREs Syst. Biol. Med.* 2013, *5*, 73–82, doi:10.1002/wsbm.1198

5 AI-Enhanced Vaccine Development against Emerging Infectious Diseases

Ashish Singh Chauhan, Pallavi Chand, Arvind Raghav, and Arif Nur Muhammad Ansori

5.1 INTRODUCTION

Emerging infectious diseases (EIDs) are now a significant and increasing threat to the health of the world. They are widely known as conditions that either started in a population recently or were there before but are now rising in how often and where they occur [1]. Examples of this classification are pathogens found in animals that infect humans, and old diseases that return due to different factors such as antimicrobial resistance, decreased immunity, or changes in human living and natural conditions [2].

EIDs are grounded in concepts from epidemiology, ecology, evolutionary biology, and global health policy. The World Health Organization (WHO) says that EIDs cause a large number of infections in people. Many species are being reduced in number due to interconnected problems, including climate change, deforestation, urbanization, increased global trade and travel, agricultural expansion, and human encroachment into natural habitats. Such environmental changes encourage pathogens to jump from animals to humans, which can result in both epidemics and pandemics [3].

This study aligns closely with several United Nations Sustainable Development Goals (SDGs), particularly **SDG 3: Good Health and Well-Being**, emphasizing the need for effective responses to infectious diseases. By leveraging artificial intelligence (AI) to accelerate vaccine development, the chapter supports global efforts to reduce the burden of emerging diseases, enhance health security, and promote equitable access to medical innovations. Furthermore, integrating AI contributes to **SDG 9: Industry, Innovation, and Infrastructure**, through advancing digital health technologies, and **SDG 17: Partnerships for the Goals**, by encouraging interdisciplinary and international collaboration in biomedical research [4].

Historically, EIDs have posed catastrophic threats to global health security. The 1918 influenza pandemic, caused by the H1N1 virus, infected approximately one-third of the worldwide population and resulted in an estimated 50 million deaths [5].

HIV/AIDS, first identified in the early 1980s, has resulted in over 40 million deaths and continues to affect nearly 38 million individuals worldwide [6]. More than 8,000 cases and 774 deaths were caused by the outbreak of severe acute respiratory syndrome (SARS-CoV-1) in the years 2002–2003, and these showed that the global measures to deal with outbreaks had major faults [7]. The 2009 H1N1 flu pandemic resulted in 151,700 to 575,400 deaths in its first year [8]. Ebola was as deadly as it was for many other outbreaks, and the infection during the West Africa epidemic of 2014 to 2016 caused more than 11,000 deaths, showing what consequences delayed detection and weak healthcare can bring [9]. While the Zika virus was present in 2015–2016, it confirmed that viruses in pregnant women can harm the brain, and it clarified that vector control is still essential (according to WHO) [10]. Recently, COVID-19, brought on by the virus SARS-CoV-2, was identified as the most serious pandemic this century, causing 770 million cases and nearly 7 million deaths at the time of writing, in early 2024. Public health ideas have been reshaped, many economies have been interrupted, and important international cooperation was required due to the pandemic [11].

In addition to harming health, EIDs result in ongoing socioeconomic costs over the long run. The COVID-19 pandemic has caused the global economy to lose US\$16 trillion, and its effects have hit low and middle-income countries (LMICs) especially hard. Such events demonstrate that pathogen emergencies can compromise global stability, which is why increased investment in surveillance, diagnostics, and countermeasures against new threats, including vaccine development, is necessary. [12].

EIDs could become a significant problem in the future. Scientists are still unaware of an estimated 1.6 million viruses in mammals and birds; more than 700,000 of these are known to infect humans. The Global Virome Project, One Health, and the WHO R&D Blueprint are among the global efforts that have helped prioritize lab studies of pathogens that might cause pandemics. Still, despite all these measures, progress on medical countermeasures, especially on vaccines, continues to slow the process down [13].

Current methods for developing vaccines, which are slow, expensive, and subject to numerous regulations, are not well-suited for addressing EIDs, which evolve rapidly. Once a new pathogen is identified, developing vaccines quickly is crucial to prevent its spread and protect public health. More attention is being given to novel technologies, with AI now playing a significant role in this

DOI: 10.1201/9781003615699-5

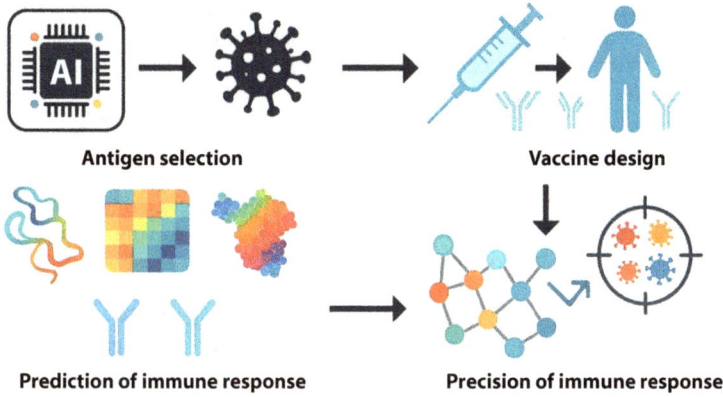

Figure 5.1 AI-driven revolution in vaccine antigen design.

area. AI approaches to vaccine research may lead to a revolution as they help combine live information, make predictions, and create logical vaccine components [14]. Even though vaccines have significantly controlled infectious diseases (Figure 5.1), the traditional development process for vaccines means they can sometimes be slow to respond to new risks. Until now, vaccine development has typically involved searching for the pathogen, selecting the antigen, conducting non-human experiments, completing clinical trials in stages, obtaining authorization from health agencies, and large-scale production. Because the process requires strict protocols, it moves slowly, resulting in an average gap of 10 to 15 years between initial discovery and making the vaccine public, and the costs can run over USD 1–2 billion for a vaccine candidate [15].

Highly time-consuming steps, such as discovering and validating antigens through trial-and-error screening and the use of lab animals, are one of the main problems in this area. An additional problem is when scientists lack complete knowledge about the biology, immune recognition, and interaction between the pathogen and the patient. Additionally, most conventional vaccine platforms struggle to elicit the desired immune response, and their effects tend to wane rapidly when confronted with rapidly evolving viruses, such as influenza, HIV, or SARS-CoV-2 [16].

The limited accuracy of both in vitro experiments and animal studies makes preclinical progress more challenging in predicting how the drug will be used in humans. When vaccine candidates are reviewed in clinical trials, the need for large numbers of participants in Phase III and the related high costs become significant hindrances. Finding participants from diverse backgrounds, monitoring their health over time, and estimating benefits using statistical analysis requires significant resources and effort. In addition, following strict guidelines and requests for documentation and evaluations in the regulatory process typically delays the completion of drug development [17].

However, another difficulty is scaling construction. Using inactivated virus or protein subunit vaccines requires careful biosafety controls, which can impact the speed of international vaccine distribution during unfolding emergencies. At the beginning of the COVID-19 crisis, it was obvious that there were supply chain limits and that vaccine manufacturing could not keep up with demand [18].

Moreover, these traditional hurdles exacerbate the problems of unequal vaccine access worldwide. Due to a shortage of both equipment and money, LMICs are more often affected by delays in getting vaccines and continue to deal with increased health problems, deaths, and economic losses [19].

The problems in current vaccine development highlight the fact that existing pipelines do not effectively respond to the need for rapid vaccines against EIDs. This has led people to look at new ways to prioritize speed, flexibility, and accuracy, and AI could play a significant role here [20].

AI has played a significant role in the life sciences due to the increasing complexity and size of biomedical information. AI supports a wide range of computer algorithms, such as those related to machine learning (ML) and deep learning (DL), enabling machines to learn from data, identify trends, and base their own decisions on the knowledge gained from that data. As biomedical research now deals mainly with multi-part, different data types, these abilities are well-suited [21, 22].

The application of AI in the biomedical sector has led to significant advancements in genomics, drug discovery, personalized medicine, and imaging for early disease diagnosis. Annotating genes involves applying AI, estimating how proteins fold (such as AlphaFold by DeepMind), helping find new drug options, and making it easier to detect diseases like cancer, cardiovascular problems, or brain disorders [23]. With AI, interpreting sequencing data has become possible in genomics, enabling the identification of mutations linked to diseases and the discovery of new relationships between genes and disease. New developments are influencing the way researchers view disease causes and effective treatment [24].

AI can significantly assist in predicting disease behavior, tracking viruses, and predicting out-breaks. Because of AI, BlueDot and HealthMap were able to predict the early signs of the virus outbreak using travel, media, and health information worldwide during the COVID-19 pandemic [25]. That is why it is now possible to integrate AI more comprehensively into pandemic prepared-ness frameworks.

AI is beneficial for more than data analysis; now, it helps scientists discover new things. Thanks to AI, generating hypotheses, designing tests, and handling lab tasks can now be done quickly, smartly, and automatically with robotic systems. Moreover, explainable AI (XAI) increases the extent to which the biomedical research community can understand and trust complex models [26].

As data becomes increasingly important in biology, the connection between AI and biotechnol-ogy becomes even more crucial. Due to its ability to process various types of data, including omics data, electronic health records, images, and sensor information, AI is essential for both precision medicine and timely health interventions. Its use in ongoing, evolving research involving algo-rithms that regularly learn from new information differs from old research models [27].

AI is a supporting device and an essential tool for progress in biomedical science today. Vaccine development for new and EIDs is a natural and timely use of CRISPR technology within biomedical science. With this merging of causes, new opportunities emerge to shorten development cycles, increase our accuracy in predicting, and quickly handle threats to global health [28].

This chapter emphasizes that the rapid recognition and control of new pathogens necessitate a shift in surveillance, diagnostics, and vaccine development approaches. AI can now handle these challenges because it can handle vast amounts of data and offer predictions and automation. AI is essential for expedited development efforts against both new and existing infectious diseases.

5.2 LIMITATIONS OF CONVENTIONAL VACCINE DEVELOPMENT PIPELINES

Developing a vaccine is a complex process that involves testing its safety, effectiveness, and quality before it is accepted. Moving a theoretical vaccination idea into a real and usable product depends heavily on the progress through each stage of development. This process is generally split into four main sections: discovery, preliminary laboratory experiments, clinical trials on people, and regula-tory acceptance following updates from what is learned [29].

a. Discovery Phase

The process begins when scientists look for antigens that can trigger a safety mechanism in the body. An important part of this process is learning about the pathogen's biology, shape, how it reproduces, what makes it cause disease, and its methods to evade the immune system. Conventionally, pathogens have been cultured, and then their potential antigens or epitopes have been identified manually using biochemical methods [30, 31].

Recent advancements in genomics, proteomics, and bioinformatics have enabled the applica-tion of reverse vaccinology, a method that screens entire pathogen genomes digitally to select antigens for vaccine development. Because of this approach, developing the MenB vaccine (Bexsero) took far less time since it was based on genomic analysis of *Neisseria meningitidis*. AI and ML are now regularly used to improve the process of finding antigens, checking their immu-nogenicity, and predicting reactions between different antigens [32].

b. Preclinical Phase

Vaccine candidates undergo preclinical evaluation following antigen identification to determine their immunogenic potential and safety profile *in vitro* (e.g., using cell lines or organoids) and *in vivo* (typically using animal models such as mice, ferrets, or non-human primates). This phase assesses multiple parameters, including antigen stability, immune activation (e.g., antibody titers, T-cell responses), toxicity, and preliminary dosing regimens [33].

Key outputs of the preclinical phase include [33, 34]:

- Selection of vaccine platforms (e.g., mRNA, viral vector, protein subunit)

- Determination of optimal adjuvants

- Preliminary formulation strategies

- Data supporting an Investigational New Drug (IND) application

While essential, this phase faces challenges such as limited translatability of animal models and variability in immune responses, which can result in promising candidates failing in subsequent human trials.

c. **Clinical Trial Phases (I–III)**

Upon successful preclinical evaluation and regulatory clearance, vaccine candidates proceed to **clinical trials**, a multi-stage process involving human participants [35]:

> **Phase I** trials focus primarily on safety, tolerability, and dose-ranging, typically involving 20–100 healthy volunteers. This phase also collects preliminary data on immune responses. **Phase II** trials expand to several hundred participants, assess immunogenicity more robustly, refine dosing regimens, and continue safety monitoring. These studies are often randomized and controlled. **Phase III** trials are large-scale studies involving thousands to tens of thousands of participants across diverse demographics and geographic regions. The primary objective is to determine the vaccine's efficacy in preventing infection or disease, typically expressed as a percentage reduction in incidence among vaccinated individuals compared to those in the placebo group. Phase III trials also monitor adverse events and rare side effects [35, 36].

The timeline for clinical trials can range from 3 to 10 years, although adaptive trial designs and emergency use authorizations (as seen with COVID-19 vaccines) can accelerate this process significantly under pandemic conditions [31].

d. **Regulatory Approval and Post-Marketing Surveillance**

Following the successful outcomes of Phase III trials, manufacturers submit a comprehensive Biologics License Application (BLA) to regulatory bodies, such as the U.S. Food and Drug Administration (FDA), the European Medicines Agency (EMA), or national regulatory authorities in other countries. These submissions include clinical data, manufacturing protocols, quality control measures, and risk assessments [37]. When the vaccine receives approval, the public may be eligible to receive it. Even so, the evaluation is not finished here. For a continued period, Phase IV focuses on detecting rare problems, long-term side effects, and the actual effectiveness of the product. In the U.S., the Vaccine Adverse Event Reporting System (VAERS) and EudraVigilance in Europe keep vaccines safe after they are approved [38].

5.2.1 Challenges Across the Pipeline

Each stage of the vaccine development pipeline presents unique scientific, logistical, and regulatory challenges. These include the following:

Long development timelines, high attrition rates – with only 6%–10% of candidates entering clinical trials reaching licensure – substantial costs ranging from US$500 million to over US$1 billion, and unequal access to resources and manufacturing capacities, especially in low-resource settings, present significant challenges in the clinical trial process [38–40].

In response to these challenges, incorporating AI and computational tools offers promising opportunities to improve efficiency, reduce failure rates, and enable data-driven decisions across all stages of the vaccine development lifecycle [41].

Traditional vaccine development methods have been effective, but they often face significant challenges when addressing diseases that can evolve rapidly. The main obstacles are long time scales, costly orders, and difficulties in determining product effectiveness during the initial stages. These constraints seriously affect global efforts to face and respond to EIDs [40].

5.2.1.1 Time Constraints

Traditionally, developing a vaccine can take considerable time, from the concept stage to licensing. Typically, developing a new medicine for patients can take 10–15 years and requires scientists to conduct research in the lab, followed by animal studies, multiple clinical studies, and, ultimately, approval from regulatory bodies. Every part of the process is necessary to ensure the vaccine is effective and safe. However, because the military pipeline is not flexible, it often causes delays, which can be particularly significant during emergencies. Although these accelerated means were approved to respond to COVID-19, this process is not typical [42].

5.2.1.2 High Development Costs

Developing a vaccine from inception to public use is expensive in terms of both time and capital. The average amount required to develop a vaccine, from development to market release, ranges widely between US$500 million and US$2 billion, depending on the method used, the target disease, and the location of trials. They encompass laboratory research, mass production activities, clinical studies, regulatory compliance, and post-approval monitoring. It is also risky because many vaccine candidates do not advance past the early research stages [43]. For this reason, investing in vaccines for diseases that offer little profit in the market is not attractive to many pharmaceutical companies.

5.2.1.3 Uncertainty in Efficacy Prediction

Traditional vaccine development often encounters difficulties early on because it is challenging to determine whether the candidate vaccine will elicit an effective immune response. Antigen selection and immune response modeling are often based on animal studies, which may not apply to humans. Because it is difficult to predict outcomes, many experiments in human trials fail. Although numerous HIV and tuberculosis vaccines show promise in experiments, the majority have experienced failure in human testing due to poor protection or some unknown risk issue [44].

Furthermore, immune system issues are complex to target with a single universal vaccine because immune reactions vary across populations due to age, genetics, other medical conditions, and prior infections. The fast mutations seen in influenza viruses and SARS-CoV-2 can lead to less effective vaccines because the changes in the pathogens happen very quickly [45].

5.2.1.4 Logistical and Manufacturing Challenges

Classic vaccine methods using inactivated or weakened viruses must usually be made in labs of level-3 (BSL-3) safety, with cells in culture and cold storage. Such needs are particularly challenging for impoverished regions, resulting in delays in releasing vaccines when the development phase is completed. Due to the inequity in accessing vaccines during the COVID-19 pandemic, there is a pressing need for vaccine solutions that are easier to develop and share globally [46].

In short, conventional vaccine manufacturing challenges – such as lengthy timelines, high costs, unreliable models, and supply chain bottlenecks – make it difficult for the world to address EIDs. Because of these challenges, new data-driven techniques, such as those offered by AI, can significantly improve how vaccines are developed while reducing costs and leading to more accurate results [47].

5.2.2 AI Intervention Opportunities at Stages

With AI used in vaccine development, there is a significant chance of overcoming the problems seen in conventional approaches. Compared to standard approaches, which rely on lengthy experiments and strict progression, AI enables a shift to research based on predictions, repetition, and data. During vaccine development, applying AI, for example, ML, DL, and natural language processing (NLP), at different stages, starting from choosing targets, optimizing clinical trials, and making decisions for regulation, can be very effective [48].

5.2.2.1 Antigen Discovery and Epitope Prediction

AI has enabled the rapid identification of potential targets for attack by pinpointing antigenic targets early in the process. Antigens were previously selected using slow and manual laboratory methods. In contrast to these approaches, ML algorithms have been utilized to accurately predict epitopes of both B and T cells from genomic, proteomic, and immunological data. For example, models such as Deep Learning Vaccine Prediction Tool (DeepVacPred) and NetMHCpan have been developed to accurately predict MHC-peptide binding and identify which parts of proteins will dominate the immune response. Such tools make it easier to spot molecule parts conserved across multiple strains of highly mutating pathogens [49].

5.2.2.2 Vaccine Design and Optimization

After identifying promising targets, AI can arrange vaccine proteins more effectively, assess their strength, and predict how they might interact with existing vaccines. VAEs and GANs from the generative model group have been utilized to develop new peptide sequences and optimize the setup of antigens. AI plays a role in developing immunization strategies by predicting the proper adjuvant mixtures and assessing their stability in several situations [50].

With the help of AI, the use of codons, RNA shape, and untranslated sequences are tested to increase both the efficiency of translation and the effectiveness of the immune response in vaccine design. Without these applications, the fast growth of COVID-19 vaccines produced by Moderna and BioNTech might not have happened [51].

5.2.2.3 AI in Preclinical Immune Modeling

AI enables the modeling of immune responses in silico, reducing reliance on animal models and enhancing the translatability of findings to human systems. Reinforcement learning (RL) and agent-based modeling can simulate host-pathogen interactions, immune signaling pathways, and cytokine dynamics, providing valuable insights into these complex systems. These models provide early insights into potential toxicity and efficacy, facilitating data-informed decision-making before clinical testing [52].

Moreover, AI-driven multi-omics integration (e.g., combining transcriptomics, proteomics, and metabolomics data) offers a systems-level view of immune responses, enabling the identification of biomarkers predictive of vaccine efficacy or adverse effects [53].

5.2.2.4 Clinical Trial Design and Patient Stratification

The clinical development phase stands to benefit considerably from AI in terms of efficiency and precision. AI models can assist in adaptive clinical trial designs, enabling real-time adjustments based on interim data analysis. ML tools are also used to stratify patients by predicting who will respond to a vaccine or experience side effects, based on genetic, demographic, and immunological profiles [54].

During the COVID-19 vaccine trials, companies employed AI to monitor patient adherence, manage site logistics, and analyze massive amounts of trial data to expedite decision-making processes. NLP tools have also been utilized to curate clinical literature and trial reports, streamlining evidence synthesis for regulatory submissions [55].

5.2.2.5 Manufacturing and Quality Control

AI technology is being used to monitor vaccine production on advanced manufacturing machines in real time. By using computer vision, predictive support, and automation in quality checks, production can run consistently, lessen batch errors, and expand more easily. AI can help detect issues in the supply chain and make vaccine distribution easier around the globe [56].

5.2.2.6 Post-Marketing Surveillance and Pharmacovigilance

When a vaccine receives approval, AI becomes crucial in monitoring its safety after it hits the market. With NLP and anomaly detection, information from electronic health records (EHRs), social media, and data from wearable devices can be leveraged to identify anomalies or reduce vaccine effectiveness. AI is being used to verify patterns of adverse events following COVID-19 vaccination by studying information from VAERS and EudraVigilance [57].

AI benefits all stages of vaccine development, including increased speed, accuracy, cost savings, and scalability. Integrating AI helps boost the effectiveness of valuable emerging infectious disease efforts by discovering new mechanisms, testing them in laboratory models, planning smart clinical trials, and closely tracking the outcomes in real-life settings. When implemented correctly, these opportunities can change the vaccine development process globally [58].

5.3 AI TECHNIQUES IN VACCINE DEVELOPMENT

AI is the field that utilizes various computing techniques to enable machines to act in ways that typically require human thought. There has been significant interest in ML and DL lately, as their implications for vaccine development are substantial. Being able to handle extensive and complicated biological data, spot fine details, and predict results significantly boosts vaccine research [59].

ML is a subset of AI and works on algorithms that improve over time through training on data. ML algorithms in vaccine research review biological, genomic, and clinical information to identify features related to immunity that an expert review might overlook. For example, supervised learning is used to build models on labeled data, so the target, such as the immunogenic quality of a peptide, is already defined. Afterward, these models can suggest which candidate antigens will generate a robust immune response. Unsupervised learning is used to study unlabeled data and discover new insights, such as identifying variants with similar antigenic properties and uncovering other important immune-related features. Still less common, RL is now being considered to mimic and improve the function of biological systems, mainly how the immune system is regulated [60].

DL, which belongs to the field of ML, utilizes artificial neural networks with multiple levels to identify more abstract features hidden within raw data. DL models can utilize a hierarchical setup to recognize the complex, nonlinear processes of biological systems. New scientific methods from DL have significantly enhanced the analysis of protein sequences and 3D structures, which are essential for vaccine development. At the start, CNNs were designed for image recognition, but later brought into biology and immunology to help find epitope-related patterns in biological sequences and immunology data [61]. Modeling sequential data is a strong point of both RNNs and their variants, improving immune memory and antigen prediction accuracy. In the past few years, transformer models, popular in NLP, have been adopted to analyze protein language and their interactions, allowing researchers to predict more advanced results, including MHC-peptide binding and the effects of viral mutations [62].

DeepMind's AlphaFold shows how DL can solve problems nearly as accurately as experts do – it predicts the form of proteins with accuracy comparable to those seen in experiments. Instead of using sums of the parts, rational vaccine design requires accurately predicting the vaccine's structure [63].

For ML and DL to work effectively, data must be easily accessible and reliable. Platforms such as the Immune Epitope Database and collections like GISAID, as well as various types of omics databases, are relied upon to develop and validate AI models. However, because the data is not the same, has some missing labels, and is not balanced, exceptional attention to preprocessing and fine-tuning is necessary for reliable results [64].

When ML and DL are applied to systems immunology and computational biology, we can truly study how hosts interact with pathogens. Adopting a multidisciplinary approach enables the design of vaccines tailored to specific populations, the administration of rapid vaccinations for recent illnesses, and the testing of vaccine impact and safety on a computer before actual use. The use of AI is bringing about a significant shift in vaccine development, as it now relies on insights based on data to help vaccine creation progress much faster from research to application [65].

5.3.1 Natural Language Processing (NLP) for Biomedical Literature Mining

NLP is a vital component of AI that enables machines to interpret and utilize human language accurately. In working on vaccines, NLP is helping scientists process and learn from the increasing amount of literature, trial results, and information on public health [66].

Scientific, patent, regulatory, and epidemiological reports increase rapidly during infectious disease outbreaks. It is impossible to thoroughly review all the material by hand to identify meaningful trends, new developments, or potential vaccine targets. NLP methods transform unstructured text data into structured, valuable information for further study [67].

In biomedical research, the primary NLP task is named entity recognition (NER), which identifies and categorizes genes, proteins, pathogens, diseases, and various chemical compounds within written text. As a result, AI can create databases that cross-reference vaccine antigens with either immune reactions or types of pathogens. For instance, if NER identifies reports on mutations of the spike protein in SARS-CoV-2, it can match them with the results from studies on vaccine efficacy, which can help inform the redesign of vaccines [68].

Relation extraction goes further by detecting interactions of biological entities, such as bonds between viral proteins and host receptors, and partnerships between vaccine adjuvants and immunological effects. Because of these insights, scientists can determine how vaccines produce their effects, suggest possible new formulas, or decide which ones should be studied deeply [69].

With the use of transformer-based models like BERT and combined models like BioBERT and SciBERT, the precision and meaningfulness of biomedical text mining have improved significantly [70]. Thanks to these models, complex details and terminology from various fields can be understood, which helps with literature reviews, question answering, and examining trends. This method organizes research results quickly and can help guide decisions about vaccines and their use.

In clinical trials, NLP plays a crucial role in data mining, as robots scan trial databases and results to track vaccine outcomes, identify signs of side effects, and assess the vaccine's performance in various populations. As a result, vaccine studies are modified as needed, and safety measures are introduced so that vaccines are safe and suitable for various groups [71].

Additionally, NLP-based knowledge graphs integrate details from multiple sources and connect data from literature with data on genomics, proteomics, and clinical settings (Table 5.1). Thanks to updates made in response to new knowledge, these graphs provide current and reliable information to anyone researching vaccines [66, 71]. NLP supports vaccine advancement by organizing unstructured data into valuable, searchable, and precise knowledge. As a result, finding innovative vaccine targets and tracking and updating vaccines for rapidly changing pathogens becomes quicker and more efficient. With the rapidly increasing biomedical literature, NLP will be crucial in handling all this information [78].

TABLE 5.1 Key Biomedical Data Types and Their Integration into Knowledge Graphs for Vaccine Development

Data Type	Examples	Primary Sources	Role in Knowledge Graphs	References
Genomic Data	Viral genome sequences, host polymorphisms	NCBI GenBank, GISAID	Identifies antigenic regions; connects pathogens with host susceptibility	[72]
Proteomic Data	Viral proteins, host receptors, and epitope structures	UniProt, PDB, IEDB	Maps epitopes to protein structures; links antigen features to immune targets	[73]
Immunological Data	T-cell and B-cell responses, HLA-binding profiles	ImmPort, IEDB, VDJdb	Connects immune responses to specific epitopes and HLA alleles	[74]
Clinical Trial Data	Vaccine efficacy, safety, and immunogenicity outcomes	ClinicalTrials.gov, WHO International Clinical Trials Registry Platform	Links vaccine formulations to outcomes and demographic characteristics	[75]
Literature-Derived Data	Protein interactions, vaccine targets, and immune correlates	PubMed, CORD-19, PMC via NLP engines	Extracts relationships and hypotheses for inclusion in the graph	[76]
Epidemiological Data	Outbreak locations, variant tracking, vaccine coverage	WHO, CDC, GISAID, Our World in Data	Integrates pathogen evolution with population-level vaccine deployment	[77]

5.3.2 Structural Bioinformatics and AI-Based Protein Modeling

Using computational and structural studies, structural bioinformatics helps reveal the interactions between virus components and human immune receptors, aiding in vaccine design. Understanding how proteins are built is essential in COVID-19 vaccines, since where epitopes are located decides their availability and effect on the immune system [79]. Although very accurate, traditional methods to determine protein structure often take a long time, require much labor, and face technical issues, mainly affecting membrane proteins and transient complexes [80].

Recent improvements in AI and computational studies have significantly transformed protein structure prediction. AI technology is now as accurate and as fast as conventional computing methods. The achievement of AlphaFold from DeepMind is significant, as it accurately predicts protein structures just like experiments using only amino acid orderings [79].

AlphaFold relies on a novel neural network design that takes evolutionary, physical, and geometric clues to produce accurate protein models. This means that vaccines could be developed much more easily in the future. A good example of this is that precise models of viral surface proteins aid in detecting epitopes essential for building vaccines that benefit the majority of people. While the SARS-CoV-2 pandemic was ongoing, relying on AlphaFold, scientists could model spike proteins quickly and help the creation of mRNA and protein subunit vaccines [51, 58].

The use of AI for structural bioinformatics extends beyond epitope mapping, enabling predictions of the impact of mutations on protein shape and stability when designing antigens. This matters greatly in the context of infectious diseases that are rapidly evolving and creating immune-escape variants. Researchers can adjust vaccine formulations by predicting important changes in the virus before trouble arises. AI tools aid in designing immunogens that mimic the shapes of viral proteins, thereby enhancing the immune response. Findings from these models also influence the design of nanoparticle vaccines, ensuring the antigen is presented to the immune system in the proper form (Figure 5.2) [49].

Applying structural bioinformatics and AI methods, such as molecular dynamics and immune system models, allows us to understand vaccine antigenicity and immunogenicity at the systems level. Combining multiple scientific fields enables the team to test vaccine candidates more efficiently on computers, utilizing fewer laboratory resources [81].

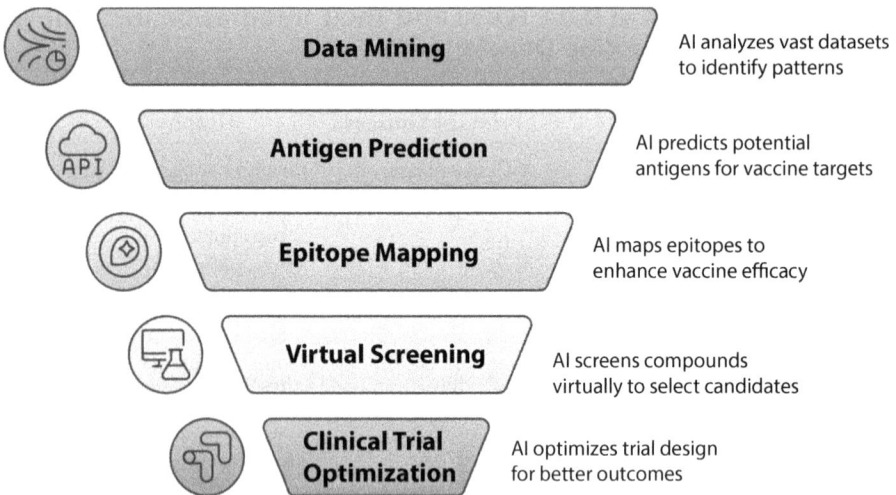

Figure 5.2 AI integration across the vaccine development pipeline.

5.3.3 Data Integration and Knowledge Graphs

Scientists must combine genomics, proteomics, and other immune, clinical, and epidemiological data to create vaccines for new infectious diseases (Table 5.1). The differences in structure and source of these datasets make it challenging for researchers to identify important and valuable results. Knowledge graphs based on AI unify all these data types to be easily used [81].

A knowledge graph combines information from buildings (nodes) and the surrounding area (edges). When creating vaccines, researchers may rely on viral proteins, the body's immune protection, epitopes, adjuvants, and important scientific information to determine how the vaccine works in people. Edges represent relationships in biology or medicine, such as "epitope as part of a protein," "vaccine plus adjuvant," or "efficacy demonstrated in a clinical trial report" [82].

Unlike other databases, knowledge graphs enable flexible question-asking, reveal more contextual reasons, and facilitate predictions about new connections. As a result, they are invaluable during pandemics, where new developments and studies must be constantly followed. Graph neural networks and other AI algorithms are used to study the structure of these graphs and find patterns that help predict the response to antigen targets or estimate the effectiveness of different vaccines [81].

Knowledge graphs played a crucial role during the COVID-19 pandemic by linking research data from articles with molecular and clinical data to accelerate efforts in the field. Researchers used these tools to discover how SARS-CoV-2 proteins connect with hosts, find immune targets inside these proteins, and identify vaccination candidates [7, 11].

Table 5.1 summarizes key data types used in vaccine research, their primary sources, and how they are incorporated into knowledge graph frameworks:

Because this data is brought together in the same computational system, experts can analyze all aspects of vaccine impact and how they work. Furthermore, these graphs can be updated repeatedly as new data appears, so researchers always use the latest and most complete understanding of the immune system [75].

All in all, AI is transforming vaccine research by using knowledge graphs. It transforms disconnected biomedical data into well-connected knowledge systems that help discover, collect, and analyze evidence quickly and predict outcomes. As biomedical data expands in scale and complexity, knowledge graphs will become more important in tomorrow's vaccine science [76].

5.4 AI APPLICATIONS IN VACCINE DISCOVERY AND DESIGN

5.4.1 Antigen and Epitope Prediction

Antigen and epitope prediction at the start of vaccine design is crucial for combating EIDs, as time is a critical factor in this process. Identifying protective antigens using traditional methods is challenging, time-consuming, and often yields incomplete results. AI has changed this field by using ML, DL, and computational immunology to quickly and reliably predict which vaccine targets should be chosen [83].

The field of immunology has recently seen reverse vaccinology stand out as a leading AI-supported method. This method reverses the conventional approach by first sequencing the pathogen's genome and using computers to forecast proteins that will trigger effective immune responses. AI improves reverse vaccinology by merging genomics, proteomics, and immunology information to choose antigens likely to act safely and competently on most populations [84].

The latest reverse vaccinology approaches utilize support vector machines (SVM), random forests (RF), and deep neural networks powered by AI to assess B-cell and T-cell epitopes, MHC affinities, and antigen processing methods. Specifically, VaxiJen uses automated cross-covariance transformation of proteins and model-based methods to predict antigenicity correctly more than 85% of the time. Similarly, NetMHCpan uses AI to forecast peptide–MHC class I lattice with excellent results for several HLA alleles [85].

You can see how AI predicts antigens in the vaccine development for COVID-19. Experts relied on DeepVacPred, a tool trained on information about SARS-CoV-2 proteins, to identify crucial parts of the spike protein that are recognized by the immune system. The model successfully identified immunogens in 90% of cases, which helped experts locate them more quickly after the genome became available. With ensemble learning, EpitopePredict evaluates the immune response of Ebola, Zika, and Lassa virus proteins for their binding to MHC and immunogenicity [48].

AI-based antigen prediction models also incorporate **population-level HLA allele frequency data**, which enables personalized vaccine design by ensuring epitope coverage across diverse ethnic and geographic populations (Figure 5.2) [84]. This approach enhances vaccine efficacy and equity, particularly in LMICs disproportionately affected by EIDs [84, 85].

Furthermore, AI facilitates **structural epitope prediction** by combining sequence-based predictions with 3D structural data of antigens. Algorithms such as **BepiPred** and **AlphaFold-based predictors** offer a high-resolution view of conformational B-cell epitopes, which are often missed by linear sequence analysis alone [86].

5.4.2 Immunogenicity Prediction

A central challenge in vaccine development is accurately predicting the immunogenic potential of candidate antigens – i.e., their capacity to elicit a protective immune response. Traditional laboratory-based assays, although informative, are often time-consuming and resource-intensive and may not scale efficiently when screening large antigen libraries. AI, particularly through ML and DL approaches, has emerged as a powerful tool for predicting immunogenicity in silico, thereby enhancing both the speed and reliability of vaccine candidate evaluation [87].

AI models trained on immunological datasets can identify features associated with high immunogenicity, such as peptide binding affinity to MHC molecules, T-cell receptor recognition motifs, and epitope structural conformations. For instance, deep convolutional neural networks (CNNs) and recurrent neural networks (RNNs) have been successfully applied to capture nonlinear, high-dimensional patterns that determine immunogenic outcomes, which are difficult to detect using traditional statistical models [88].

One prominent example is DeepImmuno, an AI framework trained on experimentally validated epitope datasets, capable of predicting T-cell immunogenicity based on peptide–MHC interaction features. DeepImmuno achieved prediction accuracies exceeding 90% in benchmark studies, significantly outperforming traditional position-specific scoring matrix (PSSM) models. Another tool, iPred, uses ML classifiers to distinguish between immunogenic and non-immunogenic peptides across multiple pathogens, providing real-time predictions during antigen screening [89].

Additionally, AI enables the prediction of how a population will respond to immunization by incorporating human leukocyte antigen diversity into the prediction process. Thanks to NetMHCIIpan, we can predict the binding of MHC class II to several HLA types, a requirement for identifying in many people helper T-cell epitopes that help develop effective pandemic vaccines [90].

Recently, knowledge from NLP and advanced transfer learning has made immunogenicity models more interpretable and adaptable to different situations. Thanks to these methods, models developed to predict influenza may also be able to predict infection for new viruses such as SARS-CoV-2 or the Nipah virus [88, 89].

In the context of new infectious diseases, utilizing AI for immunogenicity prediction is highly promising. For example, during the COVID-19 pandemic, immunoinformatics tools incorporating AI methodologies were rapidly deployed to evaluate the spike protein's immunogenic domains, facilitating the prioritization of epitopes for inclusion in mRNA and vector-based vaccines [90].

5.4.3 Allergenicity and Toxicity Screening

Safety is paramount in the design and development process of vaccine candidates. Two major concerns in this context are allergenicity – the potential to trigger allergic reactions – and toxicity, which refers to harmful biological effects. AI has become increasingly instrumental in performing high-throughput in silico screening to predict these adverse effects, thereby reducing the reliance on animal testing and expediting the preclinical safety assessment [91].

AI models utilize a range of sequence-based and structure-based features to assess the allergenic or toxic properties of peptide or protein-based vaccine components. One tool is the Allergenicity Prediction Tool (AllerTOP) v.2.0, which utilizes ML algorithms, such as k-nearest neighbor (k-NN), trained on datasets of known allergens and non-allergens. The tool has demonstrated an overall prediction accuracy of 88%–89%, making it valuable for early-stage screening [92].

Similarly, AlgPred 2.0 integrates motif-based analysis, IgE epitope prediction, and SVM learning. It was trained on experimentally validated allergens and non-allergens, achieving an average AUC (Area Under the Curve) of ~0.92 in independent benchmark tests. The Toxicity Prediction Tool (ToxinPred) utilizes SVM models for toxicity prediction, classifying peptides into toxic and non-toxic categories based on their physicochemical properties. These tools can screen thousands of candidates simultaneously, facilitating the elimination of unsafe constructs before experimental validation [93].

With the emergence of novel delivery platforms, such as lipid nanoparticles and viral vectors, AI tools are also being expanded to predict formulation-specific safety risks. Integration with toxicity databases and cross-species toxicogenomic data further enhances the predictive power of these models (as shown in Table 5.2) [91, 92].

5.4.4 Optimization of Vaccine Candidates

The optimization of vaccine candidates – especially mRNA-based vaccines – requires careful tuning of several sequence and structural parameters to ensure high translational efficiency, stability, and immunogenicity. AI and DL models have shown remarkable capability in addressing these complex, multi-objective optimization problems [99].

TABLE 5.2 Selected AI Tools in Vaccine Design: Models, Applications, and Efficacy

Tool/ Platform	Model Type	Application	Performance Metrics	Notable Output	References
Vaxign-ML	RF, SVM, Gradient Boosting	Reverse vaccinology for antigen prediction	Accuracy: 91.3%, AUC: 0.942	Identified membrane proteins for *Coxiella burnetii*, *M. tuberculosis*	[94]
DeepVacPred	CNN	Epitope prediction for subunit vaccines	Accuracy: ~94.2% for MHC class I binding	Designed multi-epitope vaccines against SARS-CoV-2	
Moderna (mRNA-1273)	RNN, RL	mRNA vaccine sequence optimization	Clinical efficacy: ~94.1% (Phase III trial)	Enabled vaccine design within 48 hours of SARS-CoV-2 genome publication	[95]
AllerTOP v.2.0	k-Nearest Neighbor (k-NN)	Allergenicity screening	Accuracy: 85%–88%	Classified allergenic proteins with high reliability	[96]
ToxinPred	SVM	Toxicity prediction for peptides	Sensitivity: 93.4%	Filtered toxic peptides during preclinical screening	[97]
Zika AI Vaccine	ANN, Docking Tools, Immune Simulation	Multi-epitope vaccine design	In silico efficacy: Elevated IgG, IFN-γ, T-helper cell levels	Complete AI-driven design: epitope prediction, docking, simulation	[98]

One of the most advanced platforms in this domain is LinearDesign, developed by Baidu Research, which utilizes a deep RL algorithm to design highly stable mRNA sequences with minimal hindrance from secondary structure. In published benchmarks, mRNA sequences designed by LinearDesign showed up to two times improvement in expression efficiency compared to standard codon-optimized sequences in *in vitro* translation systems [100].

Another platform, mRNAid, utilizes transformer-based architectures, similar to those used in NLP, to optimize mRNA sequences by balancing codon usage, GC content, and the inclusion of immunogenic motifs. The tool demonstrated 30%–40% improvement in protein expression across various constructs, including SARS-CoV-2 spike protein variants [91].

AI also plays a key role in secondary structure prediction and RNA–protein interaction modeling, which is essential for stabilizing the mRNA molecule and enhancing translation. Tools like EternaFold and SPOT-RNA apply deep neural networks trained on SHAPE-seq and DMS-seq data to predict stable and translationally favorable RNA folds with high fidelity. EternaFold, for example, achieved a Pearson correlation coefficient >0.9 when compared to experimental structure data [93, 99].

Beyond mRNA vaccines, AI is also being used to optimize DNA and recombinant subunit vaccines, particularly by predicting immunodominant regions, linker sequences, and adjuvant placements that enhance delivery and immune activation.

5.5 CASE STUDIES AND REAL-WORLD APPLICATIONS

AI has progressed from being a mere possibility in vaccine research to being put into actual use. We note a few examples of how AI has contributed to the development of vaccines for various diseases, particularly EIDs. They demonstrate that AI is key in speeding up discovery, designing more efficiently, and improving public health preparedness.

5.5.1 SARS-CoV-2 (COVID-19): Moderna's AI-Driven mRNA Vaccine

The COVID-19 pandemic served as a pivotal proving ground for the development of AI-enabled vaccines. Advanced AI systems facilitated Moderna's rapid development of the mRNA-1273 vaccine [95].

- **AI Integration**: Moderna employed AI and ML algorithms to optimize mRNA sequence structures, codon usage, and immunogenicity. NLP tools were also utilized to rapidly synthesize insights from the growing corpus of COVID-19 literature.

- **Design Timeline**: The AI-optimized mRNA vaccine candidate was finalized within 48 hours of genome publication.

- **Clinical Outcome**: The vaccine exhibited a Phase III efficacy of **94.1%** against symptomatic COVID-19.

- **Significance**: This case exemplifies how AI can compress vaccine design timelines from years to weeks while maintaining high efficacy.

5.5.2 Ebola Virus Vaccine: Structural Modeling and AI-Augmented Design

Following the Ebola outbreaks in West Africa (2014–2016), researchers leveraged AI to accelerate the development of safe and effective vaccines [101].

- **AI Use**: Structure-based ML models were employed to predict conserved B-cell and T-cell epitopes, thereby enhancing the development of multi-epitope subunit vaccines.

- **Tools**: Models trained on known viral protein structures using ensemble learning methods and validated using AI-guided immune simulations (e.g., C-ImmSim).

- **Impact**: These AI-generated vaccine constructs demonstrated high antigenicity and population coverage in silico and were further validated in preclinical trials.

5.5.3 Zika Virus: AI-Powered Multi-epitope Vaccine Design

The emergence of the Zika virus prompted researchers to explore AI for epitope prediction and immune modeling [102],

- **Modeling Strategy**: Artificial neural networks (ANNs), trained on data from the Zika virus envelope protein, were used to predict cytotoxic and helper T-cell epitopes.

- **Validation**: Docking and immune response simulations confirmed robust interaction with MHC molecules and strongly predicted immunogenicity.

- **Outcome**: The AI-designed vaccine construct was advanced to *in vivo* preclinical testing stages.

5.5.4 Influenza: Predictive Modeling for Universal Vaccine Development

Given the rapid mutation rates of influenza viruses, AI has been pivotal in identifying conserved antigenic regions across multiple strains [103].

- **Platform**: DeepFlu (Deep Learning Influenza Epitope Prediction Tool), a DL-based framework, was developed to predict conserved T-cell epitopes across influenza subtypes.

- **AI Architecture**: CNNs trained on global influenza proteome data.

- **Performance**: Achieved over **93% accuracy** in identifying cross-reactive T-cell epitopes with population coverage predictions using HLA-binding profiles.

- **Application**: Informed the rational design of prototype universal influenza vaccines.

5.5.5 Tuberculosis: Vaxign and Vaxign-ML (Machine Learning-Enhanced Vaccine Design Platform) in Antigen Prediction

Tuberculosis (TB) remains a global health challenge. The Vaxign and Vaxign-ML platforms were used to identify new TB vaccine targets [104].

- **AI Models**: Hybrid ensemble models including RFs and SVMs.

- **Outcomes**: Predicted 18 novel vaccine candidates with high antigenicity and population coverage, with five progressing to preclinical validation stages.

5.6 CHALLENGES, LIMITATIONS, AND ETHICAL CONSIDERATIONS

Although AI can significantly enhance the development of vaccines for EIDs, its use presents various challenges and ethical concerns that must be addressed with care.

A primary obstacle in designing vaccines using AI is the quality and accessibility of biomedical data. "The more thorough and correct data AI is given, the more powerful it is," according to Dr. Daphne Koller, an expert in computational biology. Currently, many data collections lack equal representation, may be inconsistent, and are biased against minority populations and different kinds of pathogens. As a result, AI models cannot be trusted in diverse groups and may, by chance, pass on differences in vaccine performance [105].

Understanding how AI models arrive at their results is also a significant concern. According to Professor Yoshua Bengio, XAI plays a crucial role in healthcare. Uncertainty about decision-making in deep neural networks delays the use and approval of new treatments. Many advocate for simple AI systems because those in charge of approvals and clinical teams want to understand each step [106].

Dr. Margaret Hamburg, who led the U.S. FDA, has noted that the existing rules are not yet prepared to evaluate vaccines that utilize AI in their development. She supports the use of adaptive laws that validate algorithms, ensure proof of data origin, and employ models that become smarter over time. Applying these models would ensure that innovation speed is safe for patients and accepted by the public [107].

Editing the genome is not just about technology or rules; it also involves important ethical considerations. When the training data lacks diversity globally, there is a real risk of algorithmic bias. Her key point is that unless AI utilizes various datasets and incorporates everyone into its design, it could exacerbate health disparities rather than alleviate them. With proactive approaches for data gathering and model teaching, it becomes possible to ensure fair access and outcomes for everyone [108].

Because patient data is increasingly being used for AI, strict rules for privacy and security must be in place. Data anonymization, safe ways to exchange data, and respecting regulations like GDPR and HIPAA are necessary to respect people's rights and allow science to develop [109].

Some operational issues are caused by the requirement for integrated teamwork and issues related to accessing advanced infrastructure worldwide. According to Professor Eric Topol, integrating AI into vaccine development requires the collaboration of experts in coding, the science of the immune system, medical treatments, and policy, as well as improved access to computer resources in areas where they are currently less available.

In short, researchers all agree that AI will transform vaccine development. However, to fully reap its benefits, we must address data constraints, improve model transparency, update rules from health-care agencies, preserve high ethical standards, and encourage worldwide cooperation. Addressing these issues as we utilize AI in global public health is crucial for benefiting the world's population.

5.7 FUTURE PROSPECTS

Using AI in vaccine development is transforming how we address global health challenges related to EIDs. AI-enhanced vaccine research is poised to rapidly develop along exciting new pathways, building on the accomplishments and addressing the challenges that must be overcome.

Future versions of AI will rely on more advanced algorithms that enable them to understand biology more effectively and make more accurate predictions. XAI and causal inference models will help explain complex immunology and ensure clear vaccine candidate selections. Also, using genetic, protein, RNA, and metabolite analysis in one experiment helps better model interactions between a pathogen and the host, permitting more reliable epitope prediction and assessment of immunogenicity [110].

Future AI systems will likely be integrated with real-time epidemiological surveillance networks, leveraging significant data streams from clinical, environmental, and social media sources to detect emerging pathogen variants rapidly. This capability will facilitate the dynamic design of vaccines and the rapid reconfiguration of vaccine formulations in response to viral mutations, thereby mitigating the lag between pathogen evolution and vaccine efficacy. AI-driven personalized vaccinology will advance by incorporating individual genetic, immunological, and environmental data to tailor vaccine formulations and dosing regimens. Precision public health initiatives, empowered by AI, will optimize vaccine distribution and deployment strategies based on predictive models of population immunity, susceptibility, and logistical constraints, thereby enhancing vaccine coverage and impact [111].

Integrating AI and synthetic biology will enable the faster design of new vaccines, including mRNA, viral vector vaccines, and those utilizing nanoparticles. Relying on AI for antigen design and delivery system improvement will increase vaccine stability, immune effect, and safety. Moreover, AI could help create universal vaccines aimed at conserved details across many pathogen types, tackling issues related to their variation [112].

Advancing AI for vaccines will require nations to share data, agree on how AI is applied, and adjust their regulatory rules. Promoting open-access platforms and collaborating with countries worldwide will be vital to ensure that everyone can utilize AI and that vaccines are shared equitably. Changes in ethical rules and regulations should follow the advancement of AI technologies. Some steps to take include reinforcing ethics and promoting transparency, fairness, and inclusivity in the creation of AI models. To ensure vaccines stay safe, regulators need adaptive rules that support AI so vaccine approval can happen faster [110, 111].

5.8 CONCLUSION

A continuing threat from EID to global health necessitates inventive, rapid, and effective vaccine development. While designers rely on scientific methods to make traditional vaccines, the process tends to be slow, costly, and complex to change for unknown pathogens. Introducing AI to develop vaccines marks a new direction, as it increases the efficiency, precision, and affordability of the research process.

We have demonstrated how AI contributes to resolving the most significant challenges in vaccine development, ranging from antigen and epitope prediction to optimizing clinical trials. These AI tools, such as ML, DL, and neural networks, have demonstrated remarkable capabilities in supporting faster antigen search, more accurate prediction of immunogenicity, and smoother processing of large datasets. Examples from AI development of vaccines against SARS-CoV-2 and similar diseases confirm their usefulness in the real world.

Nevertheless, AI faces challenges when it comes to vaccines. Matters related to data quality, model understanding, readiness for regulations, and ethical fairness need to be organized and addressed. In addition, the differences in AI infrastructure and knowledge worldwide mean that we must try harder to distribute new technology evenly.

The future of utilizing AI in vaccine research appears promising. AI is expected to revolutionize preventive healthcare when combined with high-throughput omics technologies, real-time surveillance systems, and personalized medicine. The primary ways to benefit from AI in combating infectious diseases are investing in new research, fostering global collaboration, and developing new regulations.

In short, AI provides essential support for the future development of vaccines. If ethically and adequately integrated with global health practices, this technology could enhance our ability to manage new infectious diseases, thereby securing and improving public health and biomedical research.

ABBREVIATION

AI	Artificial Intelligence
AllerTOP	Allergenicity Prediction Tool
CNN	Convolutional Neural Network
DeepVacPred	Deep Learning Vaccine Prediction Tool
EIDs	Emerging Infectious Diseases
LMICs	Low- and Middle-Income Countries
NLP	Natural Language Processing
RF	Random Forest
RL	Reinforcement Learning
RNN	Recurrent Neural Network
SVM	Support Vector Machine
ToxinPred	Toxicity Prediction Tool
Vaxign-ML	Machine Learning-Enhanced Vaccine Design Platform
WHO	World Health Organization

REFERENCES

[1] Behera JK, Mishra P, Jena AK, Behera B, Bhattacharya M. Human health implications of emerging diseases and the current situation in India's vaccine industry. *Science in One Health* 2023;2:100046. https://doi.org/10.1016/J.SOH.2023.100046

[2] Antimicrobial resistance. 2023. https://www.who.int/news-room/fact-sheets/detail/antimicrobial-resistance (accessed May 27, 2025).

[3] Wegner GI, Murray KA, Springmann M, Muller A, Sokolow SH, Saylors K, et al. Averting wildlife-borne infectious disease epidemics requires a focus on socio-ecological drivers and a redesign of the global food system. *EClinicalMedicine* 2022;47:101386. https://doi.org/10.1016/J.ECLINM.2022.101386

[4] THE 17 GOALS | Sustainable Development n.d. https://sdgs.un.org/goals (accessed May 27, 2025).

[5] Tosh PK, Ovsyannikova IG, Barry MA, Poland GA, Sambhara S, Gray GC. Vaccine development: The development of avian influenza vaccines for human use. *Encyclopedia of Microbiology*, 3rd ed. 2009:775–95. https://doi.org/10.1016/B978-012373944-5.00212-1

[6] Pitchumoni CS, Brun A. HIV and AIDS. *Geriatric Gastroenterology* 2024:659–66. https://doi.org/10.1007/978-1-4419-1623-5_71

[7] V'kovski P, Gultom M, Kelly JN, Steiner S, Russeil J, Mangeat B, et al. Disparate temperature-dependent virus-host dynamics for SARS-CoV-2 and SARS-CoV in the human respiratory epithelium. *PLoS Biol* 2021;19:e3001158. https://doi.org/10.1371/JOURNAL.PBIO.3001158

[8] Hancock K, Veguilla V, Lu X, Zhong W, Butler EN, Sun H, et al. Cross-reactive antibody responses to the 2009 pandemic H1N1 influenza virus. *New England Journal of Medicine* 2009;361:1945–52. https://doi.org/10.1056/NEJMOA0906453

[9] Kamorudeen RT, Adedokun KA, Olarinmoye AO. Ebola outbreak in West Africa, 2014–2016: Epidemic timeline, differential diagnoses, determining factors, and lessons for future response. *J Infect Public Health* 2020;13:956–62. https://doi.org/10.1016/J.JIPH.2020.03.014

[10] Zika virus 2022. https://www.who.int/news-room/fact-sheets/detail/zika-virus (accessed May 27, 2025).

[11] COVID-19 epidemiological update – 15 July 2024 n.d. https://www.who.int/publications/m/item/covid-19-epidemiological-update-edition-169 (accessed May 27, 2025).

[12] Cutler DM, Summers LH. The COVID-19 Pandemic and the $16 Trillion Virus. *JAMA* 2020;324:1495–6. https://doi.org/10.1001/JAMA.2020.19759

[13] Leifels M, Khalilur Rahman O, Sam I-C, Cheng D, Chua FJD, Nainani D, et al. The one health perspective to improve environmental surveillance of zoonotic viruses: Lessons from COVID-19 and outlook beyond. *ISME Communications* 2022;2:1–9. https://doi.org/10.1038/s43705-022-00191-8

[14] Ghattas M, Dwivedi G, Lavertu M, Alameh MG. Vaccine technologies and platforms for infectious diseases: Current progress, challenges, and opportunities. *Vaccines* 2021;9:1490. https://doi.org/10.3390/VACCINES9121490

[15] Pollard AJ, Bijker EM. A guide to vaccinology: From basic principles to new developments. *Nature Reviews Immunology* 2020;21:83–100. https://doi.org/10.1038/s41577-020-00479-7

[16] Shamsusah NA, Mohd Fadli MEI, Emrizal R, Hanifah SA, Firdaus-Raih M. Challenges in the detection of emerging novel pathogens and evolving known pathogens. In NHM Al-Hardan, A Jalar, MAA Hamid; MF Raih (Eds.) *Field-Effect Transistor Biosensors for Rapid Pathogen Detection* 2024:73–92. https://doi.org/10.1039/BK9781837673421-00073

[17] Gomez PL, Robinson JM. Vaccine manufacturing. In SA Plotkin, WA Orenstein, PA Offit, KM Edwards (Eds.). *Plotkin's Vaccines* 7th ed. 2017:51-60.e1. https://doi.org/10.1016/B978-0-323-35761-6.00005-5

[18] Sandbrink JB, Koblentz GD. Biosecurity risks associated with vaccine platform technologies. *Vaccine* 2022;40:2514–23. https://doi.org/10.1016/J.VACCINE.2021.02.023

[19] Santangelo OE, Provenzano S, Di Martino G, Ferrara P. COVID-19 vaccination and public health: Addressing global, regional, and within-country inequalities. *Vaccines* 2024;12:885. https://doi.org/10.3390/VACCINES12080885

[20] Kaushik R, Kant R, Christodoulides M. Artificial intelligence in accelerating vaccine development – Current and future perspectives. *Frontiers in Bacteriology* 2023;2:1258159. https://doi.org/10.3389/FBRIO.2023.1258159

[21] Sarker IH. Machine learning: Algorithms, real-world applications and research directions. *SN Comput Sci* 2021;2:1–21.

[22] Athanasopoulou K, Daneva GN, Adamopoulos PG, Scorilas A. Artificial intelligence: The milestone in modern biomedical research. *BioMedInformatics* 2022;2:727–44. https://doi.org/10.3390/BIOMEDINFORMATICS2040049

[23] Serrano DR, Luciano FC, Anaya BJ, Ongoren B, Kara A, Molina G, et al. Artificial intelligence (AI) applications in drug discovery and drug delivery: Revolutionizing personalized medicine. *Pharmaceutics* 2024;16:1328. https://doi.org/10.3390/PHARMACEUTICS16101328

[24] Vilhekar RS, Rawekar A, Vilhekar RS, Rawekar A. Artificial intelligence in genetics. *Cureus* 2024;16. https://doi.org/10.7759/CUREUS.52035

[25] Gawande MS, Zade N, Kumar P, Gundewar S, Weerarathna IN, Verma P. The role of artificial intelligence in pandemic responses: From epidemiological modeling to vaccine development. *Molecular Biomedicine* 2025;6:1–25. https://doi.org/10.1186/S43556-024-00238-3

[26] Bin Rashid A, Kausik MDAK. AI revolutionizing industries worldwide: A comprehensive overview of its diverse applications. *Hybrid Advances* 2024;7:100277. https://doi.org/10.1016/J.HYBADV.2024.100277

[27] Holzinger A, Keiblinger K, Holub P, Zatloukal K, Müller H. AI for life: Trends in artificial intelligence for biotechnology. *N Biotechnol* 2023;74:16–24. https://doi.org/10.1016/J.NBT.2023.02.001

[28] Kapustina O, Burmakina P, Gubina N, Serov N, Vinogradov V. User-friendly and industry-integrated AI for medicinal chemists and pharmaceuticals. *Artificial Intelligence Chemistry* 2024;2:100072. https://doi.org/10.1016/J.AICHEM.2024.100072

[29] Brisse M, Vrba SM, Kirk N, Liang Y, Ly H. Emerging concepts and technologies in vaccine development. *Front Immunol* 2020;11:583077.

[30] Vaillant AAJ, Sabir S, Jan A. *Physiology, Immune Response.* StatPearls 2024.

[31] Arora G, Joshi J, Mandal RS, Shrivastava N, Virmani R, Sethi T. Artificial intelligence in surveillance, diagnosis, drug discovery and vaccine development against covid-19. *Pathogens* 2021;10:1048.

[32] Masignani V, Pizza M, Moxon ER. The development of a vaccine against Meningococcus B using reverse vaccinology. *Front Immunol* 2019;10:441439.

[33] Shanmugaraj B, Khorattanakulchai N, Panapitakkul C, Malla A, Im-Erbsin R, Inthawong M, et al. Preclinical evaluation of a plant-derived SARS-CoV-2 subunit vaccine: Protective efficacy, immunogenicity, safety, and toxicity. *Vaccine* 2022;40:4440–52. https://doi.org/10.1016/J.VACCINE.2022.05.087

[34] Ahmed S, Khan S, Imran I, Al Mughairbi F, Sheikh FS, Hussain J, et al. Vaccine development against COVID-19: Study from pre-clinical phases to clinical trials and global use. *Vaccines* 2021;9:836. https://doi.org/10.3390/VACCINES9080836

[35] Singh K, Mehta S. The clinical development process for a novel preventive vaccine: An overview. *J Postgrad Med* 2016;62:4–11. https://doi.org/10.4103/0022-3859.173187

[36] Kandi V, Vadakedath S, Kandi V, Vadakedath S. Clinical trials and clinical research: A comprehensive review. *Cureus* 2023;15. https://doi.org/10.7759/CUREUS.35077

[37] Franco P, Jain R, Rosenkrands-Lange E, Hey C, Koban MU. Regulatory pathways supporting expedited drug development and approval in ICH member countries. *Ther Innov Regul Sci* 2023;57:484–514.

[38] Di Pasquale A, Bonanni P, Garçon N, Stanberry LR, El-Hodhod M, Tavares Da Silva F. Vaccine safety evaluation: Practical aspects in assessing benefits and risks. *Vaccine* 2016;34:6672–80. https://doi.org/10.1016/J.VACCINE.2016.10.039

[39] Jungbluth S, Gunn A, Bandara S, Yamey G, D'Alessio F, Depraetere H, et al. Pipeline analysis of a vaccine candidate portfolio for diseases of poverty using the portfolio-to-impact modelling tool. *F1000Research* 2020;8:1066. https://doi.org/10.12688/f1000research.19810.2

[40] Kennedy RB, Ovsyannikova IG, Palese P, Poland GA. Current challenges in vaccinology. *Front Immunol* 2020;11:541543.

[41] Vaghasiya J, Khan M, Milan Bakhda T. A meta-analysis of AI and machine learning in project management: Optimizing vaccine development for emerging viral threats in biotechnology. *Int J Med Inform* 2025;195:105768. https://doi.org/10.1016/J.IJMEDINF.2024.105768

[42] How vaccines are developed and approved for use | Vaccines & immunizations | CDC 2024 https://www.cdc.gov/vaccines/basics/how-developed-approved.html (accessed May 28, 2025).

[43] Kumraj G, Pathak S, Shah S, Majumder P, Jain J, Bhati D, et al. Capacity building for vaccine manufacturing across developing countries: The way forward. *Hum Vaccin Immunother* 2022;18. https://doi.org/10.1080/21645515.2021.2020529

[44] Madewell ZJ, Dean NE, Berlin JA, Coplan PM, Davis KJ, Struchiner CJ, et al. Challenges of evaluating and modelling vaccination in emerging infectious diseases. *Epidemics* 2021;37:100506. https://doi.org/10.1016/J.EPIDEM.2021.100506

[45] Jones CH, Hauguel T, Beitelshees M, Davitt M, Welch V, Lindert K, et al. Deciphering immune responses: A comparative analysis of influenza vaccination platforms. *Drug Discov Today* 2024;29:104125. https://doi.org/10.1016/J.DRUDIS.2024.104125

[46] Choy RKM, Bourgeois AL, Ockenhouse CF, Walker RI, Sheets RL, Flores J. Controlled human infection models to accelerate vaccine development. *Clin Microbiol Rev* 2022;35:1–163. https://doi.org/10.1128/CMR.00008-21

[47] Karp CL, Lans D, Esparza J, Edson EB, Owen KE, Wilson CB, et al. Evaluating the value proposition for improving vaccine thermostability to increase vaccine impact in low and middle-income countries. *Vaccine* 2015;33:3471–9. https://doi.org/10.1016/J.VACCINE.2015.05.071

[48] El Arab RA, Alkhunaizi M, Alhashem YN, Al Khatib A, Bubsheet M, Hassanein S, et al. Artificial intelligence in vaccine research and development: An umbrella review. *Front Immunol* 2025;16:1567116. https://doi.org/10.3389/FIMMU.2025.1567116

[49] Zeng X, Bai G, Sun C, Ma B. Recent progress in antibody epitope prediction. *Antibodies* 2023;12:52. https://doi.org/10.3390/ANTIB12030052

[50] Hederman AP, Ackerman ME. Leveraging deep learning to improve vaccine design. *Trends Immunol* 2023;44:333–44. https://doi.org/10.1016/J.IT.2023.03.002

[51] Zhang H, Zhang L, Lin A, Xu C, Li Z, Liu K, et al. Algorithm for optimized mRNA design improves stability and immunogenicity. *Nature* 2023;621:396–403. https://doi.org/10.1038/s41586-023-06127-z

[52] Pappalardo F, Russo G, Corsini E, Paini A, Worth A. Translatability and transferability of in silico models: Context of use switching to predict the effects of environmental chemicals on the immune system. *Comput Struct Biotechnol J* 2022;20:1764–77. https://doi.org/10.1016/J.CSBJ.2022.03.024

[53] Wang X, Fan D, Yang Y, Gimple RC, Zhou S. Integrative multi-omics approaches to explore immune cell functions: Challenges and opportunities. *IScience* 2023;26. https://doi.org/10.1016/J.ISCI.2023.106359

[54] Sedano R, Solitano V, Vuyyuru SK, Yuan Y, Hanžel J, Ma C, et al. Artificial intelligence to revolutionize IBD clinical trials: A comprehensive review. *Therap Adv Gastroenterol* 2025;18. https://doi.org/10.1177/17562848251321915

[55] Sharma A, Virmani T, Pathak V, Sharma A, Pathak K, Kumar G, et al. Artificial intelligence-based data-driven strategy to accelerate research, development, and clinical trials of COVID vaccine. *Biomed Res Int* 2022;2022:7205241. https://doi.org/10.1155/2022/7205241

[56] Bhat VN, Bharati S, Bothiraja C, Sangshetti J, Gaikwad V. A review on intervention of AI in pharmaceutical sector: Revolutionizing drug discovery and manufacturing. *Intelligent Pharmacy* 2025. https://doi.org/10.1016/J.IPHA.2025.04.001

[57] Desai MK. Artificial intelligence in pharmacovigilance – Opportunities and challenges. *Perspect Clin Res* 2024;15:116–21. https://doi.org/10.4103/PICR.PICR_290_23

[58] Ghosh A, Larrondo-Petrie MM, Pavlovic M. Revolutionizing vaccine development for COVID-19: A review of AI-based approaches. *Information* 2023;14:665. https://doi.org/10.3390/INFO14120665

[59] Asediya VS, Anjaria PA, Mathakiya RA, Koringa PG, Nayak JB, Bisht D, et al. Vaccine development using artificial intelligence and machine learning: A review. *Int J Biol Macromol* 2024;282:136643. https://doi.org/10.1016/J.IJBIOMAC.2024.136643

[60] Bravi B. Development and use of machine learning algorithms in vaccine target selection. *Npj Vaccines* 2024;9:1–14. https://doi.org/10.1038/s41541-023-00795-8

[61] Ahmed SF, Alam MS Bin, Hassan M, Rozbu MR, Ishtiak T, Rafa N, et al. Deep learning modelling techniques: Current progress, applications, advantages, and challenges. *Artificial Intelligence Review* 2023;56:13521–617. https://doi.org/10.1007/S10462-023-10466-8

[62] Meng Y, Zhang Z, Zhou C, Tang X, Hu X, Tian G, et al. Protein structure prediction via deep learning: An in-depth review. *Front Pharmacol* 2025;16:1498662.

[63] Gao W, Mahajan SP, Sulam J, Gray JJ. Deep learning in protein structural modeling and design. *Patterns* 2020;1:100142. https://doi.org/10.1016/J.PATTER.2020.100142

[64] Vita R, Blazeska N, Marrama D, Shackelford D, Zalman L, Foos G, et al. The immune epitope database (IEDB): 2024 update. *Nucleic Acids Res* 2025;53:D436–43. https://doi.org/10.1093/NAR/GKAE1092

[65] Ananya, PDC, Karthic A, Singh SP, Mani A, Chawade A, et al. Vaccine design and development: Exploring the interface with computational biology and AI. *Int Rev Immunol* 2024. https://doi.org/10.1080/08830185.2024.2374546

[66] Rojas-Carabali W, Agrawal R, Gutierrez-Sinisterra L, Baxter SL, Cifuentes-González C, Wei YC, et al. Natural language processing in medicine and ophthalmology: A review for the 21st-century clinician. *Asia-Pacific Journal of Ophthalmology* 2024;13:100084. https://doi.org/10.1016/J.APJO.2024.100084

[67] Wang J, Wang K, Li J, Jiang J, Wang Y, Mei J, et al. Accelerating epidemiological investigation analysis by using NLP and knowledge reasoning: A case study on COVID-19. *AMIA Annual Symposium Proceedings* 2021;2020:1258.

[68] Zhang Z, Chen ALP. Biomedical named entity recognition with the combined feature attention and fully-shared multi-task learning. *BMC Bioinformatics* 2022;23:1–21.

[69] Anderson LN, Hoyt CT, Zucker JD, McNaughton AD, Teuton JR, Karis K, et al. Computational tools and data integration to accelerate vaccine development: Challenges, opportunities, and future directions. *Front Immunol* 2025;16:1502484.

[70] Madan S, Lentzen M, Brandt J, Rueckert D, Hofmann-Apitius M, Fröhlich H. Transformer models in biomedicine. *BMC Medical Informatics and Decision Making* 2024;24:1–22. https://doi.org/10.1186/S12911-024-02600-5

[71] Yu W, Zheng C, Xie F, Chen W, Mercado C, Sy LS, et al. The use of natural language processing to identify vaccine-related anaphylaxis at five health care systems in the vaccine safety datalink. *Pharmacoepidemiol Drug Saf* 2020;29:182–8. https://doi.org/10.1002/PDS.4919

[72] Shu Y, McCauley J. GISAID: Global initiative on sharing all influenza data – From vision to reality. *Eurosurveillance* 2017;22:30494.

[73] Bateman A, Martin MJ, Orchard S, Magrane M, Ahmad S, Alpi E, et al. UniProt: The universal protein knowledgebase in 2023. *Nucleic Acids Res* 2023;51:D523–31. https://doi.org/10.1093/NAR/GKAC1052

[74] Bhattacharya S, Andorf S, Gomes L, Dunn P, Schaefer H, Pontius J, et al. ImmPort: Disseminating data to the public for the future of immunology. *Immunol Res* 2014;58:234–9.

[75] Tian F, Yang R, Chen Z. Safety and efficacy of COVID-19 vaccines in children and adolescents: A systematic review of randomized controlled trials. *J Med Virol* 2022;94:4644–53. https://doi.org/10.1002/JMV.27940

[76] Nicholson DN, Greene CS. Constructing knowledge graphs and their biomedical applications. *Comput Struct Biotechnol J* 2020;18:1414–28. https://doi.org/10.1016/J.CSBJ.2020.05.017

[77] Dong E, Du H, Gardner L. An interactive web-based dashboard to track COVID-19 in real time. *Lancet Infect Dis* 2020;20:533–4. https://doi.org/10.1016/S1473-3099(20)30120-1

[78] Huang L-C, Eiden AL, He L, Annan A, Wang S, Wang J, et al. Natural language processing-powered real-time monitoring solution for vaccine sentiments and hesitancy on social media: System development and validation. *JMIR Med Inform* 2024;12:e57164. https://doi.org/10.2196/57164

[79] Desai D, Kantliwala S V., Vybhavi J, Ravi R, Patel H, Patel J, et al. Review of AlphaFold 3: Transformative advances in drug design and therapeutics. *Cureus* 2024;16. https://doi.org/10.7759/CUREUS.63646

[80] Sunita, Sajid A, Singh Y, Shukla P. Computational tools for modern vaccine development. *Hum Vaccin Immunother* 2020;16:723–35. https://doi.org/10.1080/21645515.2019.1670035

[81] Olawade DB, Teke J, Fapohunda O, Weerasinghe K, Usman SO, Ige AO, et al. Leveraging artificial intelligence in vaccine development: A narrative review. *J Microbiol Methods* 2024;224:106998. https://doi.org/10.1016/J.MIMET.2024.106998

[82] Zhao T, Cai Y, Jiang Y, He X, Wei Y, Yu Y, et al. Vaccine adjuvants: Mechanisms and platforms. *Signal Transduction and Targeted Therapy* 2023;8:1–24. https://doi.org/10.1038/s41392-023-01557-7

[83] Hashempour A, Khodadad N, Bemani P, Ghasemi Y, Akbarinia S, Bordbari R, et al. Design of multivalent-epitope vaccine models directed toward the world's population against HIV-Gag polyprotein: Reverse vaccinology and immunoinformatics. *PLoS One* 2024;19:e0306559. https://doi.org/10.1371/JOURNAL.PONE.0306559

[84] Basmenj ER, Pajhouh SR, Ebrahimi Fallah A, Naijian R, Rahimi E, Atighy H, et al. Computational epitope-based vaccine design with bioinformatics approach: A review. *Heliyon* 2025;11:e41714. https://doi.org/10.1016/J.HELIYON.2025.E41714

[85] Yousaf H, Naz A, Zaman N, Hassan M, Obaid A, Awan FM, et al. Immunoinformatic and reverse vaccinology-based designing of potent multi-epitope vaccine against Marburgvirus targeting the glycoprotein. *Heliyon* 2023;9:e18059. https://doi.org/10.1016/J.HELIYON.2023.E18059

[86] Ivanisenko NV, Shashkova TI, Shevtsov A, Sindeeva M, Umerenkov D, Kardymon O. SEMA 2.0: Web-platform for B-cell conformational epitopes prediction using artificial intelligence. *Nucleic Acids Res* 2024;52:W533–9. https://doi.org/10.1093/NAR/GKAE386

[87] Oli AN, Obialor WO, Ositadimma M, Ifeanyichukwu, Odimegwu DC, Okoyeh JN, et al. Immunoinformatics and vaccine development: An Overview. *Immunotargets Ther* 2020;9:13–30. https://doi.org/10.2147/ITT.S241064

[88] Weber A, Pélissier A, Rodríguez Martínez M. T-cell receptor binding prediction: A machine learning revolution. *ImmunoInformatics* 2024;15:100040. https://doi.org/10.1016/J.IMMUNO.2024.100040

[89] Li G, Iyer B, Prasath VBS, Ni Y, Salomonis N. DeepImmuno: Deep learning-empowered prediction and generation of immunogenic peptides for T-cell immunity. *Brief Bioinform* 2021;22:1–10. https://doi.org/10.1093/BIB/BBAB160

[90] Nielsen M, Andreatta M, Peters B, Buus S. Immunoinformatics: Predicting peptide-MHC binding. *Annu Rev Biomed Data Sci* 2020;3:191–215.

[91] da Silva MK, Campos DM de O, Akash S, Akter S, Yee LC, Fulco UL, et al. Advances of reverse vaccinology for mRNA vaccine design against SARS-CoV-2: A review of methods and tools. *Viruses* 2023;15:2130. https://doi.org/10.3390/V15102130

[92] Kottarathil A, Murugan G, Rajkumar DS, Chandran AK, Elumalai V, Padmanaban R. Designing multi-epitope-based vaccine targeting immunogenic proteins of Streptococcus mutans using immunoinformatics to prevent caries. *The Microbe* 2025;7:100320. https://doi.org/10.1016/J.MICROB.2025.100320

[93] Sharma N, Patiyal S, Dhall A, Pande A, Arora C, Raghava GPS. AlgPred 2.0: An improved method for predicting allergenic proteins and mapping of IgE epitopes. *Brief Bioinform* 2021;22:1–12. https://doi.org/10.1093/BIB/BBAA294

[94] Ong E, Wang H, Wong MU, Seetharaman M, Valdez N, He Y. Vaxign-ML: Supervised machine learning reverse vaccinology model for improved prediction of bacterial protective antigens. *Bioinformatics* 2020;36:3185–91. https://doi.org/10.1093/BIOINFORMATICS/BTAA119

[95] Baden LR, El Sahly HM, Essink B, Kotloff K, Frey S, Novak R, et al. Efficacy and safety of the mRNA-1273 SARS-CoV-2 vaccine. *New England Journal of Medicine* 2021;384:403–16. https://doi.org/10.1056/NEJMOA2035389

[96] Dimitrov I, Bangov I, Flower DR, Doytchinova I. AllerTOP v.2 – A server for in silico prediction of allergens. *J Mol Model* 2014;20:1–6.

[97] Gupta S, Kapoor P, Chaudhary K, Gautam A, Kumar R, Raghava GPS. In silico approach for predicting toxicity of peptides and proteins. *PLoS One* 2013;8:e73957. https://doi.org/10.1371/JOURNAL.PONE.0073957

[98] Mittal A, Sasidharan S, Raj S, Balaji SN, Saudagar P. Exploring the Zika genome to design a potential multiepitope vaccine using an immunoinformatics approach. *Int J Pept Res Ther* 2020;26:2231–40.

[99] Jin L, Zhou Y, Zhang S, Chen SJ. mRNA vaccine sequence and structure design and optimization: Advances and challenges. *Journal of Biological Chemistry* 2025;301:108015. https://doi.org/10.1016/J.JBC.2024.108015

[100] Liu X, Wang S, Sun Y, Liao Y, Jiang G, Sun BY, et al. Unlocking the potential of circular RNA vaccines: A bioinformatics and computational biology perspective. *EBioMedicine* 2025;114:105638. https://doi.org/10.1016/J.EBIOM.2025.105638

[101] Ma R, Lu L, Suo L, Zhangzhu J, Chen M, Pang X. Evaluation of the adequacy of measles laboratory diagnostic tests in the era of accelerating measles elimination in Beijing, *China. Vaccine* 2019;37:3804–9. https://doi.org/10.1016/J.VACCINE.2019.05.058

[102] Ezzemani W, Windisch MP, Altawalah H, Guessous F, Saile R, Benjelloun S, et al. Design of a multi-epitope Zika virus vaccine candidate – an *in-silico* study. *J Biomol Struct Dyn* 2023;41:3762–71. https://doi.org/10.1080/07391102.2022.2055648

[103] Zan A, Xie ZR, Hsu YC, Chen YH, Lin TH, Chang YS, et al. DeepFlu: A deep learning approach for forecasting symptomatic influenza A infection based on pre-exposure gene expression. *Comput Methods Programs Biomed* 2022;213:106495. https://doi.org/10.1016/J.CMPB.2021.106495

[104] Ong E, Wang H, Wong MU, Seetharaman M, Valdez N, He Y. Vaxign-ML: Supervised machine learning reverse vaccinology model for improved prediction of bacterial protective antigens. *Bioinformatics* 2020;36:3185–91. https://doi.org/10.1093/BIOINFORMATICS/BTAA119

[105] Koller D. Behind the tech with Kevin Scott. Podcast. 2025. https://www.microsoft.com/en-us/behind-the-tech/ask-me-anything-with-microsoft-cto-kevin-scott (accessed May 28, 2025).

[106] Press, G. Deep learning pioneer Yoshua Bengio says AI is not magic and Intel AI experts explain why and how. n.d. https://www.forbes.com/sites/gilpress/2019/09/20/deep-learning-pioneer-yoshua-bengio-says-ai-is-not-magic-and-intel-ai-experts-explain-why-and-how (accessed May 28, 2025).

[107] Ex-FDA leader joins new initiative aiming to protect vaccine use. n.d. https://www.fiercebiotech.com/biotech/ex-fda-leader-joins-new-initiative-designed-fight-unfortunate-reality-vaccine-landscape (accessed May 28, 2025).

[108] Li F-F. *Governing AI through acquisition and procurement.* 2023. https://doi.org/10.24963/ijcai.2017/654

[109] Ohno-Machado L. Sharing data for the public good and protecting individual privacy: Informatics solutions to combine different goals. *Journal of the American Medical Informatics Association* 2013;20:1–1. https://doi.org/10.1136/AMIAJNL-2012-001513

[110] Kumar P, Chaudhary B, Arya P, Chauhan R, Devi S, Parejiya PB, et al. Advanced artificial intelligence technologies transforming contemporary pharmaceutical research. *Bioengineering* 2025;12:363. https://doi.org/10.3390/BIOENGINEERING12040363

[111] Zhao AP, Li S, Cao Z, Hu PJH, Wang J, Xiang Y, et al. AI for science: Predicting infectious diseases. *Journal of Safety Science and Resilience* 2024;5:130–46. https://doi.org/10.1016/J.JNLSSR.2024.02.002

[112] Imani S, Li X, Chen K, Maghsoudloo M, Jabbarzadeh Kaboli P, Hashemi M, et al. Computational biology and artificial intelligence in mRNA vaccine design for cancer immunotherapy. *Front Cell Infect Microbiol* 2024;14:1501010.

6 AI Technologies and Molecular Diagnostics for Managing Infectious Diseases

Jothi Dheivasikamani Abidharini, Mani Manoj, Gunasekaran Arthi, Rajamanikkam Ramya, Arul Sampath Kumar, Asirvatham Alwin Robert, Jeyabal Philomenathan Antony Prabhu, and Arumugam Vijaya Anand

6.1 NEXT-GENERATION MOLECULAR DIAGNOSTIC TECHNOLOGIES FOR INFECTIOUS DISEASES

The diseases are spreading fast and diagnosis is highly needed in this industrialized world, and emphasizes how critical it is to diagnose different kind of diseases quickly and accurately in order to manage patient health and successfully stop their spread. Increased international travel, population growth, and environmental changes are some of the reasons that lead to the rapid spread of illnesses (Velayuthaprabhu and Archunan, 2005; Velayuthaprabhu et al., 2007; Alagendran et al., 2010; Velayuthaprabhu et al., 2011; Velayuthaprabhu et al., 2013; Velayuthaprabhu et al., 2016; Varghese et al., 2020; Lee et al., 2021; Mohd, Balasubramanian, et al., 2021; Mohd, Kumar, et al., 2021; Gundappa et al., 2022; Meyyazhagan et al., 2022; Ramya et al., 2023; Sangeetha, Anand, & Begum, 2022; Sangeetha, Nargis Begum, et al., 2022; Paranitharan et al., 2025).

Ultra-sensitive digital polymerase chain reaction (PCR), and specifically droplet digital PCR (ddPCR), is now a cost-effective tool for precise quantification of pathogens in a vast array of applications, ranging from aquaculture and clinical diagnostics to environmental monitoring. ddPCR divides a reaction into thousands of nano-droplets, which allows absolute quantification of DNA or RNA targets without the need for standard curves, thus enhancing the diagnostic sensitivity. In aquaculture, ddPCR has demonstrated sensitivity levels comparable to quantitative PCR (qPCR), with greater accuracy in the detection of pathogens in various species of fish (Sumon et al., 2024). In the clinical field, ddPCR has demonstrated excellent sensitivity and specificity in the detection of *Escherichia coli* bloodstream infections, surpassing the conventional blood cultures in time-to-positivity and agreement with clinical outcome – for example, septic shock (Kitagawa et al., 2025). Further, it is more effective than qPCR in the detection of RNA viruses such as SARS-CoV-2, Influenza A, and RSV-A, with the added advantage of high linear correlation over various concentration ranges (Pushparaj et al., 2022; Arumugam et al., 2020; Meyyazhagan et al., 2020; Shanmugam et al., 2020; Kuchi Bhotla et al., 2020; Kuchi Bhotla et al., 2021; Casto et al., 2024). The ultra-sensitivity of ddPCR has also enabled the successful detection of Plasmodium species in malaria samples with low parasitaemia (Srisutham et al., 2017) and *Carassius auratus* herpesvirus (CaHV) in fish, with lower detection limits and greater positivity rates compared to qPCR (Zhao et al., 2024).

Similarly, CRISPR-Cas-based diagnostic tools, including SHERLOCK and DETECTR, are revolutionizing the rapid, point-of-care diagnosis of infectious diseases. These tools utilize CRISPR-Cas enzymes to detect and cut target nucleic acid sequences with attomolar specificity and high specificity, thereby enabling equipment-minimal detection (Kostyusheva et al., 2022; Mustafa & Makhawi, 2021). With pooled sensitivity and specificity of 94% and 97%, respectively, for SARS-CoV-2 detection, CRISPR diagnostics are strong alternatives to conventional PCR-based diagnostics (application of CRISPR diagnostics post-pandemic) (Shahin et al., 2024). Their ability to be adapted to user-friendly formats, including lateral flow assays, and introduction of streamlined protocols like CRISPR-FDS – no RNA isolation needed and results within 15 minutes – make them extremely well-positioned to be deployed in the field (Uzay & Dinçer, 2022).

At the same time, real-time nanopore sequencing, as represented by the MinION of Oxford Nanopore Technologies, is revolutionizing pathogen diagnosis through portable, fast, and whole metagenomic profiling. Sequencing is now feasible on-site with few infrastructural demands, detecting full-length antimicrobial resistance (AMR) genes and facilitating real-time taxonomic identification (Bloemen et al., 2023). In clinical practice, nanopore sequencing is extremely sensitive and specific in direct pathogen detection from whole blood and blood cultures, and greatly reduces diagnostic turnaround times (Govender et al., 2025; Zhou et al., 2021). Furthermore, it accurately predicts AMR patterns equivalent to conventional microbiological testing (Liu, Xu, et al., 2023). Notably, nanopore-targeted sequencing (NTS) has a performance that is comparable to

DOI: 10.1201/9781003615699-6

metagenomic next-generation sequencing (mNGS), but surpasses conventional blood culture methods, thereby emerging as an important critical care diagnostic tool (Han et al., 2024).

6.2 ADVANCED AI ARCHITECTURES FOR MULTI-OMICS DATA INTEGRATION AND INTERPRETATION

Deep learning models have emerged as key tools for multi-omics integration of genomics, transcriptomics, proteomics, and metabolomics data, hence significantly enhancing diagnostic precision for infectious diseases. By integrating the unique perspectives of each omics layer, genetic susceptibility elicited from genomics, gene expression patterns from transcriptomics, protein-protein interactions seen with proteomics, and metabolic alterations elicited by metabolomics, such models enable comprehensive systems-level understanding of disease pathophysiology (Ward et al., 2021). Such integrative analyses enable deep learning algorithms to detect complex biological interactions typical of single-omics analyses, leading to enhanced disease classification and risk stratification (Sanches et al., 2024). Multi-omics integration is the key to precision medicine as it combines individual omics profiles with phenotypic features, elucidating genotype-phenotype relationships and enabling personalized diagnostic and therapeutic interventions (Zaghlool & Attallah, 2022). Integration of multi-omics data with deep learning can also facilitate biomarker discovery and identification of disease-linked genes, while reducing dependence on time-consuming methods like PCR, thus enabling quicker diagnostics and cost-effectiveness.

The convergence of autoencoders, graph neural networks (GNNs), and transformer architectures has greatly revolutionized predictive modeling in infectious disease research, especially in host-pathogen interaction (HPI) and microbe-disease or medicine-microbe prediction analysis. Autoencoders, especially when combined with GNNs, display efficacy in learning concise representations of intricate biological networks, thus improving microbe-disease and medicine-microbe prediction accuracy. For example, the SARMDA model utilizes an adversarially regularized autoencoder to achieve state-of-the-art predictive capacity on different datasets like HMDAD and Disbiome, accurately capturing the topological and semantic characteristics of microbe-disease networks (He et al., 2024). Similarly, the MKAN-MMI model utilizes masked graph autoencoders and self-supervised learning techniques to address data sparsity challenges and improve the robustness of medicine-microbe interaction predictions (Ye et al., 2024). On the other hand, transformer-based models such as TEST-Net have been greatly successful in predictive modeling of the spatio-temporal dynamics of infectious disease outbreaks, utilizing attention mechanisms to boost the accuracy of long-range predictions (Chen et al., 2024). Moreover, the ReGraFT model utilizes recurrent neural networks (RNN) and transformers to improve predictions of epidemic trajectories by taking into account inter-regional interdependencies, thus leading to major improvements in predictive accuracy (Kim et al., 2024). Overall, such hybrid AI approaches prove to be crucial in demystifying the intricate biological and epidemiological connections related to infectious diseases (Yakimovich, 2021). Figure 6.1 illustrates AI-driven molecular diagnostics for infectious disease detection and biomarker identification.

AI-based approaches play a crucial role in the discovery of immune response signatures and disease biomarkers using advanced tools such as mass spectrometry (MS), single-cell RNA sequencing (scRNA-seq), and machine learning (ML) algorithms. The marriage of AI with MS-based profiling allows for in-depth molecular characterization by analyzing complex proteomic data, thereby discovering significant biomarkers of infection-related pathways (Liu et al., 2024). Furthermore, the application of ML to scRNA-seq data has revealed variations in gene expression profiles between immune cell subsets and enables discrimination between mild and severe infectious diseases, e.g., sepsis and COVID-19, thereby facilitating early diagnosis and treatment planning (Sganzerla Martinez et al., 2024). The application of AI in precision medicine is also exemplified by its application in multi-omics data analysis in response prediction of immunotherapy and prognosis (Chang et al., 2025). The novel concept of transferring transcriptomic signatures gene sets with predictive features across various datasets and populations emphasizes the application of AI in the design of universally applicable diagnostic markers, as demonstrated in tuberculosis and COVID-19-related research (di Iulio et al., 2021). Despite persistent challenges with data heterogeneity and model interpretability, AI-aided biomarker discovery continues to revolutionize infectious disease diagnostics by generating rapid, reproducible, and highly accurate information (Srivastava et al., 2025).

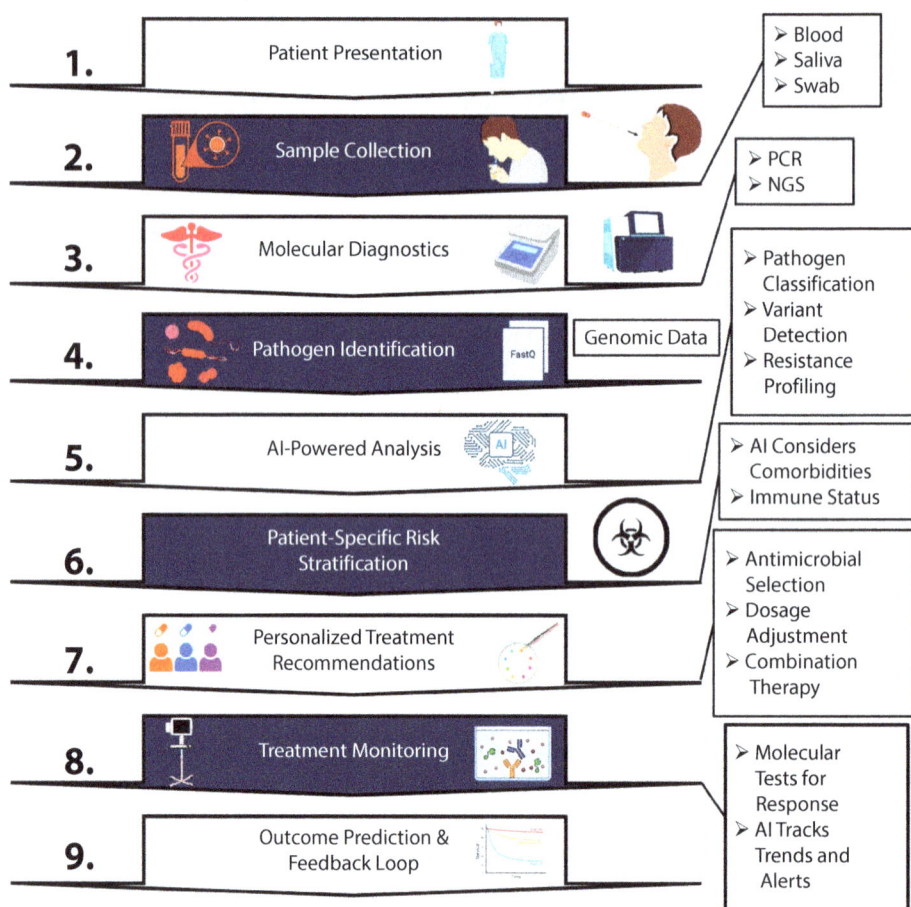

Figure 6.1 AI-models-driven molecular diagnostics of infectious disease.

6.3 AI-ENABLED REAL-TIME SURVEILLANCE, OUTBREAK PREDICTION, AND PATHOGEN EVOLUTION MODELING

Big data analysis is central to the understanding and tackling of global health issues through the integration of data from epidemiological, environmental, social media, and travel datasets. In epidemiological contexts, surveillance data such as case numbers, mortality rates, and test results can be utilized to model disease trends, locate outbreak hotspots, and predict future epidemics. In addition, environmental conditions such as climate, air quality, and land use control the spread of diseases, specifically vector-borne diseases such as malaria and dengue. Integration of these datasets supports the identification of spatio-temporal trends, thus the potential for timely intervention actions (Khoury & Ioannidis, 2014; Harris et al., 2015). Social media and mobile data contain real-time, user-generated proxies that facilitate early detection of shifts in public opinion, symptom conversation, and behavioral response. Social media such as Twitter, Facebook, and Google Trends have been useful in monitoring public health responses to pandemics such as COVID-19 and influenza. Conversely, anonymized mobile phone data and travel data have been instrumental in monitoring human mobility and transmission modeling, as observed in the Ebola and COVID-19 responses (Wesolowski et al., 2014; Abd-Alrazaq et al., 2020). The effectiveness and ethics of such analytics are dependent on good data governance, interoperability, and privacy protection. Integration of heterogeneous datasets requires ML methods that are capable of handling heterogeneity, noise, and missingness without compromising ethical considerations such as consent and equity (Mooney & Pejaver, 2018). Figure 6.2 illustrates the use of AI in the infectious disease management lifecycle, from surveillance to prediction.

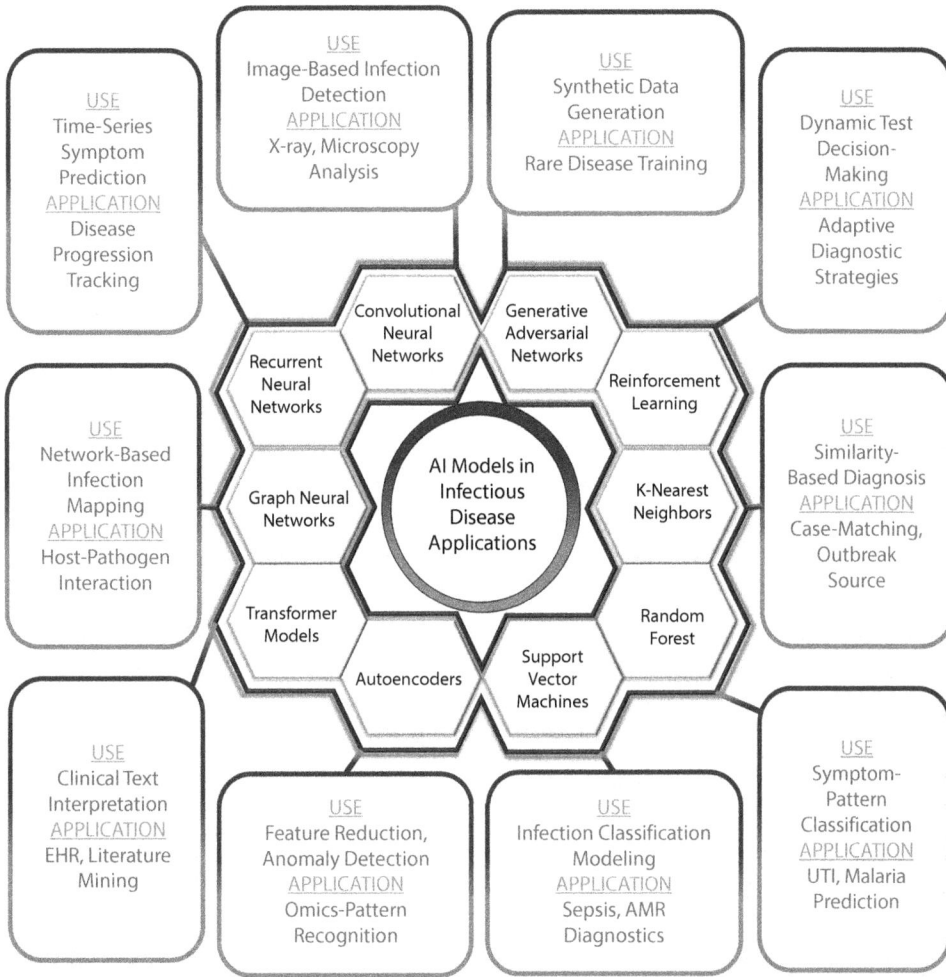

Figure 6.2 AI models in infectious disease applications.

Deep learning architectures for spatio-temporal data such as convolutional long short-term memory networks, GNNs, and transformers have been successful in predicting infectious diseases such as COVID-19, dengue, and influenza by learning large, heterogeneous datasets and generalizing complex spatial and temporal patterns (Hsieh et al., 2021; Kapil et al., 2021). Reinforcement learning enhances these architectures by enabling the ability to simulate dynamic decision-making processes with multiple intervention strategies, such as lockdowns and vaccination campaigns, to obtain optimal results. Agents using reinforcement learning can be trained to learn strategies that maximize the alleviation of infections based on economic and social considerations (Kompella et al., 2021). The combination of spatio-temporal deep learning and reinforcement learning offers strong decision-support systems; however, issues persist regarding model interpretability, data resolution, and ethical deployment. Explainable artificial intelligence-based approaches and domain-knowledge-informed architectures are being researched today to overcome the black-box nature of deep learning models and to increase stakeholder trust.

Concurrently, AI increasingly mimics viral evolution and predicts the development of variants, especially in fast-evolving pathogens like SARS-CoV-2 and influenza viruses. RNN, transformer models, and probabilistic models trained on genomic information enable the prediction of mutation hotspots, protein structure evolution, and immune evasion (Senior et al., 2020). Sequence-to-function prediction models, based on evolutionary principles like epistasis and fitness landscapes, are used to model the evolution of variants under immune and therapeutic pressure. Generative models like variational autoencoders and generative adversarial networks also make informed predictions for variant emergence, with structural predictions further improved by using tools like AlphaFold (Jumper et al., 2021). AI-facilitated platforms like Nextstrain and outbreak.info track global sequencing data in real time to identify newly emerging variants, while reinforcement learning is being developed to model evolutionary outcomes under various intervention designs. However, the credibility of these models depends on biological plausibility, fast data acquisition, and interdisciplinarity to make AI tools valid, interpretable, and relevant to global surveillance and response (Wang et al., 2022; Vatti et al., 2021).

6.4 EXPLAINABLE AI (XAI) AND TRUSTWORTHY MODELS IN MOLECULAR INFECTIOUS DISEASE DIAGNOSTICS

Explainable artificial intelligence (XAI) in healthcare is crucial to making ML models interpretable and transparent, particularly for clinicians working under life-or-death time pressures. In molecular infectious disease diagnosis, this is especially the case, as decisions need to be rapid and accurate. Among the most relevant interpretability methods are feature importance visualizations like SHAP (Shapley Additive Explanations) and LIME (Local Interpretable Model-Agnostic Explanations), which analyze the contribution of each input variable pathogen gene expression, host biomarkers, or patient metadata to a predictive outcome (Lundberg & Lee, 2017; Ribeiro et al., 2016). These visualization tools enable clinicians to trust AI-facilitated diagnostics by showing how variables such as viral load, mutation profiles, or patient symptoms contribute to end decisions.

Natively interpretable models such as decision trees, logistic regression, and attention-based neural networks are widely used across many applications. Accuracy might be slightly lower for such models than for more complex "black-box" alternatives, but they provide reasoning paths that are well-mapped to clinical experience and microbiological knowledge. Hybrid models, which leverage interpretable elements of deep learning models such as the incorporation of attention layers into convolutional neural networks for image-based pathogen detection or genomics-aware text analysis, can detect biologically relevant features (Caruana et al., 2015). Model confidence scoring and interactive dashboards are necessary for real-time clinical deployment. These interfaces provide predictions, as well as uncertainty estimates, to enable clinicians to evaluate algorithmic output in the context of professional experience. Open integration of explainability modules with electronic health records (EHRs) not only induces adoption but also facilitates continuous clinical feedback (Tonekaboni et al., 2019). XAI thus not only increases transparency but also clinical safety and accountability.

One of the main challenges in using AI for diagnostics is the "black-box" effect, with uninterpretable model outputs. In molecular diagnostics, this lack of transparency can produce fear-based resistance to the reliability and fairness of AI predictions. One approach is to use inherently interpretable models such as generalized additive models (GAMs) and sparse linear models, which enable direct insight into feature effects on predictions (Rudin, 2019). Building in explainability during model development using SHAP values, or attention mechanisms during training prevents model behavior from deviating from established clinical principles and maintains fairness across heterogeneous patient populations (Doshi-Velez & Kim, 2017). Standardized documentation and transparent validation are also crucial in addressing the black-box problem. Model cards and datasheets for datasets are tools that record model design, use case intentions, and performance metrics across population subgroups. Validation needs to go beyond retrospective datasets and include prospective clinical trials and real-world assessments, further generating trust in model recommendations (Mitchell et al., 2019; Sendak et al., 2020).

AI diagnostic use in the clinic requires careful regulation. Regulators like the U.S. FDA and the European EMA emphasize strict validation and transparency of AI-based SaMD. The regulations focus on real-world performance oversight, explanation, and reproducibility, particularly for adaptive algorithms (Zhu et al., 2022). Ethically speaking, the use of AI in molecular diagnostics must adhere to the principles of beneficence, non-maleficence, justice, and autonomy, preventing algorithms from perpetuating biases or exacerbating health inequities. Clinician control and seeking informed consent are important to ensuring trust (Floridi et al., 2018; Mittelstadt et al., 2016). Global efforts – for example, the WHO's six principles of AI in health – also ensure responsible development and deployment by ensuring inclusivity, accountability, and sustainability. In addition, the Good Machine Learning Practice (GMLP) guidelines aim to harmonize AI control across jurisdictions, allowing for the development of ethical and scientifically sound AI solutions in infectious disease diagnostics (WHO, 2024).

6.5 AI-DRIVEN DISCOVERY AND MONITORING OF ANTIMICROBIAL RESISTANCE MECHANISMS

AMR mechanisms optimization of ML algorithms for effective detection of novel resistance genes and mutations from genomic data has become an essential driver in the battle against infectious disease. New approaches, like the use of unitig-based pan-genome analysis, employ compressed de Bruijn graphs to detect unique sequence patterns and thus enable detection of novel antibiotic resistance genes unreported previously (Do et al., 2024). In *Mycobacterium tuberculosis*, artificial intelligence-driven systems that map sequence and structural mutation characteristics in high-throughput resistance genes have achieved average predictive accuracies of 85%, an indication of the significance of ML in drug resistance prediction (Jamal et al., 2020). Deep feature selection approaches, like the DNP-AAP model, employ whole-genome single-nucleotide polymorphism (SNP) data to offer information on genetic determinants of AMR, detecting established, as well as novel, determinants in *Neisseria gonorrhoeae* (Shi et al., 2019). Additionally, integrative ML models that integrate homology modeling with molecular docking provide functional information on new resistance genes by selecting candidates based on predicted interactions with antimicrobials (Sunuwar & Azad, 2022). In *M. tuberculosis*, non-parametric classification trees and gradient-boosted models have been applied to predict first-line drug resistance, an indication of the predictive gain when co-occurring resistance markers are employed (Deelder et al., 2019).

The use of AI is significant in connecting resistance genotypes with drug susceptibility phenotypes, thus improving the control of infectious diseases. Artificial neural networks (ANNs) have been utilized successfully to forecast HIV-1 resistance phenotypes from genotypic data with high accuracy and robust generalizability to various subtypes, underpinning individualized treatment regimens in extensively treated patients (Pasomsub et al., 2010). For multidrug-resistant *Acinetobacter baumannii*, deep neural networks (DNN) trained on whole-genome sequencing and gene expression data are projected to achieve antimicrobial susceptibilities of up to 98.64% accuracy, showing the potential of AI in real-time surveillance and AMR control (Jia et al., 2024). For multidrug-resistant *M. tuberculosis*, genetic determinants of underlying phenotypic testing have demonstrated better performance than traditional phenotypic tests in predicting clinical outcomes, such as sputum smear and culture conversion, thus qualifying as dependable biomarkers for predicting treatment (Che et al., 2021). In addition, ML algorithms have been used to identify unknown genetic determinants of resistance in strains that lack known resistance genes but possess resistant phenotypes, overcoming a key limitation in the detection of AMR (Sunuwar & Azad, 2022). In general, these AI-based approaches provide greater insight into mechanisms of resistance and guide the design of targeted therapeutic interventions and novel drug designs, vital approaches in combating the growing global threat of AMR.

Generative AI models have also proven to be powerful tools in the design of novel antimicrobials and peptide therapeutics that are urgently required to fight AMR. The APEX Generative Optimization (APEX GO) platform couples transformer-based variational autoencoders and Bayesian optimization to generate and optimize antimicrobial peptides (AMPs), identifying new sequences with high activity against Gram-negative pathogens and establishing their efficacy in preclinical infection models (Torres et al., 2024). Long- short-term memory (LSTM) networks have been used effectively to design short AMP sequences against bacteria like *Escherichia coli* with high classification accuracy and proving the potential of deep learning for AMP innovation (Wang et al., 2021). Also, generative models coupled with molecular simulations enable the prediction of AMP-like sequences prioritized for membrane interaction potential, hence allowing the discovery of broad-spectrum antibacterial peptides (Ferrell et al., 2020). The AMPTrans-lstm model broadens the design space by coupling LSTM and transformer architectures to generate innovative, diverse AMPs for specific antimicrobial functions (Mao et al., 2023). Further, the multi-CGAN (conditional generative adversarial network) architecture proves to have improved capability in generating AMPs with multiple desired features, which is better than current deep learning approaches and increasing training datasets required in AMP prediction tasks (Yu et al., 2023). Collectively, these generative AI approaches mark a revolutionary step in the discovery of next-generation antimicrobials that can overcome issues of resistance.

6.6 SYNERGIZING AI WITH CRISPR-BASED DIAGNOSTIC PLATFORMS

CRISPR guide RNA (crRNA) design optimization for improving the sensitivity and specificity of assays used for infectious disease diagnosis has been greatly enhanced by the incorporation of ML and deep learning strategies. For example, the easy design system uses the application of a convolutional neural network to design crRNA for the CRISPR/Cas12a detection system with high predictive capability, with a Spearman's ρ of 0.812. Of interest, easy design has shown excellent generalizability by being able to design crRNAs for pathogens beyond its training dataset, including monkeypox virus and enterovirus 71. The system is made available through an interactive web server, allowing for fast translation into clinical diagnostic pipelines (Huang et al., 2024). Likewise, the ADAPT platform uses a DNN to individualize CRISPR-based diagnostics by targeting viral genomic variation, resulting in increased sensitivity and specificity. The strategy has been applied comprehensively to 1,933 vertebrate-infecting viral species with reduced detection limits compared to standard diagnostic protocols (Metsky et al., 2020). In extension of these advances, the CRISPR-Cas12f1 system has been engineered for SNP detection, allowing for high-specificity detection of viral or bacterial mutations, crucial for the detection of drug-resistant strains (Gao et al., 2023). Improving assay sensitivity, the design of a multimodal universal reporter for CRISPR/Cas12a assays has achieved a tenfold reduction in detection limits, with flexible readout capacities that enable rapid, specific pathogen detection such as staphylococcal enterotoxin A (de Dieu Habimana et al., 2023).

Adaptive AI-enabled CRISPR diagnostics are on the rise as a very potent solution to match the fast-changing pace of pathogen mutations, particularly infectious diseases. CRISPR/Cas systems, particularly Class 1 and Class 2 effectors, have played a pivotal role in the development of rapid, sensitive, and multiplexed nucleic acid diagnostics for a broad range of pathogens, including viruses, bacteria, and fungi. These platforms enable mutation profiling to identify viral variants and drug-resistant bacteria, a critical feature for point-of-care diagnostics in resource-scarce settings (Yang, Zhang, et al., 2023). The small CRISPR-Cas12f1 nuclease exhibits excellent sensitivity and specificity for the detection of SNPs, which speaks volumes about its efficacy for accurate pathogen mutation diagnostics (Gao et al., 2023). Deep learning-aided design tools such as EasyDesign extrapolate further optimization of guide RNAs for enhanced detection efficiency even for pathogens not included in the training dataset (Huang et al., 2024). AI-enabled amplification-free CRISPR diagnostic approaches are also being pursued, with potential for streamlined workflows to improve accessibility and speed of testing (Xia et al., 2024).

Portable AI-CRISPR multiplex biosensors constitute a groundbreaking frontier for on-site, rapid infectious disease diagnosis. CRISPR/Cas systems like Cas12a and Cas13a were reengineered into biosensors that are highly specific and sensitive for nucleic acid detection. Nanotechnology advancements enhance the performance of the biosensors through inkjet-printed nanostructured electrodes, which introduce higher stability and sensitivity of the signal, well suited for point-of-care applications (Carota et al., 2024; Gupta et al., 2023). Microfluidic integration enables multiplexed detection, essential for the simultaneous diagnosis of multiple pathogens. A portable microfluidic CRISPR-Cas13a, for instance, can identify ten distinct viruses precisely at high speed, with potential in resource-limited settings (Zhang et al., 2024). Engineered CRISPR RNA (en-crRNA) enhances multiplex and visual detection of respiratory pathogens, expanding clinical utility (Gu et al., 2024). These minimal-technical-expertise portable biosensors enable decentralized testing critical for early infection detection and outbreak control. Much remains to be enhanced in scaling AI-CRISPR diagnostic platforms for mass deployment. Improvements in robustness, reduction in manufacturing and operational costs, and compatibility with a variety of clinical environments are required factors critical to global health impact and equitable access (He et al., 2023). Figure 6.3 illustrates the integration of CRISPR-Cas diagnostics with AI technologies for rapid and multiplexed infectious disease detection. Further innovation at the interface of AI and CRISPR technologies has tremendous potential to revolutionize infectious disease diagnosis and enhance pandemic preparedness.

6.7 PERSONALIZED INFECTIOUS DISEASE MANAGEMENT POWERED BY MOLECULAR DIAGNOSTICS AND AI

The convergence of AI and molecular diagnostics has revolutionized personalized disease treatment for infectious diseases. The most potent innovation is AI-aided patient stratification that integrates pathogen genotypes and host immune profiles to stratify patients into clinically relevant subgroups. These stratifications improve precision medicine approaches by personalizing treatment protocols, predicting the severity of disease, and enhancing prognostic accuracy (Chang et al., 2021; Gong et al., 2020). Sophisticated ML techniques, like clustering algorithms, support vector machines (SVM), and DNN, are now utilized as a matter of routine to process high-dimensional, heterogeneous data. These data typically include viral or bacterial genomic sequences, along with host immune markers like cytokine levels, HLA types, and T-cell receptor repertoires. The ability of

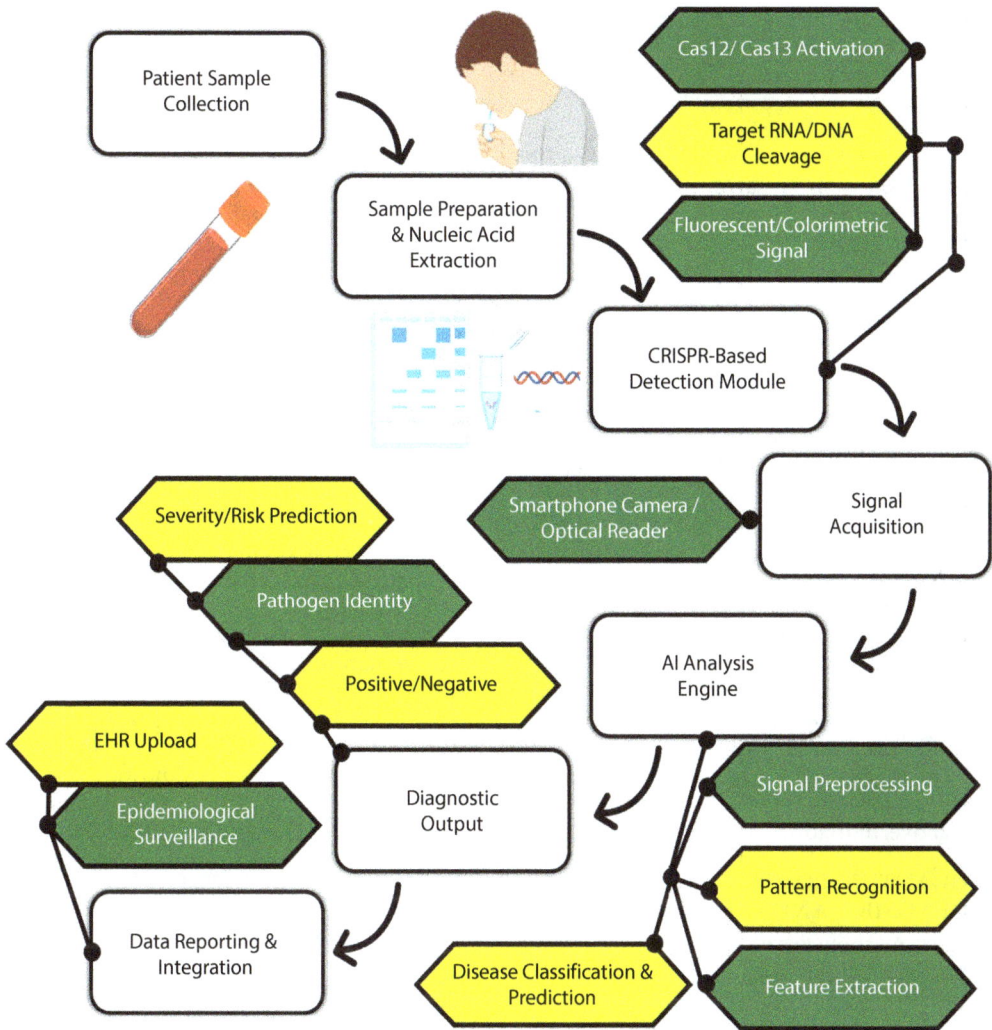

Figure 6.3 CRISPR + AI diagnostic system for infectious disease detection.

AI to combine multiple data modalities makes it especially suited to detect latent patterns related to disease outcomes.

These multimodal models, which integrate transcriptomic (e.g., peripheral blood mononuclear cell RNA-seq), proteomic, and single-cell sequencing data with pathogen-specific genomic data, have shown strong performance for predicting immune response subtypes. For example, multi-omics integration with autoencoders and GNNs has identified patient clusters with different responses to different pathogen lineages, such as SARS-CoV-2 variants (Arunachalam et al., 2020). These models do not simply stratify patients by clinical outcomes such as hospitalization risk, disease severity, or risk for long COVID but also predict vaccine responsiveness and treatments. Also useful is the use of XAI tools such as SHAP values and attention mechanisms, which improve the interpretability of complex ML models. Transparency is key to clinician trust and regulatory approval, particularly in the identification of important predictive features such as pathogen mutations or immune biomarkers.

But to guarantee equity and generalizability, these models need to be trained and validated on representative, population-level datasets. Data collection or model training bias can worsen healthcare disparities unless they are addressed sufficiently. Interdisciplinary collaboration between immunologists, computational scientists, and clinicians is required to guarantee biologically sensible and clinically interpretable output. Apart from stratification, AI is at the core of predictive modeling of response to treatment, disease progression, and risk of relapse. ML algorithms like random forests, SVMs, and deep learning models are increasingly being applied to examine clinical, genomic, and imaging data. In oncology and infectious disease, in turn, these tools predict responsiveness to therapy (e.g., to antivirals or antibiotics) and assist in personalized regimen selection (Kourou et al., 2015; Rajkomar et al., 2019).

Temporal modeling methods, specifically RNNs and LSTM networks, are powerful at predicting disease progression by learning from longitudinal data like EHRs, biomarker trajectories, and wearable sensor streams. These models allow intervention to be proactive by detecting at-risk patients for deterioration or relapse, for instance, organ failure in sepsis or viral rebound in HIV (Miotto et al., 2018). In leukemia and breast cancer, for instance, relapse forecasting models now use multi-omics data, immune profiling, and psychosocial factors to predict minimal residual disease (MRD) and recurrence. Explainable AI not only makes predictions accurate but also interpretable, promoting clinical adoption (Lundberg and Lee, 2017; Zhao et al., 2020). Perhaps most exciting is the combination of wearable biosensors and AI algorithms. These sensors, such as smartwatches, ECG patches, continuous glucose monitors, and sweat sensors, enable real-time, non-invasive monitoring of physiological markers like heart rate, skin temperature, and oxygen saturation. AI models analyze these high-frequency data to identify early evidence of infection, metabolic derangement, or cardiovascular events even before clinical symptoms develop (Heikenfeld et al., 2018; Shcherbina et al., 2017).

AI-augmented wearables during the COVID-19 pandemic identified pre-symptomatic infection using abnormal physiological patterns. In cardiology, too, such models forecast atrial fibrillation or imminent heart failure. Reinforcement learning and subject-to-subject customized baseline modeling improve prediction even further by adapting to inter-subject heterogeneity (Quer et al., 2021; Tison et al., 2018). Challenges remain despite the promise, however. Data privacy and cybersecurity take precedence, with sensitive data under perpetual surveillance requiring safeguarding. Access to the technologies on an equitable basis and prevention of algorithmic bias are also key to averting disparities. Interoperability with EHRs and rigorous clinical validation are also prerequisites for real-world use. Advances in sensor technologies, edge computing, and federated learning in the future are likely to drive global adoption of AI-augmented, wearable-based diagnostics. Together, the technologies are transforming infectious disease management, shattering the reactive paradigm to proactive, individualized, and preventive care (Kassal et al., 2018; Piwek et al., 2016).

6.8 QUANTUM COMPUTING AND AI SYNERGIES IN INFECTIOUS DISEASE GENOMICS AND DIAGNOSTICS

Quantum machine learning (QML) promises enormous benefits to improve the efficiency of genome data analysis of infectious diseases through the leveraging of quantum computing's unique computational paradigms. Quantum nature of systems – superposition and entanglement – narrowly enable parallel computing of complex, high-dimensional biomedical data sets that lie beyond the reach of conventional computer power. This innovation opens the door to quantum-inspired algorithms that can potentially bypass the complexity inherent in genomic and proteomic data, including protein folding prediction, disease biomarker identification, and real-time diagnostics (Gowda et al., 2024).

QML models like the Pegasos Quantum Support Vector Classifier and quantum neural networks have yielded promising results in genomic data classification with high sensitivity and accuracy. Overfitting due to quantum feature mapping variability, however, still needs to be addressed to provide maximum robustness and generalizability (Singh & Pokhrel, 2025). Remarkably, quantum computing's capability of significantly reducing DNA analysis processing times and resource consumption is at the core of driving AI-driven genomics forward, enabling quicker, more accurate insights required for healthcare and biotechnology innovation (Abdellah et al., 2023). The intersection of quantum computing and AI exponentially amplifies the computational horsepower required for big data analysis, molecular simulations, and predictive modeling, all of which are pivotal to precision medicine and infectious disease diagnosis (Anbazhagu et al., 2025). Quantum-aided AI can rapidly detect disease biomarkers and enable personalized treatment pathways by efficiently processing intricate molecular diagnostic information, such as multimodal molecular profiles obtained through MS (Liu et al., 2024; Yenduri et al., 2025). This synergy not only improves data quality and usability but also enables in-depth modeling of organic molecular processes, improving drug design and predicting treatment interactions with unparalleled accuracy.

In infectious disease research, quantum-inspired algorithms like quantum genetic algorithms have been identified to be superior in predicting RNA secondary structures crucial to virus dynamics and infection pathways (Gazi, 2023). Additionally, quantum computing expedites large-scale genomic data processing, thus supporting rapid vaccine development and personalized medicine strategies (Arya & Verma, 2024). While these benefits exist, several challenges persist, including quantum hardware limitations, the need for special quantum software, and the concern for data security. These challenges must be overcome through interdisciplinary research, as exemplified by initiatives such as the DOE-NIH roundtable, which supports multidisciplinary collaborations and FAIR (Findable, Accessible, Interoperable, Reusable) data principles (McWeeney et al., 2023). Overall, the intersection of quantum computing and AI provides a paradigm-shifting solution to infectious disease genomics and diagnostics. Through improved biomarker discovery to support proactive diagnosis and personalized healthcare solutions, these technologies have the potential to drive the pace of medical innovation, despite the technical and ethical hurdles that lie ahead (Flöther et al., 2024; Allami & Yousif, 2023).

6.9 ETHICAL, LEGAL, AND SOCIAL IMPLICATIONS (ELSI) OF AI-ENABLED MOLECULAR DIAGNOSTICS

The incorporation of AI in molecular diagnostics brings revolutionary functionality, as well as deep ethical, legal, and social issues. Bias in AI can result from non-representative data, flawed algorithm design, or biased deployment strategies, leading to systems that fail for underrepresented populations. For example, facial recognition software has demonstrated much higher error rates for darker-skinned populations due to a lack of diverse data (Mehrabi et al., 2021). To counter such imbalances, data used in AI model training must be representative – and representative of the population the technology is intended to serve. Fairness-aware learning methods, re-weighting, and adversarial debiasing provide fair treatment of individuals by sensitive attributes like race, gender, or age. Testing models using fairness metrics like demographic parity and equal opportunity, and ongoing post-deployment monitoring, is crucial to ensure ethical integrity. Of equal concern are legal concerns, notably data privacy. Molecular diagnostics entail sensitive genomic and health information that, if poorly managed, can lead to discrimination or identity theft. Compliance with legislation such as HIPAA and GDPR requires stringent conditions for anonymization, secure data storage, and informed consent, ensuring ethical management of data (Shabani & Marelli, 2019; Phillips et al., 2020). Cybersecurity measures like encryption, access controls, and secure transmission protocols fend off cyberattacks. Techniques like federated learning and differential privacy enable AI models to be trained on decentralized datasets without exposing personal data, weighing data utility against privacy (Kaissis et al., 2020).

Socially, equal access to AI technologies among global populations, particularly in low- and middle-income countries, is important, where healthcare infrastructure and data assets might be scarce. Collective global action, culturally adapted model development, and local capacity investment can reduce global disparities (McKee et al., 2021). Transparency and accountability across the AI lifecycle through open-sourced models, XAI, strict validation, and third-party audits are important to maintaining public trust. Ethical governance frameworks and international standards compliance, such as the WHO's AI ethics guidance, also enable AI deployment to respect human rights (World Health Organization, 2021). All in all, the ELSI of AI in molecular diagnostics need to be addressed by a multidisciplinary, inclusive, and adaptive approach that prioritizes fairness, privacy, accountability, and global equity.

6.10 THE FUTURE OF AI-DRIVEN AUTONOMOUS MOLECULAR DIAGNOSTIC LABORATORIES

Automated molecular diagnostic workflows are revolutionizing the diagnosis of infectious diseases by combining robotics, AI, and cloud computing to enhance efficiency, accuracy, and accessibility. These closed systems streamline each step of the diagnostic workflow from sample preparation to result interpretation, thus restricting human error and contamination risk. For example, open-source modular automation tools like Opentrons offer modular and cost-effective workflows, such as in a low-cost *Neisseria meningitidis* diagnostic assay using paper microarrays (Urrutia Iturritza et al., 2024). Established automated platforms like the COBAS AMPLICOR illustrate how sophisticated molecular biotechnology can provide accurate nucleic acid amplification and detection while optimizing labor efficiency and assay reliability (Miyachi, 1998). The integration of new technologies into molecular assays has not only reduced the risk of contamination but also allowed them to function in standard diagnostic laboratories, making high-quality molecular diagnosis more accessible (Kessler et al., 1999).

Recent advances in molecular diagnostic devices PCR, isothermal amplification, and high-throughput sequencing have dramatically boosted the sensitivity, specificity, and potential for automation needed for the rapid and accurate detection of pathogens (Liu, Jin, et al., 2023). Perhaps the most dramatic example of this is the combination of robotic pipetting modules with centrifugal molecular diagnostic platforms, enabling fully automated detection of respiratory viruses from nasopharyngeal swabs and showing the potential for point-of-care use (Phan et al., 2023). Collectively, these technologies showcase the revolutionary potential of automation to molecular diagnostics, enabling rapid, reliable, and scalable solutions to infectious disease management.

In the backdrop of overcoming global health challenges, the establishment of self-learning diagnostic systems capable of evolving with newly discovered pathogens with minimal human intervention is becoming increasingly vital. The integration of next-generation sequencing (NGS) and deep learning approaches is an exemplary illustration of this progress. MG2Vec, based on transformer architecture, derives strong features directly from raw metagenomic sequences to identify pathogens from uncurated clinical samples, illustrating its versatility to a wide range of pathogens and its ability to perform extensive metagenomic analysis (Narayanan et al., 2022). Similarly, mobile health (mHealth) diagnostic platforms, including SPyDERMAN, employ adversarial neural networks to quickly design image classifiers from smartphone photographs of microfluidic chips, thus enabling quick adaptation for emerging pathogens such as SARS-CoV-2 and enhancing epidemic preparedness (Shokr et al., 2020).

Pathogen-agnostic detection platforms, inspired by the human innate immune system – e.g., the PEGASUS platform – employ pattern recognition receptors to detect conserved pathogen-associated molecular patterns. This enables the development of highly sensitive diagnostic assays capable of detecting a broad range of pathogens, even in low-resource settings (Mukundan, 2023). In addition, nano-bio-analytical systems facilitate point-of-care rapid diagnosis of viral illnesses and antibiotic-resistant bacterial infections without prior knowledge of the pathogen, and overcoming considerable limitations inherent in conventional molecular diagnostics (Yang, Fang, et al., 2023).

Global AI-powered networks for real-time infectious disease surveillance and rapid data sharing represent a transformative approach to public health, leveraging advanced technologies to enhance outbreak detection and response. The integration of AI with blockchain technology offers a secure and efficient method for disease monitoring, enabling encrypted health data to be shared among stakeholders while mitigating risks of data breaches (Tripathi & Rathore, 2025). AI algorithms, when combined with mobile health data, provide unprecedented insights into disease dynamics, allowing for real-time monitoring and proactive interventions by analyzing diverse datasets such as demographic information and social media activity (Olaboye et al., 2024). The use of next-generation networking technologies, including federated learning and mobile edge computing, further enhances the capability of AI-enabled systems to discover epidemics, facilitate remote monitoring, and enable rapid health authority responses (Otoum et al., 2021). These systems can analyze vast amounts of real-time data from unconventional sources like social media and EHRs, predicting outbreaks earlier than traditional methods and offering timely alerts to healthcare providers (Giri & Gupta, 2024). However, the implementation of these technologies must address challenges such as data privacy, interoperability, and algorithm bias to ensure reliable and ethical surveillance systems. By continuing to explore and refine these AI-driven methodologies, public health authorities can significantly improve their ability to anticipate, monitor, and respond to emerging health threats in an increasingly interconnected world (Tripathi & Rathore, 2025).

Combined, these technologies promise an AI-future of autonomous molecular diagnostic labs that not only speed and improve the accuracy of diagnostics but also can dynamically adapt to detect new and as-yet-unknown pathogens. These systems have the potential to greatly enhance global health security by facilitating rapid response to infectious disease outbreaks with minimal human intervention.

6.11 CONCLUSION

This chapter highlights the revolutionary potential of synergizing molecular diagnostic innovations with cutting-edge AI and quantum computing technologies to address the multidimensional needs of infectious disease control. The accelerated innovation in ultra-sensitive detection technologies, AI-driven multi-omics analysis, and real-time monitoring of pathogens promises a diagnostic future that is not only faster and more accurate but also highly personalized. Quantum computing is seen as a powerful tool to address computational bottlenecks, enabling unprecedented data processing power critical for next-generation genomic analysis. The emergence of fully automated, self-learning molecular diagnostic labs promises to dramatically enhance diagnostic throughput and flexibility, which are critical to a timely response to emerging global health threats. However, the realization of this vision relies on cross-disciplinary collaboration, robust ethical frameworks, and rigorous data governance to ensure equitable and responsible deployment. Overall, these convergent technologies promise a new generation of infectious disease diagnostics and therapeutics that can provide improved patient outcomes and greater public health resilience at the global scale.

ACKNOWLEDGMENT

All the authors are thankful to their respective universities and institutions for their valuable supports.

DATA AVAILABILITY

The authors confirm that the data supporting the findings of this study are available within the chapter.

ABBREVIATIONS

AI	Artificial Intelligence
AMPs	Antimicrobial Peptides
AMR	Antimicrobial Resistance
ANNs	Artificial Neural Networks
CaHV	*Carassius auratus* Herpesvirus
crRNA	CRISPR RNA
ddPCR	Droplet Digital PCR
DNN	Deep Neural Networks
EHRs	Electronic Health Records
ELSI	Ethical, Legal, and Social Implications
en-crRNA	Engineered CRISPR RNA
FAIR	Findable, Accessible, Interoperable, Reusable
GAMs	Generalized Additive Models
GMLP	Good Machine Learning Practice
GNNs	Graph Neural Networks
HPI	Host-Pathogen Interaction
LIME	Local Interpretable Model-Agnostic Explanations
LSTM	Long Short-Term Memory
M. tuberculosis	*Mycobacterium tuberculosis*
mHealth	Mobile Health
ML	Machine Learning
mNGS	Metagenomic Next-Generation Sequencing
MRD	Minimal Residual Disease
MS	Mass Spectrometry
NGS	Next-Generation Sequencing
NTS	Nanopore-Targeted Sequencing
PCR	Polymerase Chain Reaction
QML	Quantum Machine Learning
qPCR	Quantitative PCR
RNNs	Recurrent Neural Networks
scRNA-seq	Single-Cell RNA Sequencing
SHAP	Shapley Additive Explanations
SNP	Single-Nucleotide Polymorphism
SVM	Support Vector Machines
WHO	World Health Organization
XAI	Explainable Artificial Intelligence

REFERENCES

Abd-Alrazaq, A., Alhuwail, D., Househ, M., Hamdi, M., & Shah, Z. (2020). Top concerns of tweeters during the COVID-19 pandemic: Infoveillance study. *Journal of Medical Internet Research*, 22(4), e19016.

Abdellah, T., Merazka, F., & Zahra, H. F. (2023). Quantum computing for DNA analysis and AI in genomics. In *2023 20th ACS/IEEE International Conference on Computer Systems and Applications (AICCSA)* (pp. 1–6). IEEE.

Alagendran, S., Archunan, G., Prabhu, S.V., Orozco, B.E., & Guzman, R.G. (2010). Biochemical evaluation in human saliva with special reference to ovulation detection. *Indian Journal of Dental Research*, 21, 165–168.

Allami, R. H., & Yousif, M. G. (2023). Integrative AI-driven strategies for advancing precision medicine in infectious diseases and beyond: A novel multidisciplinary approach. *arXiv Preprint arXiv:2307.15228*.

Anbazhagu, U. V., Priyadharshini, K., Deepajothi, S., Haripriya, M. P., & Banu, E. A. (2025). Quantum computing in healthcare using AI-driven medical technologies. In *Modern superhypersoft computing trends in science and technology* (pp. 367–398). IGI Global Scientific Publishing.

Arumugam, V. A., Thangavelu, S., Fathah, Z., Ravindran, P., Sanjeev, A. M. A., Babu, S., Arun, M., Mohd. Iqbal, Y., Khan, S., Ruchi, T., Megha Katare, P., Ranjit, S., Chandra, R., & Dhama, K. (2020). COVID-19 and the world with co-morbidities of heart disease, hypertension and diabetes. *Journal of Pure and Applied Microbiology*, *14*(3), 1623–1638.

Arunachalam, P. S., Wimmers, F., Mok, C. K. P., Perera, R. A., Scott, M., Hagan, T., & Pulendran, B. (2020). Systems biological assessment of immunity to mild versus severe COVID-19 infection in humans. *Science*, *369*(6508), 1210–1220.

Arya, R., & Verma, A. (2024). Quantum computing and healthcare. *International Journal of Advanced Research in Science, Communication and Technology*, *4*(2), 584–588.

Bloemen, B., Gand, M., Vanneste, K., Marchal, K., Roosens, N. H., & De Keersmaecker, S. C. (2023). Development of a portable on-site applicable metagenomic data generation workflow for enhanced pathogen and antimicrobial resistance surveillance. *Scientific Reports*, *13*(1), 19656.

Carota, A. G., Bonini, A., Urban, M., Poma, N., Vivaldi, F. M., Tavanti, A., & Di Francesco, F. (2024). Low-cost inkjet-printed nanostructured biosensor based on CRISPR/Cas12a system for pathogen detection. *Biosensors and Bioelectronics*, *258*, 116340.

Caruana, R., Lou, Y., Gehrke, J., Koch, P., Sturm, M., & Elhadad, N. (2015). Intelligible models for healthcare: Predicting pneumonia risk and hospital 30-day readmission. In *Proceedings of the 21st ACM SIGKDD International Conference on Knowledge Discovery and Data Mining* (pp. 1721–1730).

Casto, M., Hicks, R., Brazeau, D., & Karamchi, M. (2024). B-228 Performance comparison of three viral RNA pathogens in MK buffered solution for diagnostic based detection using quantitative PCR vs droplet digital PCR. *Clinical Chemistry*, *70*(Supplement_1), hvae106-587.

Chang, L., Liu, J., Zhu, J., Guo, S., Wang, Y., Zhou, Z., & Wei, X. (2025). Advancing precision medicine: The transformative role of artificial intelligence in immunogenomics, radiomics, and pathomics for biomarker discovery and immunotherapy optimization. *Cancer Biology & Medicine*, *22*(1), 33–47.

Chang, W., Wan, C., Zang, Y., Zhang, C., & Cao, S. (2021). Supervised clustering of high-dimensional data using regularized mixture modeling. *Briefings in Bioinformatics*, *22*(4), bbaa291.

Che, Y., Yang, T., Lin, L., Xiao, Y., Jiang, F., Chen, Y., & Zhou, J. (2021). Comparative utility of genetic determinants of drug resistance and phenotypic drug susceptibility profiling in predicting clinical outcomes in patients with multidrug-resistant *Mycobacterium tuberculosis*. *Frontiers in Public Health*, *9*, 663974.

Chen, K., Liu, Y., Ji, T., Yang, G., Chen, Y., Yang, C., & Zheng, Y. (2024). TEST-Net: Transformer-enhanced Spatio-temporal network for infectious disease prediction. *Multimedia Systems*, *30*(6), 312.

de Dieu Habimana, J., Mukama, O., Amissah, O. B., Sun, Y., Karangwa, E., Liu, Y., & Li, Z. (2023). A rationally designed CRISPR/Cas12a assay using a multimodal reporter for various readouts. *Analytical Chemistry*, *95*(31), 11741–11750.

Deelder, W., Christakoudi, S., Phelan, J., Benavente, E. D., Campino, S., McNerney, R., & Clark, T. G. (2019). Machine learning predicts accurately *Mycobacterium tuberculosis* drug resistance from whole genome sequencing data. *Frontiers in Genetics*, *10*, 922.

di Iulio, J., Bartha, I., Spreafico, R., Virgin, H. W., & Telenti, A. (2021). Transfer transcriptomic signatures for infectious diseases. *Proceedings of the National Academy of Sciences*, *118*(22), e2022486118.

Do, D. T., Yang, M. R., Vo, T. N. S., Le, N. Q. K., & Wu, Y. W. (2024). Unitig-centered pan-genome machine learning approach for predicting antibiotic resistance and discovering novel resistance genes in bacterial strains. *Computational and Structural Biotechnology Journal*, *23*, 1864–1876.

Doshi-Velez, F., & Kim, B. (2017). Towards a rigorous science of interpretable machine learning. *arXiv Preprint arXiv:1702.08608*.

Ferrell, J. B., Remington, J. M., Van Oort, C. M., Sharafi, M., Aboushousha, R., Janssen-Heininger, Y., & Li, J. (2020). A generative approach toward precision antimicrobial peptide design. *bioRxiv*, 2020-10.

Floridi, L., Cowls, J., Beltrametti, M., Chatila, R., Chazerand, P., Dignum, V., & Vayena, E. (2018). AI4People – An ethical framework for a good AI society: Opportunities, risks, principles, and recommendations. *Minds and Machines*, *28*, 689–707.

Flöther, F. F., Blankenberg, D., Demidik, M., Jansen, K., Krishnakumar, R., Laanait, N., & Utro, F. (2024). How quantum computing can enhance biomarker discovery for multi-factorial diseases. *arXiv Preprint arXiv:2411.10511*.

Gao, P., Yang, M., Chen, Y., Yan, J., Han, M., Deng, H., & Long, Q. (2023). A spacer design strategy for CRISPR-Cas12f1 with single-nucleotide polymorphism mutation resolution capability and its application in the mutations diagnosis of pathogens. *Journal of Medical Virology*, 95(10), e29189.

Gazi, S. Z. (2023). Quantum theory for infectious diseases: Insights, prospects, and challenges. *Access Microbiology*, 1, 000593.

Giri, P. A., & Gupta, M. K. (2024). Transforming disease surveillance through artificial intelligence. *Indian Journal of Community Medicine*, 49(5), 663–664.

Gong, D., Wang, M., Dong, H., & Xu, L. (2020). Multi-omics data integration and machine learning for disease classification. *Frontiers in Genetics*, 11, 626766.

Govender, K. N., Street, T. L., Sanderson, N. D., Leach, L., Morgan, M., & Eyre, D. W. (2025). Rapid clinical diagnosis and treatment of common, undetected, and uncultivable bloodstream infections using metagenomic sequencing from routine blood cultures with Oxford Nanopore. *medRxiv*, 2025-01.

Gowda, D., Patil, H. Y., Abidin, S., Panda, R. A., & Suneetha, S. (2024). Quantum machine learning for biomedical data analysis. In *Quantum innovations at the nexus of biomedical intelligence* (pp. 180–205). IGI Global.

Gu, X., Pan, A., Wu, L., Zhang, J., Xu, Z., Wen, T., & Qin, Y. (2024). Multiplexed detection of respiratory pathogens using a portable device combining a CREM strategy. *Chemical Science*, 15(44), 18411–18418.

Gundappa, M., Arumugam, V.A., Hsieh, H.L., Balasubramanian, B., & Shanmugam, V. (2022). Expression of tissue factor and TF-mediated integrin regulation in HTR-8/SVneo trophoblast cells. *Journal of Reproductive Immunology*, 150, 103473.

Gupta, J., Althomali, R. H., Chunata, D. M. G., Abdullaev, S. S., Yeslam, H. E., Sarsembenova, O., & Alkhayyat, S. (2023). Portable biosensors based on the CRISPR/Cas system for detection of pathogen bacteria: Up-to-date technology and future prospects. *Microchemical Journal*, 194, 109268.

Han, D., Yu, F., Zhang, D., Hu, J., Zhang, X., Xiang, D., & Zheng, S. (2024). Molecular rapid diagnostic testing for bloodstream infections: Nanopore targeted sequencing with pathogen-specific primers. *Journal of Infection*, 88(6), 106166.

Harris, K. M., Spacek, J., Bell, M. E., Parker, P. H., Lindsey, L. F., Baden, A. D., & Burns, R. (2015). A resource from 3D electron microscopy of hippocampal neuropil for user training and tool development. *Scientific Data*, 2(1), 1–19.

He, L., Zou, Q., Dai, Q., Cheng, S., & Wang, Y. (2024). Adversarial regularized autoencoder graph neural network for microbe-disease associations prediction. *Briefings in Bioinformatics*, 25(6), bbae584.

He, Y., Hu, Q., San, S., Kasputis, T., Splinter, M. G. D., Yin, K., & Chen, J. (2023). CRISPR-based biosensors for human health: A novel strategy to detect emerging infectious diseases. *TrAC Trends in Analytical Chemistry*, 168, 117342.

Heikenfeld, J., Jajack, A., Rogers, J., Gutruf, P., Tian, L., Pan, T., & Wang, J. (2018). Wearable sensors: Modalities, challenges, and prospects. *Lab on a Chip*, 18(2), 217–248.

Hsieh, K., Wang, Y., Chen, L., Zhao, Z., Savitz, S., Jiang, X., & Kim, Y. (2021). Drug repurposing for COVID-19 using graph neural network and harmonizing multiple evidence. *Scientific Reports*, 11(1), 23179.

Huang, B., Guo, L., Yin, H., Wu, Y., Zeng, Z., Xu, S., & Wang, X. (2024). Deep learning enhancing guide RNA design for CRISPR/Cas12a-based diagnostics. *Imeta*, 3(4), e214.

Jamal, S., Khubaib, M., Gangwar, R., Grover, S., Grover, A., & Hasnain, S. E. (2020). Artificial intelligence and machine learning based prediction of resistant and susceptible mutations in *Mycobacterium tuberculosis*. *Scientific Reports*, 10(1), 5487.

Jia, H., Li, X., Zhuang, Y., Wu, Y., Shi, S., Sun, Q., & Ruan, Z. (2024). Neural network-based predictions of antimicrobial resistance phenotypes in multidrug-resistant *Acinetobacter baumannii* from whole genome sequencing and gene expression. *Antimicrobial Agents and Chemotherapy*, 68(12), e01446-24.

Jumper, J., Evans, R., Pritzel, A., Green, T., Figurnov, M., Ronneberger, O., & Hassabis, D. (2021). Highly accurate protein structure prediction with AlphaFold. *Nature*, 596(7873), 583–589.

Kaissis, G. A., Makowski, M. R., Rückert, D., & Braren, R. F. (2020). Secure, privacy-preserving and federated machine learning in medical imaging. *Nature Machine Intelligence*, 2(6), 305–311.

Kapil, P., Soni, S., & Kumar, A. (2021). Spatiotemporal deep learning models for forecasting COVID-19 cases. *Neural Computing and Applications*, 33, 15365–15377.

Kassal, P., Kim, J., Kumar, R., et al. (2018). Smart bandages and wearable biosensors: The future of wound care. *Advanced Healthcare Materials*, 7(1), 1700883.

Kessler, H. H., Pierer, K., & Santner, B. I. (1999). Molecular diagnostics of infectious diseases: Advances in automation. *LaboratoriumsMedizin [Journal of Laboratory Medicine]*, 23(4), 207–215.

Khoury, M. J., & Ioannidis, J. P. (2014). Big data meets public health. *Science*, 346(6213), 1054–1055.

Kim, M., Kim, J. H., & Jang, B. (2024). Forecasting epidemic spread with recurrent graph gate fusion transformers. *IEEE Journal of Biomedical and Health Informatics*, 29(2), 1546–1559.

Kitagawa, H., Kojima, M., Tadera, K., Kogasaki, S., Omori, K., Nomura, T., & Ohge, H. (2025). Clinical diagnostic performance of droplet digital PCR for pathogen detection in patients with *Escherichia coli* bloodstream infection: A prospective observational study. *BMC Infectious Diseases*, 25(1), 22.

Kompella, R., Asher, M. J., Subramanian, D., & Pappas, G. J. (2021). Reinforcement learning for adaptive epidemic control. *Nature Communications*, 12(1), 4940.

Kostyusheva, A., Brezgin, S., Babin, Y., Vasilyeva, I., Glebe, D., Kostyushev, D., & Chulanov, V. (2022). CRISPR-Cas systems for diagnosing infectious diseases. *Methods*, 203, 431–446.

Kourou, K., Exarchos, T. P., Exarchos, K. P., Karamouzis, M. V., & Fotiadis, D. I. (2015). Machine learning applications in cancer prognosis and prediction. *Computational and Structural Biotechnology Journal*, 13, 8–17.

Kuchi Bhotla, H., Balasubramanian, B., Meyyazhagan, A., Pushparaj, K., Easwaran, M., Pappusamy, Alwin Robert, Arumugam, V. A., Tsibizova, V., Msaad Alfalih, A., Aljowaie, R.M., Saravanan, M., & Di Renzo, G. C. (2021). Opportunistic mycoses in COVID-19 patients/survivors: Epidemic inside a pandemic. *Journal of Infection and Public Health*, 14(11), 1720–1726.

Kuchi Bhotla, H., Kaul, T., Balasubramanian, B., Easwaran, M., Arumugam, V.A., Pappusamy, M., Muthupandian, S., & Meyyazhagan, A. (2020). Platelets to surrogate lung inflammation in COVID-19 patients. *Medical Hypotheses*, 143, 110098.

Lee, T.H., Liu, P.S., Wang, S.J., Tsai, M.M., Shanmugam, V., & Hsieh, H.L. (2021). Bradykinin, as a reprogramming factor, induces transdifferentiation of brain astrocytes into neuron-like cells. *Biomedicine*, 9(8), 923.

Liu, J., Bao, C., Zhang, J., Han, Z., Fang, H., & Lu, H. (2024). Artificial intelligence with mass spectrometry-based multimodal molecular profiling methods for advancing therapeutic discovery of infectious diseases. *Pharmacology & Therapeutics*, 263, 108712.

Liu, Q., Jin, X., Cheng, J., Zhou, H., Zhang, Y., & Dai, Y. (2023). Advances in the application of molecular diagnostic techniques for the detection of infectious disease pathogens. *Molecular Medicine Reports*, 27(5), 104.

Liu, Y., Xu, Y., Xu, X., Chen, X., Chen, H., Zhang, J., & Chen, J. (2023). Metagenomic identification of pathogens and antimicrobial-resistant genes in bacterial positive blood cultures by nanopore sequencing. *Frontiers in Cellular and Infection Microbiology*, 13, 1283094.

Lundberg, S. M., & Lee, S. I. (2017). A unified approach to interpreting model predictions. *10.48550/arXiv.1705.07874*.

Mao, J., Guan, S., Chen, Y., Zeb, A., Sun, Q., Lu, R., & Cao, D. (2023). Application of a deep generative model produces novel and diverse functional peptides against microbial resistance. *Computational and Structural Biotechnology Journal*, 21, 463–471.

McKee, M., van Schalkwyk, M. C. I., & Stuckler, D. (2021). AI in public health: Pitfalls and promises. *The Lancet Public Health*, 6(8), e452–e453.

McWeeney, S., Perciano, T., Susut, C., Chatterjee, L., Fornari, M., Biven, L., & Siwy, C. (2023). *Quantum computing for biomedical computational and data sciences: A joint DOE-NIH roundtable*. USDOE Office of Science (SC); Advanced Scientific Computing Research (ASCR); National Institutes of Health (NIH).

Mehrabi, N., Morstatter, F., Saxena, N., Lerman, K., & Galstyan, A. (2021). A survey on bias and fairness in machine learning. *ACM Computing Surveys (CSUR)*, 54(6), 1–35.

Metsky, H. C., Welch, N. L., Pillai, P. P., Haradhvala, N. J., Rumker, L., Mantena, S., & Sabeti, P. C. (2020). Designing viral diagnostics with model-based optimization. *bioRxiv*, 2020-11.

Meyyazhagan, A., Balasubramanian, B., Easwaran, M., Alagamuthu, K. K., Shanmugam, S., Kuchi Bhotla, H., Pappusamy, M., Arumugam, V. A., Thangaraj, A., Kaul, T., Keshavarao, S., & Cacabelos, R. (2020). Biomarker study of the biological parameter and neurotransmitter levels in autistics. *Molecular and Cellular Biochemistry*, 474(1–2): 277–284.

Meyyazhagan, A., Pushparaj, K., Balasubramanian, B., Kuchi Bhotla, H., Pappusamy, M., Arumugam, V. A., Easwaran, M., Pottail, L., Mani, P., Tsibizova, V., & Di Renzo, G. C. (2022). COVID-19 in pregnant women and children: Insights on clinical manifestations, complexities, and pathogenesis. *International Journal of Gynecology & Obstetrics*, 156(2), 216–224.

Miotto, R., Wang, F., Wang, S., Jiang, X., & Dudley, J. T. (2018). Deep learning for healthcare: Review, opportunities and challenges. *Briefings in Bioinformatics*, 19(6), 1236–1246.

Mitchell, M., Wu, S., Zaldivar, A., Barnes, P., Vasserman, L., Hutchinson, B., & Gebru, T. (2019, January). Model cards for model reporting. In *Proceedings of the Conference on Fairness, Accountability, and Transparency* (pp. 220–229).

Mittelstadt, B. D., Allo, P., Taddeo, M., Wachter, S., & Floridi, L. (2016). The ethics of algorithms: Mapping the debate. *Big Data & Society, 3*(2), 2053951716679679.

Miyachi, H. (1998). Molecular diagnostics by automated systems. *Rinsho byori. The Japanese Journal of Clinical Pathology, 46*(5), 413–419.

Mohd, Y., Balasubramanian, B., Meyyazhagan, A., Kuchi Bhotla, H., Shanmugam, S. K., Ramesh Kumar, M. K., Pappusamy, M., Alagamuthu, K. K., Keshavarao, S., & Arumugam, V. A. (2021). Extricating the association between the prognostic factors of colorectal cancer. *Journal of Gastrointestinal Cancer, 52*(3), 022–1028.

Mohd, Y., Kumar, P., Kuchi Bhotla, H., Meyyazhagan, A., Balasubramanian, B., Ramesh Kumar, M. K., Pappusamy, M., Alagamuthu, K. K., Orlacchio, A., Keshavarao, S., Sampathkumar, P., & Arumugam, V. A. (2021). Transmission jeopardy of adenomatosis polyposis coli and methylenetetrahydrofolate reductase in colorectal cancer. *Journal of the Renin-Angiotensin-Aldosterone System, 2021*, 7010706.

Mooney, S. J., & Pejaver, V. (2018). Big data in public health: Terminology, machine learning, and privacy. *Annual Review of Public Health, 39*(1), 95–112.

Mukundan, H. (2023). Pathogen agnostic biodetection at the point of need. In *Electrochemical Society meeting abstracts 244* (No. 63, p. 3015). The Electrochemical Society, Inc.

Mustafa, M. I., & Makhawi, A. M. (2021). Retracted: Sherlock and DETECTR: CRISPR-Cas systems as potential rapid diagnostic tools for emerging infectious diseases. *Journal of Clinical Microbiology, 59*(3). https://doi.org/10.1128/JCM.00745-20

Narayanan, S., Aakur, S. N., Ramamurthy, P., Bagavathi, A., Ramnath, V., & Ramachandran, A. (2022). Scalable pathogen detection from next generation DNA sequencing with deep learning. *arXiv preprint arXiv:2212.00015*.

Olaboye, J. A., Maha, C. C., Kolawole, T. O., & Abdul, S. (2024). Innovations in real-time infectious disease surveillance using AI and mobile data. *International Medical Science Research Journal, 4*(6), 647–667.

Otoum, S., Al Ridhawi, I., & Mouftah, H. T. (2021). Preventing and controlling epidemics through blockchain-assisted AI-enabled networks. *IEEE Network, 35*(3), 34–41.

Paranitharan, N., Kataria, S., Arumugam, V. A., Hsieh, H., Muthukrishnan, S., & Velayuthaprabhu, S. (2025). Integrin α1 upregulation by tf:fviia complex promotes cervical cancer migration through par2-dependent mek1/2 activation. *Biochemical and Biophysical Research Communications, 742*, 151151.

Pasomsub, E., Sukasem, C., Sungkanuparph, S., Kijsirikul, B., & Chantratita, W. (2010). The application of artificial neural networks for phenotypic drug resistance prediction: Evaluation and comparison with other interpretation systems. *Japanese Journal of Infectious Diseases, 63*(2), 87–94.

Phan, V. M., Nguyen, H. Q., Bui, K. H., Van Nguyen, H., & Seo, T. S. (2023). Combination of a robotic solution pipetting device with a centrifugal molecular diagnostic platform for fully automatic respiratory infectious virus detection from nasopharyngeal swab samples. *Sensors and Actuators B: Chemical, 394*, 134362.

Phillips, M., Molnár-Gábor, F., Korbel, J. O., Thorogood, A., Joly, Y., Chalmers, D., & Knoppers, B. M. (2020). Genomics: Data sharing needs an international code of conduct. *Nature, 578*(7793), 31–33.

Piwek, L., Ellis, D. A., Andrews, S., & Joinson, A. (2016). The rise of consumer health wearables: Promises and barriers. *PLoS Medicine, 13*(2), e1001953.

Pushparaj, K., Kuchi Bhotla, H., Arumugam, V. A., Pappusamy, M., Easwaran, M., Liu, W. C., Issara, U., Rengasamy, K. R. R., Meyyazhagan, A., & Balasubramanian, B. (2022). Mucormycosis (black fungus) ensuing COVID-19 and comorbidity meets – Magnifying global pandemic grieve and catastrophe begins. *Science of the Total Environment, 805*, 150355.

Quer, G., Radin, J. M., Gadaleta, M., Baca-Motes, K., Ariniello, L., Ramos, E., & Steinhubl, S. R. (2021). Wearable sensor data and self-reported symptoms for COVID-19 detection. *Nature Medicine, 27*(1), 73–77.

Rajkomar, A., Dean, J., & Kohane, I. (2019). Machine learning in medicine. *New England Journal of Medicine, 380*(14), 1347–1358.

Ramya, S., Poornima, P., Jananisri, A., Geofferina, I. P., Bavyataa, V., Divya, M., Priyanga, P., Vadivukarasi, J., Sujitha, S., Elamathi, S., Anand, A. V., & Balamuralikrishnan, B. (2023). Role of hormones and the potential impact of multiple stresses on infertility. *Stress, 3*(2), 454–474.

Ribeiro, M. T., Singh, S., & Guestrin, C. (2016). "Why should I trust you?" Explaining the predictions of any classifier. In *Proceedings of the 22nd ACM SIGKDD International Conference on Knowledge Discovery and Data Mining* (pp. 1135–1144).

Rudin, C. (2019). Stop explaining black box machine learning models for high stakes decisions and use interpretable models instead. *Nature Machine Intelligence, 1*(5), 206–215.

Sanches, P. H. G., de Melo, N. C., Porcari, A. M., & de Carvalho, L. M. (2024). Integrating molecular perspectives: Strategies for comprehensive multi-omics integrative data analysis and machine learning applications in transcriptomics, proteomics, and metabolomics. *Biology, 13*(11), 848.

Sangeetha, T., Anand, A. V., & Begum, T.N. (2022). Assessment of inter-relationship between anemia and COPD in accordance with altitude. *Open Respiratory Medicine Journal, 16*, e187430642206270.

Sangeetha, T., Nargis Begum, T., Balamuralikrishnan, B., Arun, M., Rengasamy, K. R. R., Senthilkumar, N., Velayuthaprabhu, S., Saradhadevi, M., Sampathkumar, P., & Vijaya Anand, A. (2022). Influence of SERPINA1 gene polymorphisms on anemia and chronic obstructive pulmonary disease. *Journal of the Renin-Angiotensin-Aldosterone System 2022*, 2238320.

Sendak, M. P., D'Arcy, J., Kashyap, S., Gao, M., Nichols, M., Corey, K., & Balu, S. (2020). A path for translation of machine learning products into healthcare delivery. *EMJ Innovations, 10*. https://doi.org/10.33590/emjinnov/19-00172

Senior, A. W., Evans, R., Jumper, J., Kirkpatrick, J., Sifre, L., Green, T., & Hassabis, D. (2020). Improved protein structure prediction using potentials from deep learning. *Nature, 577*(7792), 706–710.

Sganzerla Martinez, G., Garduno, A., Toloue Ostadgavahi, A., Hewins, B., Dutt, M., Kumar, A., & Kelvin, D. J. (2024). Identification of marker genes in infectious diseases from scrna-seq data using interpretable machine learning. *International Journal of Molecular Sciences, 25*(11), 5920.

Shabani, M., & Marelli, L. (2019). Re-identifiability of genomic data and the GDPR: Assessing the re-identifiability of genomic data in light of the EU General Data Protection Regulation. *EMBO Reports, 20*(6), e48316.

Shahin, F., Ishfaq, A., Asif, I., Bilal, A., Masih, S., Ashraf, T., & Ishfaq, R. (2024). CRISPR-Cas innovative strategies for combating viral infections and enhancing diagnostic technologies: CRISPR-Cas in viral diagnostics and therapeutics. *Journal of Health and Rehabilitation Research, 4*(3), 1–4.

Shanmugam, R., Thangavelu, S., Fathah, Z., Yatoo, M. I., Tiwari, R., Pandey, M. K., Dhama, J., Chandra, R., Malik, Y. S., Dhama, K., Sah, R., Chaicumpa, W., Shanmugam, V., & Arumugam, V. A. (2020). SARS-CoV-2/COVID-19 pandemic – An update. *Journal of Experimental Biology and Agricultural Sciences, 8*(Special Issue 1), S219–S245.

Shcherbina, A., Mattsson, C. M., Waggott, D., Salisbury, H., Christle, J. W., Hastie, T., & Ashley, E. A. (2017). Accuracy in wrist-worn, sensor-based measurements of heart rate and energy expenditure in a diverse cohort. *Journal of Personalized Medicine, 7*(2), 3.

Shi, J., Yan, Y., Links, M. G., Li, L., Dillon, J. A. R., Horsch, M., & Kusalik, A. (2019). Antimicrobial resistance genetic factor identification from whole-genome sequence data using deep feature selection. *BMC Bioinformatics, 20*, 1–14.

Shokr, A., Pacheco, L. G., Thirumalaraju, P., Kanakasabapathy, M. K., Gandhi, J., Kartik, D., & Shafiee, H. (2020). Mobile health (mHealth) viral diagnostics enabled with adaptive adversarial learning. *ACS Nano, 15*(1), 665–673.

Singh, N., & Pokhrel, S. R. (2025). Modeling quantum machine learning for genomic data analysis. *arXiv Preprint arXiv:2501.08193*.

Srisutham, S., Saralamba, N., Malleret, B., Rénia, L., Dondorp, A. M., & Imwong, M. (2017). Four human Plasmodium species quantification using droplet digital PCR. *PLoS One, 12*(4), e0175771.

Srivastava, V., Kumar, R., Wani, M. Y., Robinson, K., & Ahmad, A. (2025). Role of artificial intelligence in early diagnosis and treatment of infectious diseases. *Infectious Diseases, 57*(1), 1–26.

Sumon, M. A. A., Meregildo-Rodriguez, E. D., Lee, P. T., Dinh-Hung, N., Larson, E. T., Permpoonpattana, P., & Linh, N. V. (2024). Droplet digital PCR for fish pathogen detection and quantification: A systematic review and meta-analysis. *Journal of Fish Diseases, 47*(12), e14019.

Sunuwar, J., & Azad, R. K. (2022). Identification of novel antimicrobial resistance genes using machine learning, homology modeling, and molecular docking. *Microorganisms, 10*(11), 2102.

Tison, G. H., Sanchez, J. M., Ballinger, B., Singh, A., Olgin, J. E., Pletcher, M. J., & Marcus, G. M. (2018). Passive detection of atrial fibrillation using a commercially available smartwatch. *JAMA Cardiology, 3*(5), 409–416.

Tonekaboni, S., Joshi, S., McCradden, M. D., & Goldenberg, A. (2019). What clinicians want: Contextualizing explainable machine learning for clinical end use. In *Machine Learning for Healthcare Conference* (pp. 359–380). PMLR.

Torres, M. D., Zeng, Y., Wan, F., Maus, N., Gardner, J., & de la Fuente-Nunez, C. (2024). A generative artificial intelligence approach for antibiotic optimization. *bioRxiv*, 2024-11.

Tripathi, A., & Rathore, R. (2025). AI in disease surveillance – An overview of how AI can be used in disease surveillance and outbreak detection in real-world scenarios. In R Singh, A Gehlot, N Rathour, SV Akram (Eds.) *AI in Disease Detection: Advancements and Applications*, 337–359. https://doi.org/10.1002/9781394278695.ch15

Urrutia Iturritza, M., Mlotshwa, P., Gantelius, J., Alfvén, T., Loh, E., Karlsson, J., & Gaudenzi, G. (2024). An automated versatile diagnostic workflow for infectious disease detection in low-resource settings. *Micromachines*, *15*(6), 708.

Uzay, İ. A., & Dinçer, P. (2022). CRISPR-based approaches for the point-of-care diagnosis of COVID19. *Acta Medica*, *53*(1), 1–14.

Varghese, J. E., Shanmugam, V., Rengarajan, R. L., Meyyazhagan, A., Arumugam, V. A., Al-Misned, F. A., & El-Serehy, H. A., (2020). Role of vitamin D_3 on apoptosis and inflammatory-associated gene in colorectal cancer: An *in vitro* approach. *Journal of King Saud University – Science*, *32*(6), 2786–2789.

Vatti, J. J., Nguyen, T. T., & Morishita, H. (2021). AI and bioinformatics approaches for predicting viral evolution and escape. *Frontiers in Genetics*, *12*, 732859.

Velayuthaprabhu, S., & Archunan, G. (2005). Evaluation of anticardiolipin antibodies and antiphosphatidylserine antibodies in women with recurrent abortion. *Indian Journal of Medical Sciences 59*, 347–352.

Velayuthaprabhu, S., Archunan, G., & Balakrishnan, K. (2007). Placental thrombosis in experimental anticardiolipin antibodies-mediated intrauterine fetal death. *American Journal of Reproductive Immunology*, *57*, 270–276.

Velayuthaprabhu, S., Matsubayashi, H., Archunan, G. (2016). Beta-2 GPI induced tissue factor and placental apoptosis for the pathophysiology of pregnancy loss in antiphospholipid syndrome. *International Journal of Research in Medical Sciences*, *4*(8), 3109–3113.

Velayuthaprabhu, S., Matsubayashi, H., Sugi, T., Nakamura, M., Ohnishi, Y., Ogura, T., & Archunan, G. (2013). Expression of apoptosis in placenta of experimental antiphospholipid syndrome (APS) mouse. *American Journal of Reproductive Immunology*, *69*(5), 486–494.

Velayuthaprabhu, S., Matsubayashi, H., Sugi, T., Nakamura, M., Ohnishi, Y., Ogura, T., Tomiyama, T., & Archunan, G. (2011). A unique preliminary study on fetal resorption and placental apoptosis in mice with passive immunization of anti-phosphatidylethanolamine antibodies and anti-Factor XII antibodies. *American Journal of Reproductive Immunology*, *66*, 373–384.

Wang, C., Garlick, S., & Zloh, M. (2021). Deep learning for novel antimicrobial peptide design. *Biomolecules*, *11*(3), 471.

Wang, F., Lei, X., Liao, B., & Wu, F. X. (2022). Predicting drug–drug interactions by graph convolutional network with multi-kernel. *Briefings in Bioinformatics*, *23*(1), bbab511.

Ward, R. A., Aghaeepour, N., Bhattacharyya, R. P., Clish, C. B., Gaudillière, B., Hacohen, N., & Vyas, J. M. (2021). Harnessing the potential of multiomics studies for precision medicine in infectious disease. In *Open forum infectious diseases* (Vol. 8, p. ofab483). Oxford University Press.

Wesolowski, A., Buckee, C. O., Bengtsson, L., Wetter, E., Lu, X., & Tatem, A. J. (2014). Commentary: Containing the Ebola outbreak-the potential and challenge of mobile network data. *PLoS Currents*, *6*.

World Health Organization. (2024). *Ethics and governance of artificial intelligence for health: Large multi-modal models. WHO guidance*. World Health Organization.

World Health Organization Guidance (2021). *Ethics and governance of artificial intelligence for health*. World Health Organization.

Xia, Y., Rao, R., Xiong, M., He, B., Zheng, B., Jia, Y., & Yang, Y. (2024). CRISPR-powered strategies for amplification-free diagnostics of infectious diseases. *Analytical Chemistry*, *96*(20), 8091–8108.

Yakimovich, A. (2021). Machine learning and artificial intelligence for the prediction of host–pathogen interactions: A viral case. *Infection and Drug Resistance*, 3319–3326.

Yang, D., Fang, Y., Ma, J., Xu, J., Chen, Z., Yan, C., & Zhang, F. (2023). Nano-bio-analytical systems for the detection of emerging infectious diseases. In *Surface engineering and functional nanomaterials for point-of-care analytical devices* (pp. 147–171). Springer Nature Singapore.

Yang, H., Zhang, Y., Teng, X., Hou, H., Deng, R., & Li, J. (2023). CRISPR-based nucleic acid diagnostics for pathogens. *TrAC Trends in Analytical Chemistry*, *160*, 116980.

Ye, S., Wang, J., Zhu, M., Yuan, S., Zhuo, L., Chen, T., & Gao, J. (2024). MKAN-MMI: Empowering traditional medicine-microbe interaction prediction with masked graph autoencoders and KANs. *Frontiers in Pharmacology*, *15*, 1484639.

Yenduri, L. K., Chaithra, M. H., Shahedhadeennisa, S., Annamalai, M., Rajesh, L., & Imran, S. M. (2025). Quantum computing and machine learning for transforming precision medicine and drug discovery. In *Modern super-hypersoft computing trends in science and technology* (pp. 339–366). IGI Global Scientific Publishing.

Yu, H., Wang, R., Qiao, J., & Wei, L. (2023). Multi-CGAN: Deep generative model-based multiproperty antimicrobial peptide design. *Journal of Chemical Information and Modeling*, 64(1), 316–326.

Zaghlool, S. B., & Attallah, O. (2022). A review of deep learning methods for multi-omics integration in precision medicine. In *2022 IEEE International Conference on Bioinformatics and Biomedicine (BIBM)* (pp. 2208–2215). IEEE.

Zhang, Y., Guo, Y., Liu, G., Zhou, S., Su, R., Ma, Q., & Wang, G. (2024). Portable all-in-one microfluidic system for CRISPR–Cas13a-based fully integrated multiplexed nucleic acid detection. *Lab on a Chip*, 24(14), 3367–3376.

Zhao, Y., Fu, Y. X., & Ma, J. (2020). Machine learning approaches for predicting cancer recurrence. *Current Opinion in Systems Biology*, 24, 27–35.

Zhao, Y., Xu, D., Ke, F., Zhou, Y., Li, M., & Gui, L. (2024). A droplet digital PCR method (ddPCR) for sensitive detection and quantification of Carassius auratus herpesvirus (CaHV). *Water Biology and Security*, 3(2), 100252.

Zhou, M., Wu, Y., Kudinha, T., Jia, P., Wang, L., Xu, Y., & Yang, Q. (2021). Comprehensive pathogen identification, antibiotic resistance, and virulence genes prediction directly from simulated blood samples and positive blood cultures by nanopore metagenomic sequencing. *Frontiers in Genetics*, 12, 620009.

Zhu, S., Gilbert, M., Chetty, I., & Siddiqui, F. (2022). The 2021 landscape of FDA-approved artificial intelligence/machine learning-enabled medical devices: An analysis of the characteristics and intended use. *International Journal of Medical Informatics*, 165, 104828.

7 AI-Assisted Strategies for Combating Antimicrobial Resistance and Healthcare-Associated Infections

Stuti Ghosh and Sudipto Saha

7.1 INTRODUCTION

Huge amounts of whole genome sequencing (WGS) datasets of antimicrobial resistance (AMR) bacteria and fungi clinical isolates were generated and cataloged into databases, which are a rich source to apply artificial intelligence (AI) to understand the resistance mechanisms behind it. Antibiotics, antifungals, antivirals, disinfectants, and food preservatives are examples of antimicrobial agents that either inhibit or eradicate the growth and multiplication of microorganisms (Larkin 2023). The phenomenon of AMR develops as a survival mechanism for different microbial pathogens to thrive in the presence of different antimicrobial agents (Dilnessa et al. 2022). The potential causes behind the emergence of AMR can be several, like the inability of antibiotics to penetrate bacterial membranes, changes in the targets of the antimicrobials, and a reduction in the potency of the antibiotic due to its degradation (Baker et al. 2018). In the case of poultry animals, AMR largely results from antibiotic overuse in cattle and poultry production (Shang et al. 2023). This is also a major source of healthcare-related infections in humans. The AMR pathogens like *Mycobacterium tuberculosis* (MTB), a group of six pathogenic bacteria called ESKAPE (pathogenic microbes, namely *Enterococcus, Staphylococcus, Klebsiella, Acinetobacter, Pseudomonas* and *Enterobacter*), which are responsible for causing hospital-acquired infections, are a high-risk fatality to humans (Ghosh et al. 2020). The End TB strategy by the World Health Organization (WHO) aims to reduce TB incidence and the resulting mortality rates by 90% by 2035, but this goal is under threat due to the emergence of multidrug-resistant (MDR) and extreme drug-resistant (XDR) MTB (Chakaya et al. 2020). For fungal infections, the issue of AMR has significantly affected the incidence rates for fungal disease aspergillosis, caused by *Aspergillus fumigatus* (Hui et al. 2024). AMR is currently one of the most complex issues related to public health globally, as it can be influenced by trade, finance, improper waste management, and medications (Samreen et al. 2021). Each year, around one million people die due to AMR causes worldwide. With no potential antibiotics left, standard hospital-based treatments like surgical procedures are raising infection rates and thus forming grounds for the spread of AMR (Hu et al. 2020). In this scenario, AI is assisting in combating this deadly healthcare-associated problem by using the huge amount of sequencing data of AMR pathogens' WGS datasets and surveillance data, by predicting the AMR pathogen clinical isolates, and also helping in developing novel drugs.

AI has been used to address several problems posed by AMR pathogens. It is challenging to identify the specific AMR bacteria, fungi, or other pathogens using drug susceptibility testing (DST) based methods. It is tedious and time-consuming and requires expertise. AI plays a substantial role in identifying drug-resistant bacteria or fungi using complex genomics and metagenomics datasets. AI allows for the identification of AMR genes, which can be used as an indicator for surveillance of AMR in the environment and human microbiomes. The AMR problem poses a fundamental threat to infections caused by several pathogenic species, and the rapid spread of this issue calls for novel therapeutic approaches to combat AMR. AI has been used in novel drug discovery, drug repurposing, and the identification of alternative strategies, such as antimicrobial peptide development. In this chapter, emphasis was given to AI-based applications for addressing several issues in AMR – i.e., from AMR pathogen clinical isolates prediction to elucidating novel antibiotic combinations and discovering antimicrobial peptides (AMPs) for controlling AMR.

7.2 AI/ML/DL BASICS

AI is a subset of computer science that enables machines to carry out tasks traditionally done by humans. Machine learning (ML), deep learning (DL), and natural language processing (NLP) are subsets of AI. ML, DL, and NLP enable computers to learn from data effectively without being specifically programmed. Examples of AIs that are not ML/DL are symbolic logic-rule engines, expert systems, and knowledge graphs. ML can be categorized into different types like supervised learning, unsupervised learning, semi-supervised learning, and reinforcement learning. In supervised learning, the model is trained on data with appropriate labels, whereas in the case of unsupervised learning, training is done on unlabeled data with hidden patterns (Sarker 2021). Random forest (RF), support vector machine (SVM), k-nearest neighbor (kNN), and regression-based ML models are examples of supervised learning. Unsupervised models make use of hierarchical and k-means clustering, as well as dimensionality reduction tests like principal component analysis (PCA),

DOI: 10.1201/9781003615699-7

t-distributed stochastic neighbor embedding (tSNE), and non-negative matrix factorization (NMF). ML finds huge applications in classification and regression tasks (Sarker 2021). Convolutional neural network (CNN) and recurrent neural network (RNN) are included under DL. Analysis of image data is one important application of DL (Sakagianni et al. 2023). In the case of NLP, the application mainly covers text mining and sentiment analysis. Text mining is largely used for the analysis of literature notes (Rahul et al. 2020). For classification-based models, the model evaluation is mainly based on threshold-dependent metrics like accuracy, and sensitivity/recall, whereas the threshold-independent metric is AUC-ROC. On the other hand, for regression-based models, the evaluation metrics are mainly root mean square error and R2 score (Hicks et al. 2022). A confusion matrix is usually constructed for ML model evaluation, which provides overall information on the predictions correctly performed by the model (Hicks et al. 2022). AI-based approaches find huge applications in drug development and drug repurposing. AI-based mining of genomic and metabolomic data serves as the important step toward discovery of novel antimicrobial compounds. AI-based virtual screening and molecular docking give useful insights into the probable binding potency of drugs to their targets (Singh 2024). Several applications of AI for combating AMR are shown in Figure 7.1. These include the prediction of AMR pathogen clinical isolates and the

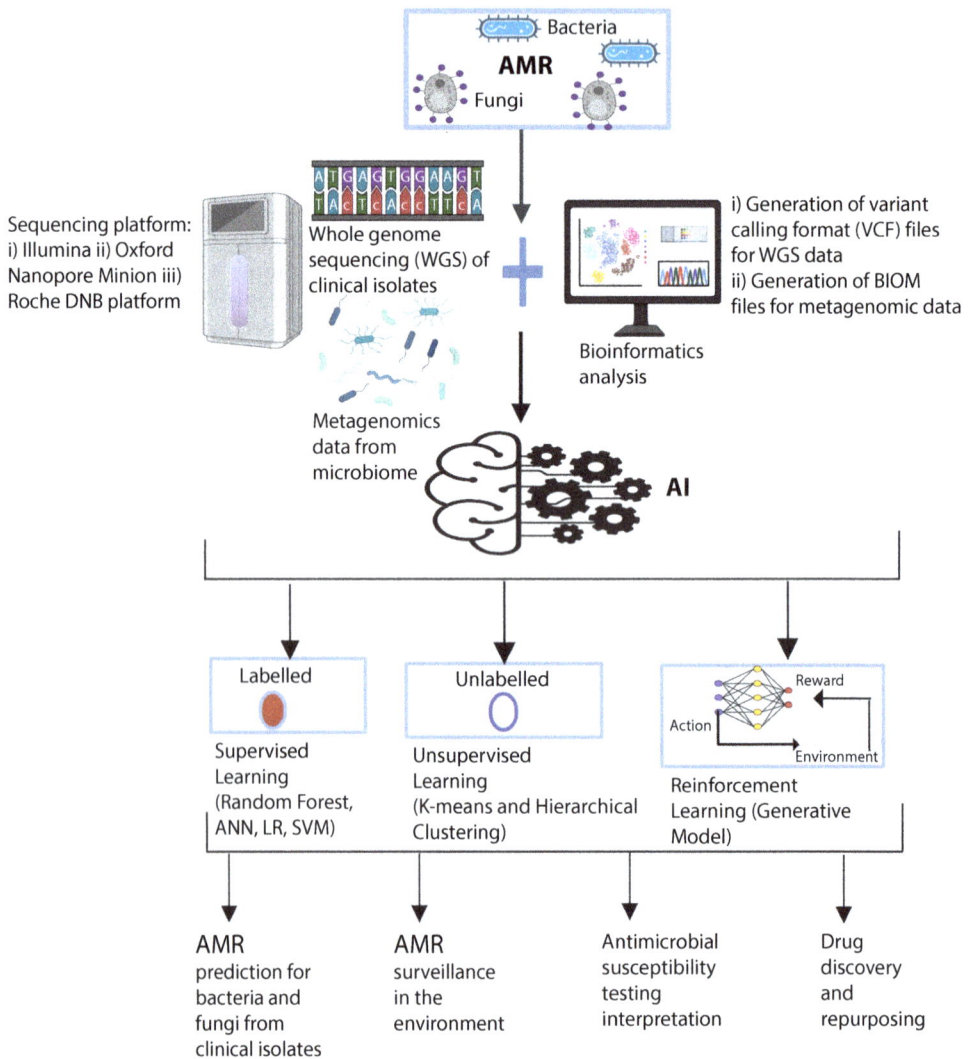

Figure 7.1 Applications of AI-based approaches to address antibiotic-resistant bacteria (ARB) in health-associated infections.

identification of AMR genes. In addition, AI is used in AMR surveillance, novel drug development, and alternative strategies like AMPs against AMR pathogens. The details of each AI-assisted application in fighting against AMR are provided in Figure 7.1.

7.3 AI-ASSISTED PREDICTION OF ARB CLINICAL ISOLATES AND IDENTIFICATION OF ARGs

AI models were developed to predict ARB and antibiotic-resistant fungi (ARF) clinical strains using mutational patterns of the genes and intergenic regions captured in the WGS datasets. Mutational information of the drug resistance-associated genes is available for ARB and ARF in several databases. These curated databases are particularly useful in studying AMR protein functions (Ren et al. 2022). There are several focused databases available for ARB like PATRIC (Davis et al. 2019), Comprehensive Antibiotic Resistance Database (CARD) (McArthur et al. 2013), and ResFinderFG v2.0 (Gschwind et al. 2023). In addition to these, there are a few databases focused on AMR in *Mycobacterium tuberculosis* (MTB), for example, MUBII-TB-DB (Flandrois et al. 2014), and Tuberculosis Drug Resistance Mutation Database (TBDReaMDB) (Sandgren et al. 2009). There is another database called DRAGdb (Ghosh et al. 2020), which provides mutation information for ARB in different bacterial pathogen groups like MTB and ESKAPE. Similarly, there are several databases for ARF, like AFRbase (Jain et al. 2023), MARDy (Nash et al. 2018), and ResFungi (Santana De Carvalho et al. 2024). Besides these, there are primary databases like NCBI where the WGS and Whole genome metagenome sequencing (WGMS) datasets of ARB and ARF are available. These datasets are processed using bioinformatics tools to generate the features, which can be used as inputs for ML/DL. In the case of WGS of ARB/ARF, the features are single-nucleotide polymorphisms (SNPs), insertions, and deletion mutations (INDELS), which captured in variant calling format (VCF). Similarly, bioinformatics tools are used to process the WGMS datasets to generate BIOM file, which includes operational taxonomic unit (OTU) information. ML models are applied to metagenomics data using OTU information (Wani et al. 2022). The nucleotide sequencing technologies are used to capture the mutational patterns in the genome of AMR pathogens, which are used in the AI/ML/DL models as discriminating features.

Several studies are showing the applications of AI-based approaches in WGS datasets for the prediction of ARB clinical isolates. For the prediction of ARB from genomic data, the input data type for ML includes mutation, insertion, and deletion of nucleotide information in the form of VCF files (Kuang et al. 2022). In a study, using WGS data of several clinical isolates of *Escherichia coli* Ren et al. employed ML models trained with classification-based algorithms such as SVM, RF, and CNN, where RF and CNN were found to perform better, to obtain insights into resistance against various antibiotics, including ceftazidime (CTZ), cefotaxime (CTX), ciprofloxacin (CIP), and gentamicin (GEN) (Ren et al. 2022). In another study, using MTB clinical isolates, different algorithms like classification-based CNN, RF, and regression-based logistic regression were used for the prediction of resistances against different anti-TB drugs like ethambutol, pyrazinamide, rifampicin, isoniazid, fluoroquinolone (Kuang et al. 2022). A study highlighted the use of the PATRIC dataset for the construction of an ML model based on classification called Ada Boost, used for the prediction of resistance against different drugs like methicillin, carbapenem, and β-lactam, in the case of several pathogenic bacteria like *Staphylococcus aureus*, *Streptococcus pneumoniae*, and *Acinetobacter baumanii* (Davis et al. 2019). Dahl et al. performed an ML-based study to find antibiotic resistance genes (ARGs) in AMR *Campylobacter jejuni* clinical isolates, which are responsible for causing foodborne disease gastroenteritis in humans. The results indicated the importance of three genes – namely, gyrA, 23S rRNA, and rpsL – in contributing to AMR in *Campylobacter jejuni* clinical isolates (Dahl et al. 2021).

A few studies have applied AI-based tools to metagenome data to predict ARB and ARGs in bacteria. AI/ML-based methods form a strong avenue for studying the interrelations between alterations of microbiome and AMR acquisition. Such insights may hugely help in elucidating novel methods for combating the problem of AMR (Rahman et al. 2018). Using microbiome data for ML-based studies is of high significance in providing targeted therapeutics for controlling AMR (Manoharan et al. 2021). There are certain platforms available for predicting AMR using metagenomic data. The most popular one in this aspect is the DL-based approach DeepARG. The utility of DeepARG remains an effective application to determine ARGs in cases of resistance against around 30 different antibiotics (Arango-Argoty et al. 2018). A structured ARG database called SARG was utilized for the construction of an online analysis platform ARGs-OAP for antibiotic resistance prediction from metagenomic datasets (Yang et al. 2016). In another study, Rahman et al. depicted the application of the RF in the prediction of ARGs in the gut microbiome of infants. Application of ML to metagenomic data from the gut microbiome was used to determine the drug resistance genes against the antibiotics cephalosporin and vancomycin (Rahman et al. 2018). Another interesting study showed the application of ML approaches to determine ARGs from the surface microbiome of the space stations. This study led to the identification of ARGs from the bacterial strains *Kalamiella piersonii, Bacillus cereus*, and *Enterobacter bugandensis* (Madrigal et al. 2022). Table 7.1 depicts the different applications of AI in the prediction of ARB using WGS and metagenomics datasets. Thus, several studies indicate the importance of ML and DL approaches to predict and interpret AMR in the case of several bacterial species. AI-based approaches can play an important role in helping clinicians make precise decisions regarding effective drug choices (Feretzakis et al. 2020).

In spite of the useful applications of ML in AMR prediction using genomic or metagenomic datasets, there are certain challenges associated with their practical implementation. A major shortcoming remains for ML applications where the resistant phenotype results from a combination of genetic features, where the AMR profile may show an intermediate phenotype – i.e., neither resistant nor susceptible (Heydari et al. 2022). Under these circumstances, class prediction based on ML approaches remains a major challenge. In these cases, ML models should be made to incorporate another category or class called the intermediate (Wang et al. 2022). The other hurdle remaining with applying ML-based approaches is the availability of a low amount of genomics data on AMR in fungi and ESKAPE pathogens (Anahtar et al. 2021).

7.4 AI-ASSISTED AMR SURVEILLANCE

AI has a role in the antimicrobial resistance surveillance network, where the ARGs are used as an indicator of pathogenic ARB, and public health records monitor the spread of AMR. The process of surveillance refers to the method of regular collection, analysis, and interpretation of public health data in order to devise appropriate strategies for health monitoring. This involves a time-dependent study of health patterns in order to reduce the risks of AMR outbreaks and ensure efficient management of the issue (Shah & Househ 2024). AMR indicators refer to certain features or genetic factors that are responsible for incorporating the capacity for resisting antibiotics into microbes. These indicators are important for AMR surveillance. They include bacterial pathogens like *Escherichia coli* (Hutinel et al. 2019), *Kbebsiella pneumoniae* (Zhou et al. 2022), and *Staphylococcus aureus* (Zhou et al. 2022). The occurrence of *E.coli* in water and sewage indicates that these are high sources for AMR spreading in the environment (Hutinel et al. 2019). The bacteria *Klebsiella pneumoniae* and *Staphylococcus aureus* belong to the ESKAPE pathogen group and contribute to a major proportion of hospital-acquired infections (Inda-Díaz et al. 2023). Thus, the presence of these pathogens in the hospital wastes and their spread throughout the environment is a strong indicator of AMR prevalence. Several AMR genes possess a drug resistance-determining region (DRDR) (Sunuwar & Azad 2022). These regions carry potential identifiers for AMR. Thus AMR markers include pathogenic microbes, ARGs disseminated in the environment, and also DRDR regions of certain AMR genes.

There are a few AMR surveillance networks available for the proper monitoring of AMR outbreaks. Constant monitoring of AMR in the form of AMR surveillance needs to be conducted in poultry, hospitals, sewage, and the entire environment. The environment serves as grounds for the sources of antimicrobial resistance genes or ARGs (Inda-Díaz et al. 2023). Monitoring of AMR in the environment is really important for sharpening our understanding of the various routes of spread for AMR-causing microbial pathogens (United Nations Environment Programme 2023). A few examples include EARS-Net or the European Antimicrobial Resistance Surveillance Network (Waterlow et al. 2024), NARMS or the National Antimicrobial Resistance Monitoring System (Karp et al. 2017), NARS-Net or the National AMR Surveillance Network in India (Walia et al. 2019). Another surveillance network called GLASS, focuses on the global AMR data (Ajulo & Awosile 2024). ML-based approaches find usage in the identification of AMR trends and healthcare-associated infections in hospital settings. This is achieved by training large datasets, which include clinical data reports, and genomic information of patients using different algorithms like RF, and SVM. Clustering Algorithms are efficient methods for the prediction of AMR outbreaks. On the other hand, DL-based approaches like CNN and RNN are quite useful for the sequential analysis of patient data. Time series analysis in AMR surveillance involves the prediction of AMR trends depending on the use of antibiotics over a certain time period. Statmodels and Prophet are the two tools useful for this purpose (Branda 2024). There are also certain AI-based tools available for real-time monitoring of AMR outbreaks. One such example includes PathogenWatch, used for the analysis of the microbial genome sequences for the efficient monitoring of probable pathogen infections (Vashisht et al. 2023). Thus, AI-based tools are successfully used for proper AMR surveillance in network-based projects to prevent AMR outbreaks.

TABLE 7.1 AI-assisted Models in the Prediction of ARB from Whole Genome and Metagenome Sequencing Data

Pathogen	Data Type	Antibiotics	Sample Size	ML Algorithm	Performance Metrics	Reference
Actinobacillus pleuropneumoniae	WGS	Ampicillin, Sulfisoxazole, Trimethoprim, Tetracycline and Enrofloxacin	96	SVM, SCM	Accuracy (%) >90	(Liu et al. 2020)
Escherichia coli	WGS	Gentamicin (GEN), Cefotaxime (CTX), Ceftazidime (CTZ), and Ciprofloxacin (CIP)	987 (Giessen data) 1509 (public dataset)	RF, SVM, LR, CNN	AUC CIP:RF (0.96) CTX:RF (0.82) CTZ:RF (0.93) GEN:RF (0.95)	(Ren et al. 2022)
Mycobacterium tuberculosis	WGS	Ethambutol (EMB), Pyrazinamide (PZA), Isoniazid (INH), Rifampicin (RIF)	10,575 MTB clinical isolates	RF, LR, CNN, Mykrobe	Accuracy (%) EMB:CNN(90) PZA:CNN(90.5) INH:CNN and Mykrobe(96.2) RIF:CNN(94.6)	(Kuang et al. 2022)
Salmonella enterica	WGS	Clavulanic Acid, Ampicillin, Cefoxitin, Amoxicillin, Ceftriaxone, Streptomycin, and Tetracycline	97	LR	Accuracy (%) 92 to 99	(Maguire et al. 2019)
Salmonella enterica	WGS	Ampicillin, Azithromycin, Amoxicillin-Clavulanicacid, Cefoxitin, Ceftiofur, Ceftriaxone, Ciprofloxacin, Chloramphenicol, Gentamicin, Kanamycin, Nalidixic Acid, Sulfisoxazole, Streptomycin, Tetracycline, Trimethoprim-Sulfamethoxazole	5278	XGBoost	Accuracy (%) 95	(Nguyen et al. 2019)
Staphylococcus aureus	WGS	Cefoxitin, Vancomycin, Daptomycin, Erythromycin, Gentamicin, Tetracycline, Ciprofloxacin, Clindamycin, Chloramphenicol, Linezolid	673	LR, SVM	Accuracy (%) >90	(Wang et al. 2021)
Kalamiella piersonii, Enterobacter bugandensis, Bacillus cereus	Metagenomic	Macrolides, Lincosamides, Tetracycline, Fluoroquinolones	24	DL		(Madrigal et al. 2022)
Enterobacteriaceae, Bifidobacterium spp., Escherichia coli, Clostridium difficile.	Metagenomic	Vancomycin, Cephalosporin	107	RF		(Rahman et al. 2018)

141

7.5 AI-ASSISTED INTERPRETATION OF ANTIMICROBIAL SUSCEPTIBILITY TESTING (AST) AND PREDICTION OF MINIMUM INHIBITORY CONCENTRATION (MIC) OF ANTIMICROBIALS

Several studies have used AI for the interpretation of antimicrobial susceptibility testing (AST). A thorough understanding of a patient's antimicrobial susceptibility test or AST profile is very crucial for the development of a precise therapy plan. AI finds huge applications in the evaluation of anti-biograms in the comprehension of the AST profile. Médecins Sans Frontières (MSF) used a combination of ML and image processing to create an AI-based application for the understanding of disk diffusion ASTs from pictures (Pascucci et al. 2021). In another study, an AI-based clustering model was developed for inhibition zone measurements in the case of antibiogram susceptibility tests. This was followed by the classification of the antibiotics using a deep-learning CNN network. The "Roboflow" online platform was used for developing a DL model (Gullu et al. 2024).

Similarly, there are reports of the application of AI regression models in predicting the minimum inhibitory concentration (MIC) of antimicrobials. A study by Pataki et al. used a logistic regression (LR) model to predict MIC for *Escherichia coli* clinical isolates (Pataki et al. 2020). MICs in Salmonella species were determined by the application of ML algorithms like RF, Multilayer Perceptron (MLP), and Deeplift (Ayoola et al. 2024). A study by Jeon et al. used ML to predict MICs for Methicillin-resistant *Staphylococcus aureus* (MRSA) isolates from the peaks obtained from the MALDI-TOF experiment (Jeon et al. 2022). ML also finds applications in assisting MIC measurements using the agar dilution method. An ML-based tool, AlgarMIC, is used in quick MIC measurements for the pathogens *Escherichia coli* and Enterobacteriaceae (Gerada et al. 2024). These AI-assisted tools are very useful for combating AMR and have direct applications in clinical settings.

7.6 AI-ASSISTED ANTIMICROBIAL DISCOVERY AND DRUG REPURPOSING APPROACHES

AI has been successfully used to discover newer potent antimicrobial drugs to fight against the emerging AMR problem. By combining biological knowledge with AI, new compounds were discovered against AMR pathogens. AI is used in several steps of drug discovery: (i) data mining of genomics, metagenomics, and proteomics data using ML-based approaches like SVM, RF, and NLP; (ii) predictive modeling, i.e., biological target identification of the potential drug molecules using ML approaches like linear regression, RF, and DL; and (iii) formation of generative models using generative adversarial network and variational autoencoder (VAE) (Bilal et al. 2025). A study by Stokes et al. showed the application of a deep neural network-based model for the identification of novel compounds with antibacterial potential (Stokes et al. 2020), where 2335 *E. coli* growth inhibitor compounds were used for the training of the neural network. This technique led to the discovery of halicin, which is a structurally unique compound with antibacterial efficacy against pathogens, including carbapenem-resistant *Mycobacterium tuberculosis* and *Enterobacteriaceae* (Stokes et al. 2020). In another study, Zheng et al. developed a novel technique to work on stationary-phase bacteria, combining a dilution-regrowth screen with ML to realistically screen chemicals across a larger chemical space. This ML-based technique led to the discovery of bactericidal compound Semapimod, showing activity against AMR pathogens like *Escherichia coli* and *Acinetobacter baumanii* bacteria (Zheng et al. 2024). Probable drug candidates with antimicrobial activities against AMR pathogens can be identified using an AI-based virtual screening approach followed by ML-powered finding of their antimicrobial potencies (Yang et al. 2019). Some of the drugs discovered using AI-based approaches include SPR206 (Zhang et al. 2020), QPX9003 (Castanheira et al. 2019), MRX-8 (Duncan et al. 2022), and Murepavadin (Sader et al. 2018), showing potencies against different AMR pathogens namely, *Acinetobacter baumanii*, *Pseudomonas aeruginosa*, *Klebsiella pneumoniae* belonging to ESKAPE group. Thus, ML-based approaches find a huge application in identifying potential drug candidates with significant antimicrobial properties. AI expedites target identification, followed by antimicrobial agent discovery, and finally preclinical and clinical development (Macesic et al. 2017).

AI also finds immense importance in drug repurposing – i.e., finding new applications for the drugs currently in use. AI-assisted drug repurposing has shown promising results in controlling AMR (Farha & Brown 2019). In the current scenario of AMR, finding novel applications of already known drugs may be a useful solution, as there is a huge shortage of novel drugs available to treat healthcare-related infections (Farha & Brown 2019). Virtual screening and molecular docking-based AI approaches may show promising results in reducing the off-target effects of drugs (Pun et al. 2023). AI-based repurposing of the drug halicin showed a promising result in its targeting a wide range of pathogens namely, *Mycobacterium tuberculosis*, *Enterobacteriaceae*, *Clostridium*, and

TABLE 7.2 AI-Assisted Novel or Repurposed Drugs against ARB

Drug	AI Model	AMR Pathogen	Reference
Halicin	Deep neural network	*Mycobacterium tuberculosis* *Enterobacteriaceae*	(Stokes et al. 2020)
Semapimod	Machine learning	*Escherichia coli* *Acinetobacter baumanii*	(Zheng et al. 2024)
Chlorambucil	AI-based virtual screening	*Escherichia coli*	(Campbell et al. 2019)
Carmustine	AI-based virtual screening	*Escherichia coli*	(Hua et al. 2022)
Streptozotocin	AI-based virtual screening	*Staphylococcus aureus*	(Alluri et al. 2023)

Acinetobacter baumannii showing AMR (Booq et al. 2021). Drug repurposing could be shown as an efficient approach in the fight for new antimicrobials to treat infections by several bacterial pathogens already showing AMR. The list includes several bacterial species like *Staphylococcus aureus*, *Streptococcus pneumoniae, Escherichia coli, Pseudomonas aeruginosa*, and *Mycobacterium tuberculosis*. The aforementioned species depict resistance against different drugs like vancomycin, cotrimoxazole, cephalosporin, carbapenem, tetracyclines, and fluoroquinolones (Serwecińska 2020). Drugs repurposed using AI approaches have shown promising results against different pathogen groups depicting AMR. The list includes several drugs like chlorambucil (Campbell et al. 2019), carmustine, and lomustine (Hua et al. 2022), showing activity against the AMR pathogen *Escherichia coli*. The list expands to streptozotocin targeting AMR pathogen *Staphylococcus aureus* (Alluri et al. 2023); busulfan showing activity against AMR pathogenic bacteria *Staphylococcus aureus, Enterococcus faecium*, and *Pseudomonas aeruginosa*; drugs mitomycin C (Cruz-Muñiz et al. 2017), doxorubicin (Lorente-Torres et al. 2024) and 5-fluoro uracil (Lorente-Torres et al. 2024) targeting the ESKAPE pathogenic group; acridine derivative 9-amino acridine showing activity against XDR pathogen *Klebsiella pneumoniae* (She et al. 2023); and the antifungal drug ciclopirox repurposed against *Klebsiella pneumoniae, Acinetobacter baumanii*, and *Escherichia coli* (Carlson-Banning et al. 2013). Table 7.2 shows a list of AI-assisted novel antimicrobials and repurposed drugs. Thus, drug repurposing is an important application of AI and can be applied to target several pathogenic bacteria exhibiting AMR. The AI-assisted novel antimicrobials and repurposed drugs are helping physicians to apply them in AMR emergency cases in hospital settings, as well as in public health.

7.7 AI-ASSISTED POTENT ANTIMICROBIAL PEPTIDES

The alternative strategy of antibiotics is the use of antimicrobial peptides (AMPs), where AI is used to develop potent AMP. These are bacterial cell membrane targeting molecules, which are small peptides with antimicrobial activities. The mode of action of the AMPs primarily involves membrane permeabilization followed by disruption of the membrane. AI has shown a promising approach in the development of novel AMPs. There are several online platforms available, namely, DL-based Deep-AmPEP30 (Yan et al. 2020), ML-based webserver IAMPE (Kavousi et al. 2020), deep network-based DeepACP (Yu et al. 2020). There can be novel AMP designing or AMP repurposing from modifications of known AMPs to generate peptides with more potent antimicrobial activities. AI-based repurposing of AMP HH2 led to the identification of DP7, which showed more potency against the *Staphylococcus aureus* pathogen. A study by Ma et al. used three NLP neural network models to design 181 novel AMPs showing potencies against different ESKAPE pathogens (Ma et al. 2022). In another study, 44 new AMPs were generated by Yoshida et al. using AI-based approaches and showing activity against AMR bacteria *Escherichia coli* (Yoshida et al. 2018). AI-based approaches were also used for the generation of ten new AMPs by Nagarajan et al. (Nagarajan et al. 2018). DL algorithms were used by Huang et al. (Huang et al. 2023) to generate around 54 novel AMPs against ESKAPE pathogens. Another study based on variational inference autoencoders (VAE) by Dean et al. led to the identification of 38 new AMPs effective against the pathogens: *Escherichia coli, Pseudomonas aeruginosa*, and *Staphylococcus aureus* (Dean et al. 2021). AI-based approaches like deep generative models followed by molecular dynamics were used to generate novel AMPs showing activity against a diverse array of pathogenic bacteria, which includes multidrug-resistant *Klebsiella pneumonia* (Das et al. 2021). Taken together, AI has a huge application in developing potent AMP against ESKAPE pathogens.

7.8 CONCLUSIONS

In spite of AMR posing a major threat to human health, AI can be used to fight this problem in several ways. AI is used in the *in silico* diagnosis of clinical isolates of ARB from WGS and metagenome sequence data, AMR surveillance, drug repurposing, and the development of potent AMPs. In the current scenario, where a huge amount of WGS datasets of bacterial clinical isolates are available, the quick development of AI-based models is highly essential for the diagnosis of AMR clinical isolates and for the design of new drug molecules with a unique mode of action. The availability of genomic and metagenomic data in the form of several databases provides favorable grounds for the application of ML approaches to predict AMR. The application of AI is also helpful for AMR surveillance in the environment and clinical settings. Thus, in controlling healthcare-related infections, AI-based strategies have shown new directions in checking the AMR problem and further understanding the underlying mechanism behind AMR acquisition.

BIBLIOGRAPHY

Ajulo S, Awosile B. 2024. Global antimicrobial resistance and use surveillance system (GLASS 2022): Investigating the relationship between antimicrobial resistance and antimicrobial consumption data across the participating countries. *PLOS ONE* **19**: e0297921.

Alluri R, Kilari EK, Pasala PK, Kopalli SR, Koppula S. 2023. Repurposing diltiazem for its neuroprotective anti-dementia role against intra-cerebroventricular streptozotocin-induced sporadic Alzheimer's disease-type rat model. *Life* **13**: 1688.

Anahtar MN, Yang JH, Kanjilal S. 2021. Applications of machine learning to the problem of antimicrobial resistance: An emerging model for translational research. *J Clin Microbiol* **59**: e01260–20.

Arango-Argoty G, Garner E, Pruden A, Heath LS, Vikesland P, Zhang L. 2018. DeepARG: A deep learning approach for predicting antibiotic resistance genes from metagenomic data. *Microbiome* **6**: 23.

Ayoola MB, Das AR, Krishnan BS, Smith DR, Nanduri B, Ramkumar M. 2024. Predicting salmonella MIC and deciphering genomic determinants of antibiotic resistance and susceptibility. *Microorganisms* **12**: 134.

Baker S, Thomson N, Weill F-X, Holt KE. 2018. Genomic insights into the emergence and spread of antimicrobial-resistant bacterial pathogens. *Science* **360**: 733–738.

Bilal H, Khan MN, Khan S, Shafiq M, Fang W, Khan RU, Rahman MU, Li X, Lv Q-L, Xu B. 2025. The role of artificial intelligence and machine learning in predicting and combating antimicrobial resistance. *Comput Struct Biotechnol J* **27**: 423–439.

Booq RY, Tawfik EA, Alfassam HA, Alfahad AJ, Alyamani EJ. 2021. Assessment of the antibacterial efficacy of halicin against pathogenic bacteria. *Antibiotics* **10**: 1480.

Branda F. 2024. The impact of artificial intelligence in the fight against antimicrobial resistance. *Infect Dis* **56**: 484–486.

Campbell O, Gagnon J, Rubin JE. 2019. Antibacterial activity of chemotherapeutic drugs against *Escherichia coli* and *Staphylococcus pseudintermedius*. *Lett Appl Microbiol* **69**: 353–357.

Carlson-Banning KM, Chou A, Liu Z, Hamill RJ, Song Y, Zechiedrich L. 2013. Toward repurposing ciclopirox as an antibiotic against drug-resistant acinetobacter baumannii, *Escherichia coli*, and *Klebsiella pneumoniae* ed. B. Adler. *PLOS ONE* **8**: e69646.

Castanheira M, Lindley J, Huynh H, Mendes RE, Lomovskaya O. 2019. 690. Activity of a novel polymyxin analog, QPX9003, tested against resistant gram-negative pathogens, including carbapenem-resistant acinetobacter, enterobacterales, and pseudomonas. *Open Forum Infect Dis* **6**: S313–S313.

Chakaya JM, Harries AD, Marks GB. 2020. Ending tuberculosis by 2030 – Pipe dream or reality? *Int J Infect Dis* **92**: S51–S54.

Cruz-Muñiz MY, López-Jacome LE, Hernández-Durán M, Franco-Cendejas R, Licona-Limón P, Ramos-Balderas JL, Martinéz-Vázquez M, Belmont-Díaz JA, Wood TK, García-Contreras R. 2017. Repurposing the anticancer drug mitomycin C for the treatment of persistent *Acinetobacter baumannii* infections. *Int J Antimicrob Agents* **49**: 88–92.

Dahl LG, Joensen KG, Østerlund MT, Kiil K, Nielsen EM. 2021. Prediction of antimicrobial resistance in clinical *Campylobacter jejuni* isolates from whole-genome sequencing data. *Eur J Clin Microbiol Infect Dis* **40**: 673–682.

Das P, Sercu T, Wadhawan K, Padhi I, Gehrmann S, Cipcigan F, Chenthamarakshan V, Strobelt H, Dos Santos C, Chen P-Y, et al. 2021. Accelerated antimicrobial discovery via deep generative models and molecular dynamics simulations. *Nat Biomed Eng* **5**: 613–623.

Davis JJ, Wattam AR, Aziz RK, Brettin T, Butler R, Butler RM, Chlenski P, Conrad N, Dickerman A, Dietrich EM, et al. 2019. The PATRIC Bioinformatics Resource Center: Expanding data and analysis capabilities. *Nucleic Acids Res* gkz943.

Dean SN, Alvarez JAE, Zabetakis D, Walper SA, Malanoski AP. 2021. PepVAE: Variational Autoencoder Framework for Antimicrobial Peptide Generation and Activity Prediction. *Front Microbiol* **12**: 725727.

Dilnessa T, Getaneh A, Hailu W, Moges F, Gelaw B. 2022. Prevalence and antimicrobial resistance pattern of *Clostridium difficile* among hospitalized diarrheal patients: A systematic review and meta-analysis ed. Y.-F. Chang. *PLOS ONE* **17**: e0262597.

Duncan LR, Wang W, Sader HS. 2022. *In vitro* potency and spectrum of the novel polymyxin MRX-8 tested against clinical isolates of gram-negative bacteria. *Antimicrob Agents Chemother* **66**: e00139-22.

Farha MA, Brown ED. 2019. Drug repurposing for antimicrobial discovery. *Nat Microbiol* **4**: 565–577.

Feretzakis G, Loupelis E, Sakagianni A, Kalles D, Lada M, Christopoulos C, Dimitrellos E, Martsoukou M, Skarmoutsou N, Petropoulou S et al. 2020. Using machine learning algorithms to predict antimicrobial resistance and assist empirical treatment. In *Studies in Health Technology and Informatics*, IOS Press https://doi.org/10.3233/SHTI200497 (Accessed May 9, 2025).

Flandrois J-P, Lina G, Dumitrescu O. 2014. MUBII-TB-DB: A database of mutations associated with antibiotic resistance in *Mycobacterium tuberculosis*. *BMC Bioinformatics* **15**: 107.

Gerada A, Harper N, Howard A, Reza N, Hope W. 2024. Determination of minimum inhibitory concentrations using machine-learning-assisted agar dilution. *Microbiol Spectr* **12**: e04209–23.

Ghosh A, Saran, N, Saha S. 2020. Survey of drug resistance associated gene mutations in *Mycobacterium tuberculosiss*, ESKAPE and other bacterial species. *Sci Rep* **10**: 8957.

Gschwind R, Ugarcina Perovic S, Weiss M, Petitjean M, Lao J, Coelho LP, Ruppé E. 2023. ResFinderFG v2.0: A database of antibiotic resistance genes obtained by functional metagenomics. *Nucleic Acids Res* **51**: W493–W500.

Gullu E, Bora S, Beynek B. 2024. Exploiting image processing and artificial intelligence techniques for the determination of antimicrobial susceptibility. *Appl Sci* **14**: 3950.

Heydari A, Kim ND, Horswell J, Gielen G, Siggins A, Taylor M, Bromhead C, Palmer BR. 2022. Co-selection of heavy metal and antibiotic resistance in soil bacteria from agricultural soils in New Zealand. *Sustainability* **14**: 1790.

Hicks SA, Strümke I, Thambawita V, Hammou M, Riegler MA, Halvorsen P, Parasa S. 2022. On evaluation metrics for medical applications of artificial intelligence. *Sci Rep* **12**: 5979.

Hu X-Y, Logue M, Robinson N. 2020. Antimicrobial resistance is a global problem – A UK perspective. *Eur J Integr Med* **36**: 101136.

Hua Y, Dai X, Xu Y, Xing G, Liu H, Lu T, Chen Y, Zhang Y. 2022. Drug repositioning: Progress and challenges in drug discovery for various diseases. *Eur J Med Chem* **234**: 114239.

Huang J, Xu Y, Xue Y, Huang Y, Li X, Chen X, Xu Y, Zhang D, Zhang P, Zhao J, et al. 2023. Identification of potent antimicrobial peptides via a machine-learning pipeline that mines the entire space of peptide sequences. *Nat Biomed Eng* **7**: 797–810.

Hui ST, Gifford H, Rhodes J. 2024. Emerging antifungal resistance in fungal pathogens. *Curr Clin Microbiol Rep* **11**: 43–50.

Hutinel M, Huijbers PMC, Fick J, Åhrén C, Larsson DGJ, Flach C-F. 2019. Population-level surveillance of antibiotic resistance in *Escherichia coli* through sewage analysis. *Eurosurveillance* **24**. https://www.eurosurveillance.org/content/10.2807/1560-7917.ES.2019.24.37.1800497 (Accessed May 21, 2025).

Inda-Díaz JS, Lund D, Parras-Moltó M, Johnning A, Bengtsson-Palme J, Kristiansson E. 2023. Latent antibiotic resistance genes are abundant, diverse, and mobile in human, animal, and environmental microbiomes. *Microbiome* **11**: 44.

Jain A, Singhal N, Kumar M. 2023. AFRbase: A database of protein mutations responsible for antifungal resistance. *Bioinformatics* **39**: btad677.

Jeon K, Kim J-M, Rho K, Jung SH, Park HS, Kim J-S. 2022. Performance of a machine learning-based methicillin resistance of *Staphylococcus aureus* identification system using MALDI-TOF MS and comparison of the accuracy according to SCCmec types. *Microorganisms* **10**: 1903.

Karp BE, Tate H, Plumblee JR, Dessai U, Whichard JM, Thacker EL, Hale KR, Wilson W, Friedman CR, Griffin PM, et al. 2017. National antimicrobial resistance monitoring system: Two decades of advancing public health through integrated surveillance of antimicrobial resistance. *Foodborne Pathog Dis* **14**: 545–557.

Kavousi K, Bagheri M, Behrouzi S, Vafadar S, Atanaki FF, Lotfabadi BT, Ariaeenejad S, Shockravi A, Moosavi-Movahedi AA. 2020. IAMPE: NMR-assisted computational prediction of antimicrobial peptides. *J Chem Inf Model* **60**: 4691–4701.

Kuang X, Wang F, Hernandez KM, Zhang Z, Grossman RL. 2022. Accurate and rapid prediction of tuberculosis drug resistance from genome sequence data using traditional machine learning algorithms and CNN. *Sci Rep* **12**: 2427.

Larkin H. 2023. Increasing antimicrobial resistance poses global threat, WHO says. *JAMA* **329**: 200.

Liu, Z., Deng, D., Lu, H., Sun, J., Lv, L., Li, S., Peng, G., Ma, X., Li, J., Li, Z., Rong, T., & Wang, G. (2020). Evaluation of machine learning models for predicting antimicrobial resistance of *Actinobacillus pleuropneumoniae* from whole genome sequences. *Frontiers in Microbiology*, 11, 474876.

Lorente-Torres B, Llano-Verdeja J, Castañera P, Ferrero HÁ, Fernández-Martínez S, Javadimarand F, Mateos LM, Letek M, Mourenza Á. 2024. Innovative strategies in drug repurposing to tackle intracellular bacterial pathogens. *Antibiotics* **13**: 834.

Ma Y, Guo Z, Xia B, Zhang Y, Liu X, Yu Y, Tang N, Tong X, Wang M, Ye X, et al. 2022. Identification of antimicrobial peptides from the human gut microbiome using deep learning. *Nat Biotechnol* **40**: 921–931.

Macesic N, Polubriaginof F, Tatonetti NP. 2017. Machine learning: Novel bioinformatics approaches for combating antimicrobial resistance. *Curr Opin Infect Dis* **30**: 511–517.

Madrigal P, Singh NK, Wood JM, Gaudioso E, Hernández-del-Olmo F, Mason CE, Venkateswaran K, Beheshti A. 2022. Machine learning algorithm to characterize antimicrobial resistance associated with the International Space Station surface microbiome. *Microbiome* **10**: 134.

Maguire F, Rehman MA, Carrillo C, Diarra MS, Beiko RG. 2019. Identification of primary antimicrobial resistance drivers in agricultural nontyphoidal *Salmonella enterica* serovars by using machine learning. *mSystems* **4**: e00211-19.

Manoharan RK, Srinivasan S, Shanmugam G, Ahn Y-H. 2021. Shotgun metagenomic analysis reveals the prevalence of antibiotic resistance genes and mobile genetic elements in full scale hospital wastewater treatment plants. *J Environ Manage* **296**: 113270.

McArthur AG, Waglechner N, Nizam F, Yan A, Azad MA, Baylay AJ, Bhullar K, Canova MJ, De Pascale G, Ejim L, et al. 2013. The comprehensive antibiotic resistance database. *Antimicrob Agents Chemother* **57**: 3348–3357.

Nagarajan D, Nagarajan T, Roy N, Kulkarni O, Ravichandran S, Mishra M, Chakravortty D, Chandra N. 2018. Computational antimicrobial peptide design and evaluation against multidrug-resistant clinical isolates of bacteria. *J Biol Chem* **293**: 3492–3509.

Nash A, Sewell T, Farrer RA, Abdolrasouli A, Shelton JMG, Fisher MC, Rhodes J. 2018. MARDy: Mycology antifungal resistance database ed. J. Kelso. *Bioinformatics* **34**: 3233–3234.

Nguyen M, Long SW, McDermott PF, Olsen RJ, Olson R, Stevens RL, Tyson GH, Zhao S, Davis JJ. 2019. Using machine learning to predict antimicrobial MICs and associated genomic features for nontyphoidal *Salmonella*. *J Clin Microbiol* **57**: e01260-18.

Pascucci M, Royer G, Adamek J, Asmar MA, Aristizabal D, Blanche L, Bezzarga A, Boniface-Chang G, Brunner A, Curel C, et al. 2021. AI-based mobile application to fight antibiotic resistance. *Nat Commun* **12**. https://www.nature.com/articles/s41467-021-21187-3 (Accessed January 14, 2025).

Pataki BÁ, Matamoros S, Van Der Putten BCL, Remondini D, Giampieri E, Aytan-Aktug D, Hendriksen RS, Lund O, Csabai I, Schultsz C, et al. 2020. Understanding and predicting ciprofloxacin minimum inhibitory concentration in *Escherichia coli* with machine learning. *Sci Rep* **10**: 15026.

Pun FW, Ozerov IV, Zhavoronkov A. 2023. AI-powered therapeutic target discovery. *Trends Pharmacol Sci* **44**: 561–572.

Rahman SF, Olm MR, Morowitz MJ, Banfield JF. 2018. Machine learning leveraging genomes from metagenomes identifies influential antibiotic resistance genes in the infant gut microbiome ed. N. Segata. *mSystems* **3**. 10.1128/msystems.00123-17

Rahul, Adhikari S, Monika. 2020. NLP based machine learning approaches for text summarization. In *2020 Fourth International Conference on Computing Methodologies and Communication (ICCMC)*, pp. 535–538, IEEE, Erode, India https://ieeexplore.ieee.org/document/9076358 (Accessed May 19, 2025).

Ren Y, Chakraborty T, Doijad S, Falgenhauer L, Falgenhauer J, Goesmann A, Hauschild A-C, Schwengers O, Heider D. 2022. Prediction of antimicrobial resistance based on whole-genome sequencing and machine learning. *Bioinformatics* **38**: 325–334.

Sader HS, Dale GE, Rhomberg PR, Flamm RK. 2018. Antimicrobial activity of murepavadin tested against clinical isolates of *Pseudomonas aeruginosa* from the United States, Europe, and China. *Antimicrob Agents Chemother* **62**: e00311–18.

Sakagianni A, Koufopoulou C, Feretzakis G, Kalles D, Verykios VS, Myrianthefs P, Fildisis G. 2023. Using machine learning to predict antimicrobial resistance – A literature review. *Antibiotics* **12**: 452.

Samreen, Ahmad I, Malak HA, Abulreesh HH. 2021. Environmental antimicrobial resistance and its drivers: A potential threat to public health. *J Glob Antimicrob Resist* **27**: 101–111.

Sandgren A, Strong M, Muthukrishnan P, Weiner BK, Church GM, Murray MB. 2009. Tuberculosis drug resistance mutation database. *PLOS Med* **6**: e1000002.

Santana De Carvalho D, Bastos RW, Rossato L, Teixeira Se Aguiar Peres N, Assis Santos D. 2024. ResFungi: A novel protein database of antifungal drug resistance genes using a hidden markov model profile. *ACS Omega* **9**: 30559–30570.

Sarker IH. 2021. Machine learning: Algorithms, real-world applications and research directions. *SN Comput Sci* **2**: 160.

Serwecińska L. 2020. Antimicrobials and antibiotic-resistant bacteria: A risk to the environment and to public health. *Water* **12**: 3313.

Shah HA, Househ M. 2024. Purpose-oriented review of public health surveillance systems: Use of surveillance systems and recent advances. *BMJ Public Health* **2**: e000374.

Shang K, Kim J-H, Park J-Y, Choi Y-R, Kim S-W, Cha S-Y, Jang H-K, Wei B, Kang M. 2023. Comparative studies of antimicrobial resistance in *Escherichia coli, Salmonella*, and *Campylobacter* isolates from broiler chickens. *Foods* **12**: 2239.

She P, Li Y, Li Z, Liu S, Yang Y, Li L, Zhou L, Wu Y. 2023. Repurposing 9-aminoacridine as an adjuvant enhances the antimicrobial effects of Rifampin against multidrug-resistant *Klebsiella pneumoniae*. *Microbiol Spectr* **11**: e04474-22.

Singh A. 2024. Artificial intelligence for drug repurposing against infectious diseases. *Artif Intell Chem* **2**: 100071.

Stokes JM, Yang K, Swanson K, Jin W, Cubillos-Ruiz A, Donghia NM, MacNair CR, French S, Carfrae LA, Bloom-Ackermann Z, et al. 2020. A deep learning approach to antibiotic discovery. *Cell* **180**: 688–702.

Sunuwar J, Azad RK. 2022. Identification of novel antimicrobial resistance genes using machine learning, homology modeling, and molecular docking. *Microorganisms* **10**: 2102.

United Nations Environment Programme. 2023. *Bracing for superbugs: Strengthening environmental action in the one health response to antimicrobial resistance*. United Nations. https://www.un-ilibrary.org/content/books/9789210025799 (Accessed May 21, 2025).

Vashisht V, Vashisht A, Mondal AK, Farmaha J, Alptekin A, Singh H, Ahluwalia P, Srinivas A, Kolhe R. 2023. Genomics for emerging pathogen identification and monitoring: Prospects and obstacles. *BioMedInformatics* **3**: 1145–1177.

Walia K, Madhumathi J, Veeraraghavan B, Chakrabarti A, Kapil A, Ray P, Singh H, Sistla S, Ohri VC. 2019. Establishing antimicrobial resistance surveillance & research network in India: Journey so far. *Indian J Med Res* **149**: 164–179.

Wang W, Baker M, Hu Y, Xu J, Yang D, Maciel-Guerra A, Xue N, Li H, Yan S, Li M, et al. 2021. Whole-genome sequencing and machine learning analysis of *Staphylococcus aureus* from multiple heterogeneous sources in china reveals common genetic traits of antimicrobial resistance. *mSystems* **6**: e01185-20.

Wang H, Jia C, Li H, Yin R, Chen J, Li Y, Yue M. 2022. Paving the way for precise diagnostics of antimicrobial resistant bacteria. *Front Mol Biosci* **9**: 976705.

Wani AK, Roy P, Kumar V, Mir TUG. 2022. Metagenomics and artificial intelligence in the context of human health. *Infect Genet Evol* **100**: 105267.

Waterlow NR, Cooper BS, Robotham JV, Knight GM. 2024. Antimicrobial resistance prevalence in bloodstream infection in 29 European countries by age and sex: An observational study. *PLOS Med* **21**: e1004301.

Yan J, Bhadra P, Li A, Sethiya P, Qin L, Tai HK, Wong KH, Siu SWI. 2020. Deep-AmPEP30: Improve short antimicrobial peptides prediction with deep learning. *Mol Ther – Nucleic Acids* **20**: 882–894.

Yang K, Swanson K, Jin W, Coley C, Eiden P, Gao H, Guzman-Perez A, Hopper T, Kelley B, Mathea M, et al. 2019. Analyzing learned molecular representations for property prediction. *J Chem Inf Model* **59**: 3370–3388.

Yang Y, Jiang X, Chai B, Ma L, Li B, Zhang A, Cole JR, Tiedje JM, Zhang T. 2016. ARGs-OAP: Online analysis pipeline for antibiotic resistance genes detection from metagenomic data using an integrated structured ARG-database. *Bioinformatics* **32**: 2346–2351.

Yoshida M, Hinkley T, Tsuda S, Abul-Haija YM, McBurney RT, Kulikov V, Mathieson JS, Galiñanes Reyes S, Castro MD, Cronin L. 2018. Using evolutionary algorithms and machine learning to explore sequence space for the discovery of antimicrobial peptides. *Chem* **4**: 533–543.

Yu L, Jing R, Liu F, Luo J, Li Y. 2020. DeepACP: A novel computational approach for accurate identification of anticancer peptides by deep learning algorithm. *Mol Ther – Nucleic Acids* **22**: 862–870.

Zhang Y, Zhao C, Wang Q, Wang X, Chen H, Li H, Zhang F, Wang H. 2020. Evaluation of the *in vitro* activity of new polymyxin B analogue SPR206 against clinical MDR, colistin-resistant and tigecycline-resistant Gram-negative bacilli. *J Antimicrob Chemother* **75**: 2609–2615.

Zheng EJ, Valeri JA, Andrews IW, Krishnan A, Bandyopadhyay P, Anahtar MN, Herneisen A, Schulte F, Linnehan B, Wong F, et al. 2024. Discovery of antibiotics that selectively kill metabolically dormant bacteria. *Cell Chem Biol* **31**: 712–728.e9.

Zhou N, Cheng Z, Zhang X, Lv C, Guo C, Liu H, Dong K, Zhang Y, Liu C, Chang Y-F, et al. 2022. Global antimicrobial resistance: A system-wide comprehensive investigation using the Global One Health Index. *Infect Dis Poverty* **11**: 92.

8 Approaches of AI and Advanced Computational Biology for Combating HIV Infections

Kazi Asraf Ali, Sabyasachi Choudhuri, Amlan Bishal, Tanushree Das, Subhrajyoti Nandy, and Chowdhury Mobaswar Hossain

8.1 INTRODUCTION

Human immunodeficiency virus (HIV) continues to be one of the most formidable infectious diseases globally, despite substantial advancements in antiretroviral therapy (ART) and public health strategies [1]. Since the beginning of the epidemic, over 85 million people have been infected, and more than 40 million have died due to HIV-related illnesses. The Joint United Nations Programme on HIV/AIDS (UNAIDS) reported that in 2023, approximately 1.3 million individuals were newly infected, and around 630,000 people died from HIV-related causes [2]. HIV's impact is multifaceted, extending beyond physical health to affect social structures, economies, and psychological well-being. Despite global initiatives to control the epidemic, several challenges persist: the emergence of drug-resistant strains, failure to develop a robust and lasting vaccine, and stigma that deters testing and treatment. These hurdles call for innovative scientific approaches that can transcend traditional methods of disease management.

Artificial intelligence (AI) and computational biology have rapidly emerged as transformative tools in the biomedical sciences. In the context of HIV, these technologies are reshaping how we understand viral behavior, host-pathogen interactions, drug resistance, and vaccine design. AI techniques, such as deep learning, natural language processing, and reinforcement learning, are now applied to mine complex datasets, simulate biological systems, predict viral mutations, and design novel therapeutics [3]. On the other hand, computational biology complements AI by providing the frameworks and models to interpret biological phenomena from genomic, proteomic, and systems biology perspectives. Together, AI and computational biology offer a powerful synergy. They enable researchers to analyze vast quantities of heterogeneous patient records and epidemiological data to high-throughput sequencing datasets – and derive actionable insights that can guide public health strategies, clinical interventions, and therapeutic development [4].

8.2 UNDERSTANDING HIV PATHOGENESIS AND IMMUNE EVASION

8.2.1 Molecular Biology of HIV: Key Mechanisms

HIV is a retrovirus belonging to the *Lentivirus* genus and primarily targets the immune system, specifically CD4+ T cells, macrophages, and dendritic cells. The virus carries two single-stranded RNA genomes, and the enzymes reverse transcriptase, integrase, and protease, which are essential for its replication cycle [5]. The lifecycle of HIV includes several key stages, which are mentioned in Table 8.1.

The key feature that makes HIV particularly difficult to eradicate is its capacity for immune evasion. The virus employs multiple strategies to escape detection and destruction by the host immune system:

1. **High Genetic Variability**: HIV's rapid replication and high mutation rates generate a diverse viral quasispecies population, enabling escape from neutralizing antibodies and cytotoxic T lymphocyte (CTL) responses.

2. **Glycan Shielding**: The HIV envelope glycoprotein (gp120) is heavily glycosylated, masking critical epitopes and preventing effective antibody binding.

3. **Latency Establishment**: HIV can integrate its genome into long-lived resting memory CD4+ T cells, establishing latent reservoirs that are invisible to the immune system and unaffected by ART.

4. **Downregulation of Host Proteins**: Viral proteins like Nef, Vpu, and Vif downregulate host molecules involved in immune recognition (e.g., MHC-I, CD4), impairing the presentation of viral peptides to immune cells.

5. **Subversion of Innate Immunity**: HIV interferes with innate immune sensing pathways and dampens the production of interferons, delaying early antiviral responses [6].

TABLE 8.1 Key Stages in the HIV Replication Cycle

Stage	Description
Attachment and Entry	HIV binds to CD4 receptors and chemokine co-receptors (CCR5 or CXCR4) on host cells.
Reverse Transcription	Viral RNA is reverse-transcribed into complementary DNA (cDNA) by reverse transcriptase.
Integration	The cDNA is integrated into the host genome via the integrase enzyme.
Transcription and Translation	Host cellular machinery is hijacked to produce viral RNA and proteins.

TABLE 8.2 Key Challenges in HIV Treatment and Management

Challenge	Impact
Latent Reservoirs	These reservoirs are not targeted by ART and pose a barrier to a complete cure.
Drug Resistance	The emergence of resistant strains undermines the long-term efficacy of ART regimens.
Lack of a Vaccine	Efforts to develop a universal vaccine are hindered by HIV's genetic variability and immune evasion tactics.
Adherence and Access	Long-term adherence to ART is challenging, particularly in low-resource settings.
Comorbidities	Chronic HIV infection is associated with an increased risk of cardiovascular, renal, and neurocognitive disorders [7].

TABLE 8.3 Outlines of the Major Obstacles Currently Hindering Effective HIV Vaccine Development

Obstacle	Description
Antigenic diversity	A high mutation rate prevents the identification of a single conserved immunogen.
Ineffective immune memory	Previous vaccine attempts failed to induce durable, broadly neutralizing antibodies.
Lack of correlates of protection	Unclear immune parameters are needed for adequate protection.
Animal model limitations	Differences in immune responses between animal models and humans limit extrapolation [8].

These mechanisms collectively allow HIV to persist in the host, evade immune clearance, and establish chronic infection.

8.2.2 Challenges in HIV Treatment and Vaccine Development

Despite significant progress in HIV therapeutics, several scientific and clinical challenges remain (Table 8.2). Moreover, the virus's ability to manipulate host immune responses means that conventional vaccine strategies – such as those that work for polio or measles – are ineffective against HIV (Table 8.3).

8.3 AI AND COMPUTATIONAL BIOLOGY TECHNIQUES IN HIV RESEARCH

AI and computational biology have become transformative tools in HIV research, enabling a deeper understanding of the virus, its evolution, and treatment responses. These technologies assist in analyzing vast genomic and clinical datasets to uncover patterns in viral mutations, predict drug resistance, and identify potential vaccine targets. Machine learning algorithms are employed to

**Applications of Artificial Intelligence and
Computational Biology in HIV Research and Precision**

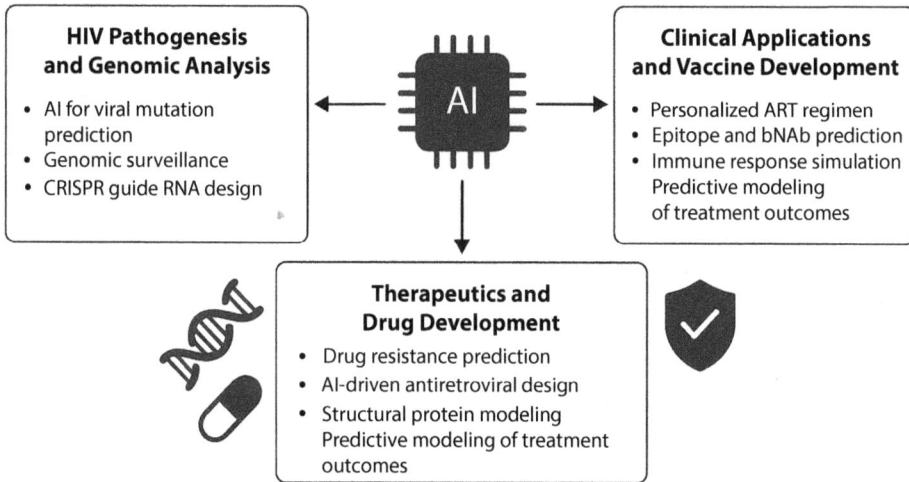

Figure 8.1 Applications of AI and computational biology in HIV research and precision medicine.

model HIV-host interactions, optimize ART, and simulate disease progression, while structural bioinformatics aids in drug design by modeling HIV proteins and their interactions (Figure 8.1). Together, AI and computational biology accelerate discovery and precision medicine approaches, offering new hope in the fight against HIV/AIDS.

8.3.1 Machine Learning and Data Mining for HIV Genomics

Genomic Sequence Analysis: Machine learning and data mining have revolutionized genomic sequence analysis in HIV research, enabling rapid, scalable, and accurate classification and surveillance of viral variants. Traditional HIV-1 subtyping and outbreak detection methods often relied on sequence alignment and phylogenetic tree construction, which are computationally intensive and can struggle with large datasets or recombination events. Recent advances leverage alignment-free approaches, such as k-mer-based encoding and natural language processing techniques, to represent viral sequences for input into supervised machine learning models, significantly reducing computational costs while maintaining high accuracy [15]. Deep learning methods, particularly convolutional neural networks (CNNs), have further enhanced outbreak detection by treating pairwise genetic distance matrices as images, allowing real-time identification of active transmission clusters with high sensitivity and specificity [16]. These machine learning frameworks accelerate subtype classification and outbreak monitoring and facilitate the discovery of functional sequence features and resistance mutations, supporting more effective public health interventions and personalized treatment strategies [17–19].

Drug Resistance Prediction: Machine learning and data mining techniques have become powerful tools for predicting HIV drug resistance, enabling clinicians to tailor ART based on the genetic makeup of viral strains. Approaches such as random forests, support vector machines, artificial neural networks, and deep learning have been applied to analyze amino acid substitutions in key HIV enzymes like protease and reverse transcriptase, achieving high predictive accuracy (often above 0.8) in classifying resistant versus susceptible variants [20–22]. Recent models utilize k-mers or multi-n-grams derived from protein sequences, and innovative methods like generative topographic mapping (GTM) provide interpretable resistance landscapes that link specific mutation patterns to resistance profiles [21, 22]. These computational tools accelerate the identification of resistance-associated mutations and support the development of web-based applications for real-time clinical decision-making, ultimately improving therapy effectiveness and reducing the risk of treatment failure due to emerging drug resistance [20, 22].

Viral Evolution and Epidemiology: Machine learning and data mining have become pivotal in studying HIV viral evolution and epidemiology, enabling rapid and scalable analysis of large genomic datasets to track transmission dynamics and outbreak events. Traditional phylogenetic methods, while informative, are computationally intensive and less suited for real-time surveillance. Recent advances employ deep learning approaches, such as CNNs [23], which analyze pairwise genetic distance matrices as images to classify epidemiological scenarios and detect active outbreaks efficiently. This method has demonstrated high accuracy and scalability, successfully mapped the start and end of documented HIV-1 outbreaks, and outperformed traditional clustering tools like HIV-TRACE in sensitivity and specificity. By bypassing the need for full phylogenetic reconstruction, these AI-driven models offer a powerful, real-time solution for monitoring HIV transmission and informing public health interventions [16].

CRISPR Target Site Identification: Machine learning and data mining are increasingly vital for identifying optimal CRISPR target sites within the highly diverse HIV genome. Computational approaches analyze large-scale viral sequence datasets to pinpoint conserved regions such as the long terminal repeat (LTR), *tat*, and *rev* genes suitable for guide RNA (gRNA) targeting, maximizing the likelihood of effectively disrupting viral replication across multiple HIV subtypes. Advanced algorithms assess sequence conservation, predict on-target efficacy, and minimize off-target effects by scoring potential gRNAs against HIV and host genomes [24–26]. For example, pipelines have been developed to align candidate gRNAs to panels of HIV-1 strains from global databases, ensuring broad-spectrum activity and reducing the risk of viral escape due to sequence variability [26]. Additionally, genome-wide unbiased methods such as GUIDE-Seq and CIRCLE-Seq- are employed to experimentally validate computational predictions and detect off-target cleavage events, further refining target site selection for safer and more effective CRISPR-based HIV therapies [25].

8.3.2 Structural Biology and Molecular Modeling of HIV Proteins

Protein Structure Prediction: Protein structure prediction has become a cornerstone in the structural biology and molecular modeling of HIV proteins, enabling researchers to elucidate three-dimensional conformations essential for understanding viral function and drug targeting. Homology modeling is widely used to predict the 3D structures of HIV proteins such as gp120 and gp41, which are challenging to crystallize experimentally. This computational approach leverages sequence similarity between target HIV proteins and known structural templates, with tools like MODELLER generating reliable models subsequently validated by quality assessment metrics such as Ramachandran plots and QMEAN scores [27]. Additionally, advanced machine learning algorithms, including support vector machines and deep learning, have been applied to classify and predict HIV protein structures based on amino acid composition and physicochemical properties, improving accuracy and efficiency [28]. These predicted structures facilitate the identification of potential drug-binding sites and support studies on mutational tolerance and resistance, thereby informing the rational design of antiretroviral therapies [29, 30].

Drug-binding simulations: Molecular modeling and drug-binding simulations are central to understanding HIV protein-ligand interactions, guiding drug design, and explaining resistance mechanisms. Recent studies have leveraged advanced computational approaches to provide detailed insights into these processes. Molecular dynamics (MD) simulations are a cornerstone for understanding how drugs bind to HIV proteins, especially protease. A 2025 study modeled the unbinding of ritonavir, a key HIV protease inhibitor, revealing multiple dissociation pathways. The most common pathway involved a broad residue-residue correlation network, including regions affected by drug resistance mutations. These findings underscore the importance of transient interactions and protein flexibility in drug binding and resistance [31]. MD simulations have also demonstrated that both antibody and inhibitor (e.g., ritonavir) binding can restrict the dynamic motions of HIV-1 protease, particularly the flap regions that regulate access to the active site. This restriction is linked to enzyme inhibition and provides a mechanistic basis for drug action and resistance [32].

Resistance Mechanism Elucidation: Molecular modeling and structural biology have elucidated intricate resistance mechanisms in HIV proteins, particularly highlighting the interplay between active-site and distal mutations. Drug resistance in HIV-1 protease, for instance, arises not only from direct mutations in the catalytic site (e.g., I50V) but also from distal mutations (e.g., A71V, L76V) that propagate dynamic changes across the protein, altering inhibitor binding while maintaining enzymatic activity [33, 34]. Accelerated MD simulations reveal that distal mutations induce conformational shifts in flap regions and dimer interfaces, reducing drug affinity through long-range allosteric effects. Epistatic interactions between mutations further stabilize resistant variants. Potts model analyses and free energy perturbation studies demonstrate that co-evolving mutations mitigate fitness costs while enhancing resistance [33, 35]. Structural models of highly resistant variants (e.g., 11-mutant protease) and inhibition assays reveal that resistance escalates with mutation accumulation, driven by interdependent adjustments in hydrogen bonding networks and hydrophobic interactions. These insights, derived from MD simulations, machine learning-based QSAR models [34], and evolutionary trajectory analyses, under score the need to target static and dynamic protein features to design robust inhibitors against rapidly evolving HIV strains.

Envelope Protein Modeling: Structural biology and molecular modeling have significantly advanced our understanding of the HIV envelope (Env) glycoprotein, a trimer comprising gp120 and gp41 subunits. MD simulations and crystallographic studies reveal that gp120 undergoes CD4-induced conformational changes, transitioning from a closed perfusion state to an open state that exposes the gp41 fusion machinery [36, 37]. These dynamic rearrangements are critical for viral entry and are targeted by broadly neutralizing antibodies (bnAbs), such as those binding the membrane-proximal external region (MPER) of gp41. MD simulations demonstrate how bnAbs like 4E10 and PGZL1 embed into lipid bilayers, forming stable interactions with phospholipids and conserved epitopes, a mechanism crucial for their neutralization potency [38]. Additionally, site-specific glycosylation analysis highlights Env's glycan shield, with ~90 N-linked glycans influencing antibody accessibility. Modeling efforts integrate glycan occupancy data, distinguishing high-mannose and complex-type glycans to inform immunogen design [39]. Structural plasticity in gp120's layered architecture comprising invariant β-sandwich cores and plastic loops enables immune evasion while maintaining gp41 interactions, as shown by cryoelectron tomography and crystal structures [37]. These insights guide the rational design of vaccines and entry inhibitors by targeting conserved, dynamic epitopes and accounting for glycan heterogeneity.

8.3.3 Bioinformatics and Systems Biology in HIV Analysis

Bioinformatics and systems biology have become essential in advancing our understanding of HIV pathogenesis, drug resistance, and therapeutic strategies. Researchers can accurately predict drug resistance mutations, track viral evolution, and identify key host-pathogen interactions by integrating high-throughput sequencing, machine learning, and multi-omics data. For example, machine learning models trained on HIV genetic and structural data can forecast resistance to protease and integrase inhibitors, aiding in personalized therapy design [21, 40, 41]. Systems biology approaches, such as transcriptomic and proteomic profiling, have revealed novel biomarkers of HIV latency and mapped the complex protein-protein interaction networks the virus exploits within host cells [42, 43]. These integrative analyses have also informed vaccine development by identifying immune correlates of protection in large cohort studies [42]. Despite challenges such as data heterogeneity and model interpretability, the synergy between bioinformatics and systems biology continues to drive innovations in HIV research, from AI-driven drug discovery to real-time monitoring of treatment adherence and viral load [21, 44].

8.4 AI FOR DRUG DISCOVERY AND DEVELOPMENT

AI is transforming drug discovery and development by enabling predictive modeling, enhancing drug design, and optimizing therapeutic strategies, particularly in ART for HIV.

a. **Predictive Modeling of Drug Resistance and Efficacy**: AI and machine learning (ML) approaches, such as random forests, support vector machines, and deep neural networks, are now widely used to predict HIV drug resistance by analyzing genomic sequences and clinical data. These models can identify complex patterns associated with resistance mutations, allowing clinicians to anticipate treatment failure and tailor ART to individual patient profiles [41, 45]. Integrating genomic and clinical data further improves predictive accuracy, providing early warnings for emerging resistance and supporting proactive treatment management [40].

b. **AI-Enhanced Drug Design and Optimization**: AI accelerates drug discovery by screening vast chemical libraries and optimizing molecular structures for improved efficacy and reduced toxicity. Deep learning and other advanced algorithms can predict new compounds' binding affinity and pharmacological properties, prioritize promising drug candidates, and suggest structural

153

modifications to overcome resistance mechanisms [46, 47]. In HIV research, AI-driven optimization techniques have been used to design novel protease and integrase inhibitors and refine existing antiretroviral agents for better patient outcomes [46].

c. **Case Studies: Antiretroviral Drug Development**: Recent case studies highlight the impact of AI in optimizing HIV treatment regimens. For example, AI models have been used to analyze patient-specific data, including viral load, CD4 count, genetic markers, and treatment history, to recommend the most effective ART combinations, resulting in improved viral suppression and fewer side effects. Collaborative projects, such as those between the University of Oxford and the Kenya Medical Research Institute, have demonstrated that AI-driven personalized treatment plans significantly enhance patient outcomes compared to standard care [46]. Furthermore, AI has been instrumental in predicting resistance to new and existing antiretrovirals, guiding drug development and clinical decision-making in real-time [30, 40]. These advances underscore the transformative potential of AI in drug discovery, resistance prediction, and personalized medicine for HIV and other infectious diseases [30, 46]. AI has revolutionized the field of drug discovery by accelerating the identification and development of therapeutic compounds. In the context of ART – particularly for HIV/AIDS – AI technologies have played a pivotal role in identifying novel drug targets, repurposing existing drugs, optimizing lead compounds, and predicting protein structures critical to viral replication. Researchers have shortened the traditionally lengthy and costly drug development process by leveraging ML, deep learning, natural language processing, and molecular modeling. Table 8.4 presents key case studies highlighting how AI has been applied to advance antiretroviral drug research and development.

8.5 AI IN HIV VACCINE DEVELOPMENT

8.5.1 Role of AI Epitope Prediction and Immune Response Simulation

By improving epitope prediction and mimicking immune responses, AI is transforming the development of HIV vaccines [47]. A crucial stage in vaccine design is epitope prediction, which entails determining certain antigenic areas that trigger immune responses [48]. The accuracy and speed needed to negotiate HIV's high mutation rates are sometimes lacking in traditional methods. AI-powered methods that predict CD8+ T-cell epitopes with previously unheard-of accuracy include MUNIS, a deep learning model created in partnership between the Ragon Institute and MIT's Jameel Clinic [49]. MUNIS efficiently detects immunogenic epitopes in various diseases, including HIV, by examining large human leukocyte antigen (HLA) ligands datasets. This method lessens the need for time-consuming experimental assays, simplifying vaccine creation [50].

Furthermore, computational technologies such as Vaccine CAD and Epi-Assembler optimize epitope alignment and make creating immunogenic consensus sequence (ICS) epitopes easier, respectively [51]. By creating representative epitopes from various viral variations and reducing junctional immunogenicity when positioning epitopes in a string-of-beads arrangement for DNA expression vectors, these techniques address the heterogeneity of HIV [52]. In preclinical research, these AI-powered tactics have demonstrated improved immune responses in transgenic mouse models, indicating their potential [53].

Researchers can more successfully create vaccinations that take into consideration HIV's genetic variety by incorporating AI into immune response modeling and epitope prediction, thereby boosting the search for an HIV vaccine that works everywhere.

8.5.2 Broadly Neutralizing Antibodies and AI-Driven Discovery

By making it easier to find broadly neutralizing antibodies (bNAbs), which are essential for successful immunization campaigns, AI is transforming the development of HIV vaccines [54]. It takes a lot of work and time to identify bNAbs using traditional approaches [55]. AI-driven methods, such as the Rapid Automatic Identification of bNAbs (RAIN), use ML algorithms to evaluate immune repertoires, making it possible to anticipate possible bNAbs quickly and accurately. The CD4-binding location of the HIV-1 envelope glycoprotein, a crucial area for viral entrance into host cells, is the target of bNAbs that RAIN has successfully discovered [56].

AI models can also forecast the evolution of viruses and their escape mutations, which helps designers create antibody combinations that reduce the likelihood of resistance [57]. Computational studies have, for example, determined the best bNAb combinations, such as PG9, PGT151, and VRC01, which lower the risk of viral rebound to less than 1% [58]. These AI-driven findings make possible the development of vaccines that elicit strong and widespread immune responses against various HIV strains [59].

TABLE 8.4 Application of AI in Advanced Antiretroviral

Case Study	AI Application	Drug/Target	AI Tools/Techniques Used	Impact/Outcome	References
BenevolentAI – HIV Drug Repurposing	Drug repurposing using AI-driven knowledge graph	Baricitinib (originally for rheumatoid arthritis)	Knowledge graphs, NLP, machine learning	Identified Baricitinib as a potential HIV therapeutic due to anti-inflammatory properties	[9]
Atomwise – Small Molecule Screening	Virtual screening and binding prediction for HIV-1 protease inhibitors	HIV-1 Protease	Deep learning-based structure prediction and docking	Accelerated screening of millions of compounds; narrowed down to promising leads	[10]
Insilico Medicine – Target Identification	AI-assisted pathway and target identification	Novel HIV targets	Deep generative models, transcriptomic data analysis	Proposed novel targets for intervention based on systems biology and omics data	[11]
IBM Watson for Drug Discovery	Literature mining and hypothesis generation	Reverse Transcriptase, Integrase	NLP, ML for text analysis	Generated new insights on drug-target interactions and resistance mechanisms	[12]
DeepMind AlphaFold	Protein structure prediction	HIV-1 Envelope Protein, Protease	Deep learning (AlphaFold)	Enabled accurate modeling of complex HIV proteins to support vaccine and inhibitor design	[13]
Exscientia – Lead Optimization	AI-guided design of HIV inhibitors	NNRTIs (Non-nucleoside Reverse Transcriptase Inhibitors)	Reinforcement learning, predictive modeling	Improved potency and selectivity of NNRTIs with fewer iterations	[14]

The identification and optimization of bNAbs are accelerated by the use of AI in HIV vaccine research, providing encouraging opportunities to create potent vaccinations to combat this enduring global health issue [60].

8.5.3 Computational Vaccine Design: Promising Approaches

The creation of HIV vaccines has been transformed by combining computational biology and AI, making it possible to create immunogens that produce broadly neutralizing antibodies (bNAbs) [61]. One significant development is the eOD-GT6 immunogen, which was produced using advanced genetic engineering and computational methods. In order to resist HIV's rapid mutation rate, this immunogen efficiently activates germline B cells, starting the maturation process toward the creation of bNAbs [62].

Computational methods have also made the development of multiepitope vaccines that target HIV-1 easier [63]. Researchers can forecast and choose epitopes that trigger strong immune responses using immunoinformatics and bioinformatics techniques, creating vaccines with improved efficacy.

Additionally, by identifying important antigens through the analysis of HLA diversity and CD8 T-cell immunological responses, AI applications in HIV research have improved vaccine formulation [64]. Deep learning algorithms that forecast the sensitivity of virus strains to monoclonal antibodies help find effective antibody combinations. Furthermore, the humoral response of individuals living with HIV to SARS-CoV-2 mRNA vaccines has been predicted using ML techniques, indicating customized booster tactics.

These computational techniques have greatly improved the accuracy and effectiveness of HIV vaccine development, which also presents encouraging opportunities for developing successful anti-virus vaccination campaigns [65].

8.6 PERSONALIZED TREATMENT AND PREDICTIVE MODELS FOR HIV MANAGEMENT

8.6.1 AI-Based Precision Medicine in HIV Care

AI transforms HIV care by enabling precision medicine techniques that customize therapies to each patient's unique profile. ML algorithms that evaluate large genomic and clinical datasets to identify drug resistance mutations can tailor ART to improve efficacy and minimize side effects. For example, models using support vector machines and random forests have proven adept at predicting resistance patterns and enhancing treatment results [60].

AI-powered predictive models have been created for clinical contexts to identify patients who may not respond to ART [66]. Using ML techniques to assess clinical information, a study with 1,577 HIV-positive participants was able to identify independent risk variables for immunological non-responders (INR) [67]. The most successful predictor was the random forest model, highlighting AI's potential to improve patient care and clinical decision-making. Additionally, by searching large datasets for possible protease inhibitors and antibodies, AI speeds up the development of new HIV treatments. The drug development process has been streamlined using deep learning and variational autoencoders to predict ligand-binding affinities and find interesting compounds [68].

AI applications have also helped with customized treatment plans. Researchers have created frameworks that combine reinforcement learning and Bayesian statistical approaches to optimize ART regimens, striking a balance between viral suppression and adverse effect reduction. Patient outcomes have significantly improved as a result of the application of these AI-driven recommendations; for example, the depression scores of those impacted have increased by 22% [69].

AI-based precision medicine is revolutionizing HIV management by giving medical professionals cutting-edge resources to forecast treatment outcomes, customize treatments, and accelerate the creation of potent medications, all of which will enhance the standard of care for people with HIV [70].

8.6.2 Predictive Models for Disease Progression and Therapy Response

The creation of predictive models for the course of HIV disease and the effectiveness of treatment has been greatly aided by developments in AI and ML [71]. These models enable individualized treatment plans, predicting patient outcomes by analyzing intricate, high-dimensional clinical data. A concordance score (c-index) of 0.83 in training sets and 0.81 in validation sets showed strong performance in a systematic review and meta-analysis assessing ML models for predicting mortality risk in people with HIV (PWH). This demonstrates how ML models can be used in healthcare settings to pinpoint high-risk individuals and enhance intervention tactics [72, 73].

Computational models have been created to forecast virological reactions to ART in resource-constrained environments without genotypic resistance testing [74]. When interpreted using rules, these models' accuracy was on par with genotyping, indicating their potential for treatment decision optimization in situations where resistance testing is not accessible.

Additionally, a study using ML developed prediction models to find PWH INR [75]. The most successful predictor was the random forest (RF) model, underscoring the significance of early INR detection for prompt intervention implementation [76].

When taken as a whole, this research demonstrates how AI and ML are revolutionizing HIV management by creating predictive models that allow for more individualized treatment plans and better patient outcomes [77].

8.6.3 Applications in Chronic HIV Management

By facilitating predictive modeling and individualized treatment plans, AI is transforming the management of chronic HIV and improving patient outcomes [78]. In order to predict the course of a disease and the likelihood of death, ML algorithms can evaluate large amounts of patient data. According to a comprehensive review and meta-analysis, these models performed well in predicting mortality risk among PWH, as evidenced by their concordance index of 0.81 in validation sets [79].

AI has been used to forecast treatment interruptions among PWH in resource-constrained environments, such as Nigeria [79]. Healthcare professionals can proactively identify patients who may be at risk of stopping ART by incorporating ML models into standard health information systems [80]. This enables prompt interventions to enhance patient outcomes and adherence.

AI also makes it easier to create frameworks for tailored therapy decision-making. To customize combination ART (cART) regimens, researchers at Johns Hopkins University have developed a two-step strategy that combines offline reinforcement learning with Bayesian statistical techniques [74]. Patients' depression levels improved by 22% as a result of this tailored approach, demonstrating AI's potential to improve quality of life by lowering side effects related to HIV therapy [81].

All things considered, incorporating AI into chronic HIV management helps medical practitioners to create customized treatment regimens, anticipate possible issues and carry out focused treatments, all of which increase the effectiveness of HIV care [82].

8.7 CHALLENGES AND ETHICAL CONSIDERATIONS

AI and advanced computational biology present both tremendous potential and significant ethical challenges for HIV research [83]. To ensure responsible and fair advancements, concerns about algorithmic bias, data privacy, and wider ethical, legal, and social implications (ELSI) must be methodically addressed when new technologies are integrated into clinical practice and biological research [84, 85].

8.7.1 Data Privacy and Security in HIV Research

Given the sensitive nature of data pertaining to HIV and the substantial stigma and discrimination surrounding the condition, security and privacy become crucial issues in AI-driven research [86]. Data breaches, unauthorized access, and re-identification have become more likely as AI applications depend more on the massive aggregation of genetic data, electronic medical records, and social determinants of health [87]. According to a global survey, over 60% of people with HIV expressed concern about the potential exploitation of personal health data in digital health research [88].

AI-enabled re-identification attacks usually outperform conventional de-identification techniques [89]. There are serious dangers for HIV genomics research, for example, as research has indicated that ML algorithms might re-identify people from anonymized genetic records with up to 95% accuracy [90]. Potential solutions are provided by privacy-preserving technologies like federated learning; Sheller et al. (2020) showed that by avoiding centralized data storage, federated learning models could maintain 98% of model performance while drastically lowering privacy risks [91]. Dynamic consent models are also strongly recommended to provide participants with more influence over how their data is used in future research projects.

8.7.2 Bias and Fairness in AI Models

Healthcare results can be significantly impacted by bias and fairness issues in AI models for HIV research [92, 93]. The underrepresentation of historically oppressed groups in clinical datasets, including women, members of ethnic minorities, and individuals from low- and middle-income countries (LMICs), is a major problem due to data representation bias [94].

These disparities are further exacerbated by algorithmic biases. As an illustration of how model architectures may perpetuate current inequalities, Obermeyer et al. (2019) showed that AI systems consistently underestimated HIV risk by about 8% for Black female patients [95]. Fairness solutions, like data resampling, re-weighting, and algorithm designs that consider fairness, are increasingly being utilized to deal with these problems [96, 97]. Adding fairness limits raised anticipated equity by 15% across a range of ethnic groups, according to Rajkumar et al. (2018), suggesting possible paths toward more inclusive AI systems. For AI to be used in HIV research and care, addressing bias and ensuring equity are not only technical problems; they are also vital ethical requirements [98].

8.7.3 Ethical, Legal, and Social Implications

There are significant ethical, legal, and sociological issues with using AI technology in HIV research that need to be addressed in advance [99, 100]. Ensuring accountability for AI-driven judgments, preserving informed consent, as well as patient autonomy, and fostering transparency – particularly with "black-box" AI models that obscure internal decision-making processes – are ethical challenges [101]. Establishing data ownership, protecting intellectual property rights, and making sure that international data protection laws such as HIPAA (Health Insurance Portability and Accountability Act) and the General Data Protection Regulation (GDPR) are followed are all legal problems [102]. Social ramifications are particularly crucial because, if fair access is not prioritized, predictive AI models may exacerbate stigma and the digital divide, resulting in greater global health disparities [103].

A recent study emphasizes how important it is to deal with these ELSI issues. According to an international survey conducted by Mozafari et al. (2023), 60% of HIV-positive patients favored direct community involvement in AI system design, while 70% expressed worry about AI transparency decision-making [104]. Significant gaps in global governance were highlighted by a 2021 World Health Organization (WHO) report that discovered that only 34% of LMICs had national regulations governing AI applications in genomics [105]. In Kenya, an AI-based HIV intervention that incorporated feedback from the local community considerably decreased stigma-related worries by 40% when compared to conventional approaches [106]. As AI plays a bigger role in HIV research and care, these incidents highlight the urgent need for ethical frameworks that guarantee inclusivity, transparency, and strict worldwide regulatory compliance.

8.8 FUTURE DIRECTIONS AND EMERGING TRENDS

Significant progress is anticipated at the intersection of HIV research, computational biology, and AI. Future advancements seek to improve therapeutic treatments, broaden our understanding of biology, and promote global health equity [107].

8.8.1 Integrating Multi-omics Data and AI for HIV Research

To fully comprehend HIV biology, multi-omics integration – including transcriptomics, proteomics, metabolomics, and genomics – is essential. When evaluating these intricate, high-dimensional datasets, AI, in particular ML and deep learning models, is essential [108, 109]. Applications include developing customized treatment plans based on multi-layered biological profiles, identifying new biomarkers for HIV progression, and analyzing viral-host interactions across biological scales [110]. Nonetheless, challenges like handling noisy or missing data and harmonizing data across several platforms continue to be significant barriers. Potential solutions are provided by cutting-edge AI techniques like graph neural networks (GNNs) and multi-modal fusion architectures, with future directions suggesting real-time integration of clinical and multi-omics data for adaptive therapeutic algorithms [111].

Recent studies have demonstrated the potential of AI in HIV multi-omics analysis. By using AI to combine transcriptome and proteome data, Wang et al. (2022) found three new host biomarkers that were more than 85% accurate in predicting HIV progression [112]. Contrasting multi-omics datasets with deep learning, Zeng et al. (2023) significantly increased the number of personalized treatment predictions compared to single-omics methods [113]. New technologies, such as GNNs, have demonstrated cutting-edge performance in handling the intricacy of multi-omics integration, which has aided the identification of new HIV therapeutic targets [114]. These developments highlight how AI-driven multi-omics techniques are improving our understanding of HIV pathogenesis at the molecular level and applying precision medicine to the management of pathogenic illnesses.

8.8.2 The Potential of AI in Eradication and Functional Cure

Among the primary objectives of current research is to achieve a functional cure for HIV, which is defined as sustained viral suppression without ongoing treatment. AI is necessary for this endeavor

[115, 116]. By facilitating the creation of vaccinations tailored to various HIV strains, AI-driven modeling of viral epitopes and immune responses is accelerating vaccine development [117]. Algorithms for ML in drug development are speeding up the procedure for finding new therapeutic targets, opportunities for drug repurposing, and optimizing latency-reversing agents (LRAs), which are essential for "shock and kill" or "block and lock" curative techniques. Additionally, AI-enhanced clinical trial optimization expedites patient recruiting, forecasts trial results, and models' long-term implications to expedite cure research timeframes. AI models additionally assist in mapping and prioritizing viral reservoir sites, boosting the targeting of hidden virus populations [118].

The revolutionary potential of AI in this field is highlighted by several significant experimental developments. Manganaro et al. (2021) found five new latency-reversing medications using in-silk-screening, and several of them advanced to early preclinical evaluation [119]. More precise and effective therapeutic solutions are now possible thanks to AI-enhanced models that have also achieved over 90% prediction accuracy in predicting HIV reservoir dynamics during small cohort studies [120]. Additionally, DeepMind's AlphaFold deep learning model has revolutionized the prediction of protein structures, particularly those of critical HIV proteins, opening up new avenues for targeted therapeutics and logical vaccine design [121]. These developments demonstrate how AI is not only improving our comprehension of HIV persistence but also accelerating the hunt for a workable and potentially comprehensive cure.

8.8.3 Prospects for Global Health Equity via AI and Computational Biology

Though careful use is necessary to avoid exacerbating already-existing gaps, AI and computational biology possess the capacity to improve global health equity in HIV research and care [122]. While open-source AI platforms enable researchers in LMICs to engage in and benefit from global scientific advances, mobile AI applications can improve diagnostic access in remote populations. The efficiency of interventions is further increased by customizing AI models to local epidemiological data [123].

Infrastructure limitations and data sovereignty issues persist in spite of these prospects. Investing in the field of digital health infrastructure, creating ethical frameworks that emphasize inclusivity and openness, and encouraging global partnerships for fair innovation sharing are some ways to deal with them [124]. Examples from the actual world highlight this potential: in Rwanda, early HIV detection among rural residents increased by 30% due to AI-powered mobile diagnostics [125]. Mollura and colleagues describe RAD-AID's "three-pronged strategy," which includes partnerships with numerous local health institutions and deployments in over 30 nations with poor and moderate incomes for integrating AI into radiology procedures in resource-poor environments. This indicates that open-source and scalable AI systems have a considerable institutional adoption rate in LMICs [126]. Further highlighting the significance of developing AI capacity globally, Masters et al. carried out a comprehensive analysis of public health programs' economic assessments (many of which leverage digital tools) and discovered a median return on investment of 14.3:1 with a median cost-benefit ratio of 8.3:1 – meaning that for every US$1 invested, roughly US$8–US$14 is saved in social and healthcare costs [127].

8.9 CONCLUSION

Integrating AI and advanced computational biology has opened transformative avenues in the fight against HIV. From unraveling the complex molecular mechanisms of HIV pathogenesis to overcoming long-standing challenges in treatment and vaccine development, these technologies are reshaping the landscape of HIV research and care. AI-driven tools enhance our understanding of viral genomics, enable precise structural modeling of HIV proteins, and support systems-level analyses that reveal novel therapeutic targets.

In drug discovery, AI accelerates identifying and optimizing antiretroviral compounds, offering predictive insights into drug resistance and efficacy. Similarly, in vaccine development, AI facilitates epitope prediction, immune response simulation, and the discovery of broadly neutralizing antibodies – key advancements in the design of effective and adaptive vaccines.

AI's potential extends to personalized treatment, where predictive models guide individualized therapy and long-term disease management. However, realizing this potential requires addressing critical challenges, including data privacy, algorithmic bias, and ethical considerations accompanying sensitive health data and predictive technologies.

Looking forward, integrating multi-omics data with AI, pursuing a functional cure, and promoting global health equity underscore the promising future of this interdisciplinary approach. As the field evolves, the synergy between AI and computational biology holds great promise in transforming HIV care, research, and, ultimately, the goal of eradication.

ABBREVIATIONS

AI	Artificial Intelligence
API	Application Programming Interface
CNN	Convolutional Neural Network
CT	Computed Tomography
DL	Deep Learning
HLA	Human Leukocyte Antigen
IoT	Internet of Things
ML	Machine Learning
mRNA	Messenger Ribonucleic Acid
NLP	Natural Language Processing

REFERENCES

[1] Bouza E, Arribas JR, Bernardino JI, Alejos B, Coiras M, Coll P, et al. Past and future of HIV infection. A document based on expert opinion. *Revista Española de Quimioterapia* 2022;35(2):131–156. doi:10.37201/req/083.2021

[2] Payagala S, Pozniak A. The global burden of HIV. *Clinics in Dermatology* 2024;42(2):119–127. doi:10.1016/j.clindermatol.2024.02.001

[3] Tarasova O, Poroikov V. Machine learning in discovery of new antivirals and optimization of viral infections therapy. *CMC* 2021;28(38):7840–7861. doi:10.2174/0929867328666210504114351

[4] Zhang Y, Luo M, Wu P, Wu S, Lee TY, Bai C. Application of computational biology and artificial intelligence in drug design. *IJMS* 2022;23(21):13568. doi:10.3390/ijms232113568

[5] Parija SC. Retrovirus. In: *Textbook of Microbiology and Immunology*. Springer Nature Singapore; 2023:881–902. doi:10.1007/978-981-19-3315-8_61

[6] Li S, Moog C, Zhang T, Su B. HIV reservoir: Antiviral immune responses and immune interventions for curing HIV infection. *Chinese Medical Journal*. December 28, 2022. doi:10.1097/CM9.0000000000002479

[7] Singh H, Chauhan CK, Choudhary K, Meena G, Phadke P, Lachyan A, et al. Navigating complexities in HIV care: Challenges, solutions, and strategies. *International STD Research & Reviews* 2023;12(2):56–62. doi:10.9734/ISRR/2023/v12i2164

[8] Derking R, Sanders RW. Structure-guided envelope trimer design in HIV-1 vaccine development: A narrative review. *Journal of the International AIDS Society* 2021;24(S7):e25797. doi:10.1002/jia2.25797

[9] Singh A. Artificial intelligence for drug repurposing against infectious diseases. *Artificial Intelligence Chemistry* 2024;2(2):100071. doi:10.1016/j.aichem.2024.100071

[10] Falls Z, Fine J, Chopra G, Samudrala R. Accurate prediction of inhibitor binding to HIV-1 protease using CANDOCK. *Frontiers in Chemistry* 2021;9:775513. doi:10.3389/fchem.2021.775513

[11] Fu C, Chen Q. The future of pharmaceuticals: Artificial intelligence in drug discovery and development. *Journal of Pharmaceutical Analysis*. February 2025:101248. doi:10.1016/j.jpha.2025.101248

[12] Martin RL, Martinez Iraola D, Louie E, Pierce D, Tagtow BA, Labrie JJ, et al. Hybrid natural language processing for high-performance patent and literature mining in IBM Watson for Drug Discovery. *IBM Journal of Research and Development* 2018;62:1–8. doi:10.1147/JRD.2018.2888975

[13] Jumper J, Evans R, Pritzel A, Green T, Figurnov M, Ronneberger O, et al. Highly accurate protein structure prediction with AlphaFold. *Nature* 2021;596(7873):583–589. doi:10.1038/s41586-021-03819-2

[14] Li J, Ye B, Gao S, Liu X, Zhan P. The latest developments in the design and discovery of non-nucleoside reverse transcriptase inhibitors (NNRTIs) for the treatment of HIV. *Expert Opinion on Drug Discovery* 2024;19(12):1439–1456. doi:10.1080/17460441.2024.2415309

[15] Solis-Reyes S, Avino M, Poon A, Kari L. An open-source k-mer based machine learning tool for fast and accurate subtyping of HIV-1 genomes. *PLoS One* 2018;13:e0206409. doi:10.1371/journal.pone.0206409

[16] Kupperman MD, Leitner T, Ke R. A deep learning approach to real-time HIV outbreak detection using genetic data. *PLoS Computational Biology* 2022;18:e1010598. doi:10.1371/journal.pcbi.1010598

[17] Ramirez-Mata AS, Ostrov D, Salemi M, Marini S, Magalis BR. Machine learning prediction and phyloanatomic modeling of viral neuroadaptive signatures in the macaque model of HIV-mediated neuropathology. *Microbiology Spectrum* 2023;11(2):e0308622. doi:10.1128/spectrum.03086-22

[18] Foglierini M, Nortier P, Schelling R, Winiger RR, Jacquet P, O'Dell S, et al. RAIN: Machine learning-based identification for HIV-1 bNAbs. *Nature Communications* 2024;15(1):5339. doi:10.1038/s41467-024-49676-1

[19] Blassel L, Zhukova A, Villabona-Arenas CJ, Atkins KE, Hué S, Gascuel O. Drug resistance mutations in HIV: New bioinformatics approaches and challenges. *Current Opinion in Virology* 2021;51:56–64. doi:10.1016/j.coviro.2021.09.009

[20] Chui CY, Chan AWE. Machine learning prediction of HIV1 drug resistance against integrase strand transfer inhibitors. April 28, 2025. doi:10.1101/2025.04.25.650610

[21] Paremskaia AI, Rudik AV, Filimonov DA, Lagunin AA, Poroikov VV, Tarasova OA. Web service for HIV drug resistance prediction based on analysis of amino acid substitutions in main drug targets. *Viruses* 2023;15(11):2245. doi:10.3390/v15112245

[22] Pikalyova K, Orlov A, Lin A, Tarasova O, Marcou M, Horvath D, et al. HIV-1 drug resistance profiling using amino acid sequence space cartography. *Bioinformatics* 2022;38(8):2307–2314. doi:10.1093/bioinformatics/btac090

[23] Ali KA, Mohin S, Mondal P, Goswami S, Ghosh S, Choudhuri S. Influence of artificial intelligence in modern pharmaceutical formulation and drug development. *Future Journal of Pharmaceutical Sciences* 2024;10(1):53. doi:10.1186/s43094-024-00625-1

[24] Ophinni Y, Inoue M, Kotaki T, Kameoka M. CRISPR/Cas9 system targeting regulatory genes of HIV-1 inhibits viral replication in infected T-cell cultures. *Scientific Reports* 2018;8(1):7784. doi:10.1038/s41598-018-26190-1

[25] Panfil AR, London JA, Green PL, Yoder KE. CRISPR/Cas9 genome editing to disable the latent HIV-1 provirus. *Frontiers in Microbiology* 2018;9:3107. doi:10.3389/fmicb.2018.03107

[26] Sullivan NT, Dampier W, Chung CH, Allen AG, Atkins A, Pirrone V, et al. Novel gRNA design pipeline to develop broad-spectrum CRISPR/Cas9 gRNAs for safe targeting of the HIV-1 quasispecies in patients. *Scientific Reports* 2019;9(1):17088. doi:10.1038/s41598-019-52353-9

[27] Elalouf A. *In-silico* structural modeling of human immunodeficiency virus proteins. *Biomedical Engineering and Computational Biology* 2023;14:11795972231154402. doi:10.1177/11795972231154402

[28] Mei J, Zhao J. Prediction of HIV-1 and HIV-2 proteins by using Chou's pseudo amino acid compositions and different classifiers. *Scientific Reports* 2018;8(1):2359. doi:10.1038/s41598-018-20819-x

[29] Ota R, So K, Tsuda M, Higuchi Y, Yamashita F. Prediction of HIV drug resistance based on the 3D protein structure: Proposal of molecular field mapping. *PLoS One* 2021;16:e0255693. doi:10.1371/journal.pone.0255693

[30] Humphris-Narayanan E, Akiva E, Varela R, Ó Conchúir S, Kortemme T. Prediction of mutational tolerance in HIV-1 protease and reverse transcriptase using flexible backbone protein design. *PLoS Computational Biology* 2012;8:e1002639. doi:10.1371/journal.pcbi.1002639

[31] Vig E, Sun J, Chang C-en A. Pathway specific unbinding free energy profiles of ritonavir dissociation from HIV-1 protease. *Biochemistry* 2025;64(4):940–952. doi:10.1021/acs.biochem.4c00560

[32] Badaya A, Sasidhar YU. Inhibition of the activity of HIV-1 protease through antibody binding and mutations probed by molecular dynamics simulations. *Scientific Reports* 2020;10(1):5501. doi:10.1038/s41598-020-62423-y

[33] Henes M, Lockbaum GJ, Kosovrasti K, Leidner F, Nachum GS, Nalivaika EA, et al. Picomolar to micromolar: Elucidating the role of distal mutations in HIV-1 protease in conferring drug resistance. *ACS Chemical Biology* 2019;14(11):2441–2452. doi:10.1021/acschembio.9b00370

[34] Wang B, He Y, Wen X, Xi Z. Prediction and molecular field view of drug resistance in HIV-1 protease mutants. *Scientific Reports* 2022;12(1):2913. doi:10.1038/s41598-022-07012-x

[35] Biswas A, Choudhuri I, Huang K, Sun Q, Sali A, Echeverria I, et al. Evolutionary sequence and structural basis for the epistatic origins of drug resistance in HIV. May 2, 2025. doi:10.1101/2025.04.30.651576

[36] Prabakaran P, Dimitrov AS, Fouts TR, Dimitrov DS. Structure and function of the HIV envelope glycoprotein as entry mediator, vaccine immunogen, and target for inhibitors. In: *Advances in Pharmacology*. Vol 55. Elsevier; 2007:33–97. doi:10.1016/S1054-3589(07)55002-7

[37] Pancera M, Majeed S, Ban YEA, Chen L, Huang C Kong L, et al. Structure of HIV-1 gp120 with gp41-interactive region reveals layered envelope architecture and basis of conformational mobility. *Proceedings of the National Academy of Sciences of the United States of America* 2010;107(3):1166–1171. doi:10.1073/pnas.0911004107

[38] Maillie CA, Golden K, Wilson IA, Ward AB, Mravic M. Ab initio prediction of specific phospholipid complexes and membrane association of HIV-1 MPER antibodies by multi-scale simulations. *eLife* 2025;12:RP90139. doi:10.7554/eLife.90139

[39] Cao L, Diedrich JK, Kulp DW, Pauthner M, He L, Park SKR, et al. Global site-specific N-glycosylation analysis of HIV envelope glycoprotein. *Nature Communications* 2017;8(1):14954. doi:10.1038/ncomms14954

[40] Kibibi WH. Harnessing artificial intelligence for HIV drug resistance prediction and personalized treatment. *NIJRMS* 2024;5:59–64. doi:10.59298/NIJRMS/2024/5.3.5964

[41] Cai Q, Yuan R, He J, Li M, Guo Y. Predicting HIV drug resistance using weighted machine learning method at target protein sequence-level. *Molecular Diversity* 2021;25(3):1541–1551. doi:10.1007/s11030-021-10262-y

[42] Zhang Z, Jin H, Zhang X, Bai M, Zheng K, Tian J, et al. Bioinformatics and system biology approach to identify the influences among COVID-19, influenza, and HIV on the regulation of gene expression. *Frontiers in Immunology* 2024;15:1369311. doi:10.3389/fimmu.2024.1369311

[43] Kibe A, Buck S, Gribling-Burrer AS, Gilmer O, Bohn P, Koch T, et al. The translational landscape of HIV-1 infected cells reveals key gene regulatory principles. *Nature Structural & Molecular Biology* January 15, 2025. doi:10.1038/s41594-024-01468-3

[44] Ebulue Charles Chukwudalu, Ekkeh Ogochukwu Virginia, Ebulue Ogochukwu Roseline, Ekesiobi Chukwunonso Sylvester. Developing predictive models for HIV Drug resistance: A genomic and AI approach. *IMSRJ* 2024;4(5):521–543. doi:10.51594/imsrj.v4i5.1119

[45] Branda F, Scarpa F. Implications of artificial intelligence in addressing antimicrobial resistance: Innovations, global challenges, and healthcare's future. *Antibiotics* 2024;13(6):502. doi:10.3390/antibiotics13060502

[46] Jin R, Zhang L. AI applications in HIV research: Advances and future directions. *Frontiers in Microbiology* 2025;16:1541942. doi:10.3389/fmicb.2025.1541942

[47] Imani S, Li X, Chen K, Maghsoudloo M, Jabbarzadeh Kaboli P, Hashemi M, et al. Computational biology and artificial intelligence in mRNA vaccine design for cancer immunotherapy. *Frontiers in Cellular and Infection Microbiology* 2025;14. doi:10.3389/fcimb.2024.1501010

[48] Basmenj ER, Pajhouh SR, Fallah AE, Naijian R, Rahimi E, Atighy H, et al. Computational epitope-based vaccine design with bioinformatics approach; a review. *Heliyon* 2025;11(1). doi:10.1016/j.heliyon.2025.e41714

[49] Wohlwend J, Nathan A, Shalon N, Crain CR, Tano-Menka R, Goldberg B, Richards E, Gaiha GD, Barzilay R. Deep learning enhances the prediction of HLA class I-presented CD8+ T cell epitopes in foreign pathogens. *Nature Machine Intelligence* 2025;7. https://www.nature.com/articles/s42256-024-00971-y. Accessed March 30, 2025

[50] Xu Y, Tan Y, Peng Z, Liu M, Zhang B, Wei K. Advancing myocarditis research: Evaluating animal models for enhanced pathophysiological insights. *Current Cardiology Reports* 2025;27(1):6. doi:10.1007/s11886-024-02182-8

[51] Khan MS, Shakya M, Verma CK. Computational modelling of a multiepitope vaccine targeting glycoprotein-D for herpes simplex virus 2 (HSV-2): An immunoinformatic analysis. *Molecular Diversity* 2025. doi:10.1007/s11030-025-11148-z

[52] Tang X, Zhang W, Zhang Z. Developing T cell epitope-based vaccines against infection: Challenging but worthwhile. *Vaccine* 2025;13(2):135. doi:10.3390/vaccines13020135

[53] Uthamacumaran A, Kiyokawa J, Wakimoto H. AI-driven hybrid ecological model for predicting oncolytic viral therapy dynamics. January 18, 2025. doi:10.48550/*arXiv*.2501.10620

[54] Gawande MS, Zade N, Kumar P, Gundewar S, Weerarathna IN, Verma P. The role of artificial intelligence in pandemic responses: From epidemiological modeling to vaccine development. *Molecular Biomedicine* 2025;6(1):1. doi:10.1186/s43556-024-00238-3

[55] Foulkes C, Friedrich N, Ivan B, Stiegeler E, Magnus C, Schmidt D, et al. Assessing bnAb potency in the context of HIV-1 envelope conformational plasticity. *PLoS Pathogens* 21(1). https://journals.plos.org/plospathogens/article?id=10.1371/journal.ppat.1012825. Accessed March 30, 2025

[56] Qi Y, Zhang S, Wang K, Ding H, Zhang Z, Anang S, et al. The membrane-proximal external region of human immunodeficiency virus (HIV-1) envelope glycoprotein trimers in A18-lipid nanodiscs. *Communications Biology*. https://www.nature.com/articles/s42003-025-07852-z. Accessed March 30, 2025

[57] Moritz U, Kraemer G, Tsui JL-H, Chang SY, Lytras S, Khurana MP, Vanderslott S. Artificial intelligence for modelling infectious disease epidemics. *Nature* 2025;638. https://www.nature.com/articles/s41586-024-08564-w. Accessed March 30, 2025

[58] Parthasarathy D, Pickthorn S, Ahmed S, Mazurov D, Jeffy J, Shukla RK, et al. Incompletely closed HIV-1CH040 envelope glycoproteins resist broadly neutralizing antibodies while mediating efficient HIV-1 entry. *NPJ Viruses*. https://www.nature.com/articles/s44298-024-00082-w. Accessed March 30, 2025

[59]   Adepoju VA, Udah DC, Onyezue OI, Adnani QES, Jamil S, Bin Ali MN. Navigating the complexities of HIV vaccine development: Lessons from the mosaico trial and next-generation development strategies. *Vaccine* 2025;13(3):274. doi:10.3390/vaccines13030274

[60]   Jin R, Zhang L. AI applications in HIV research: Advances and future directions. *Frontiers in Microbiology* 2025;16. doi:10.3389/fmicb.2025.1541942

[61]   Weerarathna IN, Doelakeh ES, Kiwanuka L, Kumar P, Arora S. Prophylactic and therapeutic vaccine development: Advancements and challenges. *Molecular Biomedicine* 2024;5(1):57. doi:10.1186/s43556-024-00222-x

[62]   Govindan R, Stephenson KE. HIV vaccine development at a crossroads: New B and T cell approaches. *Vaccine* 2024;12(9):1043. doi:10.3390/vaccines12091043

[63]   Hashempour A, Khodadad N, Bemani P, Ghasemi Y, Akbarinia S, Bordbari R, et al. Design of multivalent-epitope vaccine models directed toward the world's population against HIV-Gag polyprotein: Reverse vaccinology and immunoinformatics. *PLoS One* 2024;19(9):e0306559. doi:10.1371/journal.pone.0306559

[64]   Garrido-Mesa J, Brown MA. Antigen-driven T cell responses in rheumatic diseases: Insights from T cell receptor repertoire studies. *Nature Reviews Rheumatology* 2025;21(3):157–173. doi:10.1038/s41584-025-01218-9

[65]   Birthriya SK, Ahlawat P, and Jain AK. A comprehensive survey of social engineering attacks: Taxonomy of attacks, prevention, and mitigation strategies. *Journal of Applied Security Research*. 1–49. doi:10.1080/19361610.2024.2372986

[66]   Kandhare P, Kurlekar M, Deshpande T, Pawar A. A review on revolutionizing healthcare technologies with AI and ML applications in pharmaceutical sciences. *Drugs Drug Candidates* 2025;4(1). https://www.mdpi.com/2813-2998/4/1/9. Accessed March 30, 2025

[67]   Namale PE, Boloko L, Vermeulen M, Haigh KA, Bagula F, Maseko A, et al. Testing novel strategies for patients hospitalised with HIV-associated disseminated tuberculosis (NewStrat-TB): Protocol for a randomised controlled trial. *Trials* 2024;25(1):311. doi:10.1186/s13063-024-08119-4

[68]   Otun MO. Artificial intelligence and machine learning approaches for target-based drug discovery: A focus on GPCR-ligand interactions. *Journal of Applied Sciences and Environmental Management* 2025;29(3):737–745.

[69]   Hsu CY, Ismail SM, Ahmad I, Abdelrasheed NSG, Ballal S, Kalia R, et al. The impact of AI-driven sentiment analysis on patient outcomes in psychiatric care: A narrative review. *Asian Journal of Psychiatry* 2025;107:10F4443. doi:10.1016/j.ajp.2025.104443

[70]   Abbaoui W, Retal S, El Bhiri B, Kharmoum N, Ziti S. Towards revolutionizing precision healthcare: A systematic literature review of artificial intelligence methods in precision medicine. *Informatics in Medicine Unlocked* 2024;46:101475. doi:10.1016/j.imu.2024.101475

[71]   Li Y, Feng Y, He Q, Ni Z, Hu X, Feng X, et al. The predictive accuracy of machine learning for the risk of death in HIV patients: A systematic review and meta-analysis. *BMC Infectious Diseases* 2024;24(1):474. doi:10.1186/s12879-024-09368-z

[72]   Javaid M, Haleem A, Pratap Singh R, Suman R, Rab S. Significance of machine learning in healthcare: Features, pillars and applications. *International Journal of Intelligent Networks* 2022;3:58–73. doi:10.1016/j.ijin.2022.05.002

[73]   Ali G, Mijwil MM, Adamopoulos I, Buruga BA, Gök M, Sallam M. Harnessing the potential of artificial intelligence in managing viral hepatitis. *Mesopotamian Journal of Big Data* 2024. https://mesopotamian.press/journals/index.php/bigdata/article/view/484. Accessed March 30, 2025

[74]   Jin, W. Novel Bayesian methods for precision medicine in HIV. Dissertation, Johns Hopkins University 2022. https://jscholarship.library.jhu.edu/items/fe21857a-41d1-4150-ba02-aa28f86dc46b. Accessed March 30, 2025

[75]   Ali KA. *AI-Assisted Computational Approaches For Immunological Disorders*. Vol. 1. IGI Global; 2025.

[76]   Bomrah S, Uddin M, Upadhyay U, Komorowski M, Priya J, Dhar E, et al. A scoping review of machine learning for sepsis prediction-feature engineering strategies and model performance: A step towards explainability. *Critical Care* 2024;28(1):180. doi:10.1186/s13054-024-04948-6

[77]   Saeed A, Husnain A, Rasool S, Gill AY, Amelia. Healthcare revolution: How AI and machine learning are changing medicine. *EBSCOhost*. doi:10.59141/jrssem.v3i3.558

[78]   Ahmadi A. Digital health transformation: Leveraging AI for monitoring and disease management. *International Journal of BioLife Sciences (IJBLS)* 2024;3(1):10–24. doi:10.22034/ijbls.2024.188037

[79]   Belus JM, Johnson NE, Yoon GH, Tschumi N, Lerotholi M, Falgas-Bague I, et al. SMSs as an alternative to provider-delivered care for unhealthy alcohol use: Study protocol for Leseli, an open-label randomised

controlled trial of mhGAP-Remote vs mhGAP-Standard in Lesotho. *Trials* 2024;25(1):575. doi:10.1186/s13063-024-08411-3

[80]   Assefa H. *Predictive model to detect first-line antiretroviral therapy failure among HIV/Aids patients in Zewditu Hospital, Addis Ababa.* Thesis. St. Mary's University; 2022. http://repository.smuc.edu.et/handle/123456789/6926. Accessed March 30, 2025

[81]   Pavlopoulos A, Rachiotis T, Maglogiannis I. An overview of tools and technologies for anxiety and depression management using AI. https://www.mdpi.com/2076-3417/14/19/9068. Accessed March 30, 2025

[82]   Dybul M, Attoye T, Cherutich P, Dabis F, Deeks SG. The case for an HIV cure and how to get there – *The Lancet HIV.* https://www.thelancet.com/journals/lanhiv/article/PIIS2352-3018(20)30232-0/fulltext. Accessed March 30, 2025

[83]   Lainjo B. Artificial intelligence with machine learning and the enigmatic discovery of HIV cure. *Journal of Autonomous Intelligence* 2023;7(2). doi:10.32629/jai.v7i2.697

[84]   Izankar SV, Kumar P, Waghmare G. Evolution of artificial intelligence in biotechnology: From discovery to ethical and beyond. *AIP Conference Proceedings* 2024:080037. doi:10.1063/5.0241106

[85]   Padhi A, Agarwal A, Saxena SK, Katoch CDS. Transforming clinical virology with AI, machine learning and deep learning: A comprehensive review and outlook. *VirusDisease* 2023;34(3):345–355. doi:10.1007/s13337-023-00841-y

[86]   Garett R, Kim S, Young SD. Ethical considerations for artificial intelligence applications for HIV. *AI* 2024;5(2):594–601. doi:10.3390/ai5020031

[87]   Khatiwada P, Yang B, Lin JC, Blobel B. Patient-generated health data (PGHD): Understanding, requirements, challenges, and existing techniques for data security and privacy. *JPM* 2024;14(3):282. doi:10.3390/jpm14030282

[88]   Mootz JJ, Evans H, Tocco J, Ramon CV, Gordon P, Wainberg ML, et al. Acceptability of electronic healthcare predictive analytics for HIV prevention: A qualitative study with men who have sex with men in New York City. *Mhealth* 2020;6:11. doi:10.21037/mhealth.2019.10.03

[89]   Clunie D, Taylor A, Bisson T, Gutman D, Xiao Y, Schwarz CG, et al. Summary of the national cancer institute 2023 virtual workshop on medical image de-identification—part 2: Pathology whole slide image de-identification, de-facing, the role of AI in image de-identification, and the NCI MIDI datasets and pipeline. *Journal of Imaging Informatics in Medicine* 2024;38(1):16–30. doi:10.1007/s10278-024-01183-x

[90]   Erlich Y, Narayanan A. Routes for breaching and protecting genetic privacy. *Nature Reviews. Genetics* 2014;15(6):409–421. doi:10.1038/nrg3723

[91]   Sheller MJ, Edwards B, Reina GA, Martin J, Pati S, Kotrotsou A, et al. Federated learning in medicine: Facilitating multi-institutional collaborations without sharing patient data. *Scientific Reports* 2020;10(1):12598. doi:10.1038/s41598-020-69250-1

[92]   Garett R, Kim S, Young SD. Ethical considerations for artificial intelligence applications for HIV. *AI* 2024;5(2):594–601. doi:10.3390/ai5020031

[93]   Fletcher RR, Nakeshimana A, Olubeko O. Addressing fairness, bias, and appropriate use of artificial intelligence and machine learning in global health. *Frontiers in Artificial Intelligence* 2021;3:561802. doi:10.3389/frai.2020.561802

[94]   Woods WA, Watson M, Ranaweera S, Tajuria G, Sumathipala A. Under-representation of low and middle income countries (LMIC) in the research literature: Ethical issues arising from a survey of five leading medical journals: Have the trends changed? *Global Public Health* 2023;18(1):2229890. doi:10.1080/17441692.2023.2229890

[95]   Obermeyer Z, Powers B, Vogeli C, Mullainathan S. Dissecting racial bias in an algorithm used to manage the health of populations. *Science* 2019;366(6464):447–453. doi:10.1126/science.aax2342

[96]   Chen P, Wu L, Wang L. AI fairness in data management and analytics: A review on challenges, methodologies and applications. *Applied Sciences* 2023;13(18):10258. doi:10.3390/app131810258

[97]   Sadeghi F. Towards fairness-aware online machine learning from imbalanced data streams. August 10, 2023. doi:10.20381/RUOR-29452

[98]   Ferraro C, Hemsley A, Sands S. Embracing diversity, equity, and inclusion (DEI): Considerations and opportunities for brand managers. *Business Horizons* 2023;66(4):463–479. doi:10.1016/j.bushor.2022.09.005

[99]   Department of Computer Science, National Open University of Nigeria, C.U E, J.A. U, Department of Computer Science, National Open University of Nigeria, O.I N, Department of Computer Science, David Umahi Federal University of Health Sciences, Uburu, Ebonyi State, Nigeria., et al. Personalized risk

reduction of HIV plans with artificial intelligence: A narrative review. *KJHS* 2024;4:1–11. doi:10.59568/ KJHS-2024-4-1-01

[100] Marcus JL, Sewell WC, Balzer LB, Krakower DS. Artificial intelligence and machine learning for HIV prevention: Emerging approaches to ending the epidemic. *Current HIV/AIDS Reports* 2020;17(3):171–179. doi:10.1007/s11904-020-00490-6

[101] Al-Amiery A. The ethical implications of emerging AI technologies in healthcare. *MedMat.* April 22, 2025. doi:10.1097/mm9.0000000000000015

[102] Isibor E. Regulation of healthcare data security: Legal obligations in a digital age. 2024. doi:10.2139/ ssrn.4957244

[103] Lainjo B. The global social dynamics and inequalities of artificial intelligence. *IJISRR* 2023;5(8):4966–4974. doi:10.13140/RG.2.2.20078.79689

[104] Liu AY, Buchbinder SP. CROI 2023: Epidemiologic trends and prevention for HIV and other sexually transmitted infections. *Top Antivir Medicin* 2023;31(3):468–492.

[105] *Global Strategy on Digital Health 2020-2025.* 1st ed. World Health Organization; 2021.

[106] Pebody R. Artificial intelligence powers HIV testing services in Kenya. *The Lancet HIV* 2025;12(2):e92–e93. doi:10.1016/S2352-3018(25)00003-7

[107] Ory MG, Adepoju OE, Ramos KS, Silva PS, Vollmer Dahlke D. Health equity innovation in precision medicine: Current challenges and future directions. *Frontiers in Public Health* 2023;11:1119736. doi:10.3389/ fpubh.2023.1119736

[108] Babu M, Snyder M. Multi-omics profiling for health. *Molecular & Cellular Proteomics* 2023;22(6):100561. doi:10.1016/j.mcpro.2023.100561

[109] Mikaeloff F, Gelpi M, Benfeitas R, Knudsen AD, Vestad B, Høgh J, et al. Network-based multi-omics integration reveals metabolic at-risk profile within treated HIV-infection. *eLife* 2023;12:e82785. doi:10.7554/ eLife.82785

[110] Korla K, Chandra N. A systems perspective of signalling networks in host–pathogen interactions. *Journal of the Indian Institute of Science* 2017;97(1):41–57. doi:10.1007/s41745-016-0017-x

[111] Xu X, Li J, Zhu Z, Zhao L, Wang H, Song C, et al. A comprehensive review on synergy of multi-modal data and AI technologies in medical diagnosis. *Bioengineering* 2024;11(3):219. doi:10.3390/bioengineering11030219

[112] Hojo H, Saito T, He X, Guo Q, Onodera S, Azuma T, et al. Runx2 regulates chromatin accessibility to direct the osteoblast program at neonatal stages. *Cell Reports* 2022;40(10):111315. doi:10.1016/j.celrep.2022.111315

[113] Peng X, Dorman KS. Accurate estimation of molecular counts from amplicon sequence data with unique molecular identifiers. *Bioinformatics* 2023;39:btad002. doi:10.1093/bioinformatics/btad002

[114] Bao S, Zhao H, Yuan J, Fan D, Zhang Z, Su J, et al. Computational identification of mutator-derived lncRNA signatures of genome instability for improving the clinical outcome of cancers: A case study in breast cancer. *Briefings in Bioinformatics* 2020;21(5):1742–1755. doi:10.1093/bib/bbz118

[115] Lainjo B. Artificial intelligence with machine learning and the enigmatic discovery of HIV cure. *Journal of Autonomous Intelligence* 2023;7(2). doi:10.32629/jai.v7i2.697

[116] Addissouky TA, El Tantawy El Sayed I, Ali MM, Alubiady MH, Wang Y. Bending the curve through innovations to overcome persistent obstacles in HIV prevention and treatment. *Journal of AIDS and HIV Treatment* 2024;6(1):44–53. doi:10.33696/AIDS.6.051

[117] Olawade DB, Teke J, Fapohunda O, Weerasinghe K, Usman SO, Ige AO, et al. Leveraging artificial intelligence in vaccine development: A narrative review. *Journal of Microbiological Methods* 2024;224:106998. doi:10.1016/j.mimet.2024.106998

[118] Yingngam B. AI-driven drug discovery and repurposing for neurodegenerative disorders. In: *Machine learning for neurodegenerative disorders.* 1st ed. CRC Press; 2025:80–124. doi:10.1201/9781032661025-4

[119] Radanliev P. Artificial intelligence: Reflecting on the past and looking towards the next paradigm shift. *Journal of Experimental & Theoretical Artificial Intelligence.* February 28, 2024:1–18. doi:10.1080/0952813X. 2024.2323042

[120] Zhang M, Armendariz M, Xiao W, Rose O, Bendtz K, Livingstone M, et al. Look twice: A generalist computational model predicts return fixations across tasks and species. *PLoS Computational Biology* 2022;18:e1010654. doi:10.1371/journal.pcbi.1010654

[121] Jumper J, Evans R, Pritzel A, Green T, Figurnov M, Ronneberger O, et al. Highly accurate protein structure prediction with AlphaFold. *Nature* 2021;596(7873):583–589. doi:10.1038/s41586-021-03819-2

[122] Ogundeko-Olugbami Oluwafunmilayo, Ogundeko Oluwaseun. AI-enhanced predictive analytics systems combatting health disparities while driving equity in U.S. healthcare delivery. *World Journal of Advanced Research and Reviews* 2025;25(1):2067–2084. doi:10.30574/wjarr.2025.25.1.0298

[123] Sibiya SE, Hurchund R, Omondi B, Owira P. Artificial intelligence for digital healthcare in the low and medium income countries. *Health Technology* 2025;15(2):323–332. doi:10.1007/s12553-025-00950-2

[124] Marino J, Camiciotti L, Cheinasso F, Olivero A, Risso F. Enabling Compute and Data Sovereignty with Infrastructure-Level Data Spaces. In: *Proceedings of the 3rd Eclipse Security, AI, Architecture and Modelling Conference on Cloud to Edge Continuum. eSAAM '23.* Association for Computing Machinery; 2023:77–85. doi:10.1145/3624486.3624509

[125] *Reducing vulnerability and strengthening inclusion in Rwanda through rural development and enhancing social protection.* World Bank. https://www.worldbank.org/en/results/2019/10/29/reducing-vulnerability-and-strengthening-inclusion-in-rwanda-through-rural-development-and-enhancing-social-protection. Accessed May 8, 2025

[126] Mollura DJ, Culp MP, Pollack E, Battino G, Scheel JR, Mango VL, et al. Artificial intelligence in low- and middle-income countries: Innovating global health radiology. *Radiology* 2020;297(3):513–520. doi:10.1148/radiol.2020201434

[127] Masters R, Anwar E, Collins B, Cookson R, Capewell S. Return on investment of public health interventions: A systematic review. *Journal of Epidemiology and Community Health* 2017;71(8):827–834. doi:10.1136/jech-2016-208141

9 Machine Learning and Systems Biology for Precision Tuberculosis Treatment

Anuradha Singh

9.1 INTRODUCTION

Tuberculosis (TB), an infectious disease primarily affecting the lungs and caused by the bacterium-*Mycobacterium tuberculosis* (Mtb), remains a formidable global health challenge in the year 2025. (World Health Organization, 2025). Despite significant advancements in understanding its pathogenesis and developing effective therapies over the past century, TB continues to be a leading cause of morbidity and mortality worldwide. Recent statistics from the World Health Organization (WHO) highlight the persistent impact of this disease, with an estimated 10.8 million people falling ill with TB in 2023 and 1.25 million deaths reported in the same year (World Health Organization, 2025). Notably, TB has likely reclaimed its position as the world's leading cause of death from a single infectious agent, underscoring the urgent need for more effective control and management strategies. The intertwined epidemics of TB and HIV further complicate the landscape, with TB remaining the primary cause of death among people living with HIV (Venkatesan et al., 2025). Moreover, the escalating threat of antimicrobial resistance (AMR) in Mtb strains poses a grave danger to global public health security, demanding innovative approaches to combat this resilient pathogen (Serajian et al., 2025).

Traditional strategies for TB management have historically relied on standardized, one-size-fits-all approaches, primarily involving prolonged courses of multi-drug therapy (Jeffry and Seikka, 2025), such as the standard six-month regimen consisting of isoniazid, rifampicin, pyrazinamide, and ethambutol (often referred to as the 'intensive phase' followed by a 'continuation phase' with isoniazid and rifampicin). During the intensive phase, the four-drug combination (isoniazid, rifampicin, pyrazinamide, and ethambutol) is administered daily for two months. This initial phase aims to rapidly reduce the bacterial load in the patient's body, rendering them non-infectious and preventing the emergence of drug-resistant mutants. The inclusion of pyrazinamide is crucial in targeting semi-dormant bacilli, while ethambutol helps to prevent the development of resistance to the other three drugs, particularly rifampicin.

Following the intensive phase, the continuation phase lasts for four months and typically involves only isoniazid and rifampicin, administered daily or intermittently depending on the specific treatment guidelines and patient factors. This phase focuses on eliminating any remaining persistent bacteria to ensure a complete cure and prevent relapse of the disease. The total duration of six months is critical to achieve a bacteriological cure and minimize the risk of recurrence in individuals with drug-susceptible tuberculosis. While these regimens have been instrumental in reducing the global burden of TB, they suffer from inherent limitations that hinder optimal patient outcomes and effective disease control. One significant challenge lies in the delayed and often insensitive nature of traditional diagnostic methods, which can impede early detection and contribute to ongoing transmission within communities (Matteo et al., 2025).

Sputum smear microscopy, a cornerstone of TB diagnosis in resource-limited settings, exhibits low sensitivity, particularly in cases with low bacterial loads or among individuals co-infected with HIV (Mugenyi et al., 2024). Culture-based methods, while more sensitive, are time-consuming, often taking weeks to yield results, thus delaying the initiation of appropriate treatment. Furthermore, the empirical selection of first-line anti-tuberculosis drugs, while effective for drug-susceptible strains, fails to address the growing problem of drug-resistant TB (Mugenyi et al., 2024). The emergence and spread of multi-drug-resistant TB (MDR-TB) and extensively drug-resistant TB (XDR-TB) necessitate the use of more toxic and less effective second-line drugs, leading to prolonged treatment duration, increased adverse events, and poorer patient outcomes. These limitations underscore the urgent need to move beyond standardized approaches and embrace more personalized strategies that account for the inherent heterogeneity of the disease (Chowdhury et al., 2023).

The advent of precision medicine offers a transformative paradigm for addressing the complexities of TB management. Precision medicine, in the context of TB, aims to individualize patient care by integrating a comprehensive understanding of the specific characteristics of the host (e.g., genetic background, immunological status) and the pathogen (e.g., strain genotype, drug resistance profile, metabolic state; Lange et al., 2020). This individualized approach holds the potential to revolutionize TB diagnosis, treatment, and prevention by enabling the selection of the most appropriate

DOI: 10.1201/9781003615699-9

interventions for each patient at the right time. Key elements of precision medicine in TB include the precise identification of the Mtb strain's genotype and the patient's genetic and immunological profiles, which can inform drug susceptibility and predict the likelihood of treatment success. Critically, drug susceptibility testing (DST) plays a vital role in identifying which anti-TB drugs are effective against the patient's specific strain, thus guiding the selection of individualized treatment regimens. By tailoring treatment to the unique biological characteristics of the patient and the infecting pathogen, precision medicine promises to improve treatment outcomes, reduce drug toxicity, minimize the development and transmission of drug resistance, and, ultimately, accelerate progress toward TB elimination (Dohál et al., 2023).

The convergence of machine learning (ML) and systems biology is pivotal for advancing precision medicine in tuberculosis. Systems biology employs computational analyses to elucidate intricate biological interactions, leveraging multi-omics data to identify biomarkers indicative of disease progression, treatment response, and drug susceptibility. Complementarily, ML, a branch of artificial intelligence, offers robust analytical capabilities to interpret these extensive datasets, enabling the prediction of clinical outcomes, early detection of drug resistance, and refined patient risk stratification, thereby facilitating personalized tuberculosis management (Allami and Yousif, 2023).

This review aims to provide a comprehensive overview of the current landscape and future directions of precision medicine in tuberculosis. By critically examining the limitations of traditional TB management strategies and highlighting the transformative potential of ML and systems biology, this article will explore the emerging avenues for developing individualized diagnostic, therapeutic, and preventive interventions. Through a crisp scientific lens, we will delve into the recent advancements and challenges in leveraging these cutting-edge approaches to combat the global TB epidemic and usher in a new era of personalized care for all individuals affected by this debilitating disease.

9.2 MACHINE LEARNING IN TUBERCULOSIS

ML is rapidly emerging as a transformative force in the fight against tuberculosis. By leveraging advanced computational algorithms, ML offers powerful capabilities to analyze complex datasets inherent to TB research and clinical practice. This section will detail the diverse applications of ML in enhancing TB diagnostics, predicting drug resistance, stratifying patient risk, and ultimately contributing to the advancement of precision medicine strategies for this global health challenge.

9.2.1 Diagnostic Applications

The application of ML is poised to significantly enhance the diagnostic landscape of tuberculosis, offering solutions to overcome the limitations of traditional methods (Srivastava et al., 2025).

Leveraging the power of computational algorithms to analyze complex datasets, ML approaches are demonstrating promising results in improving the accuracy, speed, and accessibility of TB diagnosis across various domains. This section will explore the emerging diagnostic applications of ML, focusing on image analysis for chest X-rays and sputum smears, as well as the potential for early detection of active TB through the integration of clinical data and biomarkers.

9.2.1.1 Image Analysis for Enhanced TB Detection

Radiological imaging, particularly chest X-rays (CXRs), plays a crucial role in the diagnosis and monitoring of pulmonary TB. However, the interpretation of CXRs is often subjective and requires expertise that may not be readily available in resource-limited settings where the burden of TB is highest (Candemir and Antani, 2019). ML algorithms, particularly those based on deep learning architectures, offer a compelling solution to address these challenges. Trained on vast datasets of annotated CXR images, these algorithms can learn to identify subtle radiographic patterns indicative of TB with remarkable accuracy. Studies have shown that ML-powered systems can achieve diagnostic performance comparable to, and in some cases exceeding, that of experienced radiologists. This capability holds significant potential for improving the efficiency and consistency of TB screening programs, especially in high-burden countries where access to skilled radiologists is limited. Furthermore, ML algorithms can be deployed on portable digital radiography devices, enabling point-of-care TB diagnosis in remote and underserved communities, thereby facilitating earlier treatment initiation and reducing transmission (Qin et al., 2021).

Beyond CXR analysis, ML is also being explored for the automated analysis of sputum smears, a traditional cornerstone of TB diagnosis. While microscopy remains widely used, it is labor-intensive and prone to human error. ML algorithms can be trained to automatically detect the presence of acid-fast bacilli (Mtb) in digitalized sputum smear images with high sensitivity and specificity

(Chen et al., 2024). This automation not only reduces the workload of laboratory technicians but also improves the consistency and accuracy of smear microscopy results. Moreover, ML-based systems can be integrated with digital microscopy platforms, allowing for real-time analysis and remote consultation, further enhancing diagnostic capabilities in resource-constrained settings. The development of robust ML algorithms for both CXR and sputum smear analysis represents a significant step toward more efficient and accessible TB diagnostics.

9.2.1.2 Early Detection of Active TB Through Data Integration

Early detection of active TB is critical for preventing disease progression and interrupting transmission. Traditional diagnostic approaches often rely on the presence of overt clinical symptoms, which may be non-specific or absent in the early stages of the disease, particularly in vulnerable populations such as individuals with HIV, or children. ML offers the potential to overcome this limitation by integrating and analyzing diverse sources of clinical data and biomarkers to identify individuals at high risk of having or developing active TB, even before the onset of prominent symptoms (Hamna Mariyam et al., 2025).

Clinical data, including demographic information, medical history, and reported symptoms, can be leveraged by ML algorithms to develop predictive models for TB risk. By analyzing patterns in large clinical datasets, these models can identify individuals who warrant further investigation through more specific diagnostic tests. Furthermore, the integration of biomarker data, such as levels of specific proteins or metabolites in blood or other bodily fluids, can further enhance the accuracy of early TB detection. Systems biology approaches are instrumental in identifying potential biomarkers associated with early stage TB disease. ML algorithms can then be trained to recognize complex patterns of these biomarkers that are indicative of active infection, even in individuals with latent TB infection who are at risk of progressing to active disease. This capability is particularly valuable for targeted screening programs aimed at high-risk populations, allowing for timely intervention and prevention of onward transmission. The development and validation of ML-powered early detection tools, integrating clinical data and emerging biomarkers, hold significant promise for improving TB control efforts by enabling earlier diagnosis and treatment (Aamir et al., 2024).

9.2.1.3 Drug Susceptibility Testing

The effective management of tuberculosis is increasingly challenged by the emergence and spread of drug-resistant *Mycobacterium tuberculosis* strains. Rapid and accurate DST is therefore crucial for guiding appropriate treatment regimens and improving patient outcomes. ML is emerging as a powerful tool to revolutionize DST by enabling the prediction of drug resistance patterns from genomics data and by accelerating the turnaround time for obtaining DST results (Bano et al., 2025).

9.2.1.4 Predicting Drug Resistance Patterns Using Genomic Data

Whole-genome sequencing (WGS) of Mtb isolates provides a comprehensive snapshot of their genetic makeup, including mutations associated with drug resistance. However, interpreting the vast amount of genomic data to accurately predict resistance to various anti-tuberculosis drugs can be complex and time-consuming, often requiring specialized expertise. ML algorithms offer a powerful approach to overcome these challenges. By training on large datasets that link Mtb genomic sequences with their corresponding phenotypic drug susceptibility profiles (determined through traditional laboratory-based DST methods), ML models can learn intricate patterns and identify genetic markers that predict resistance or susceptibility to specific drugs (Pal and Mohanty, 2025).

Several ML techniques, including classification trees, gradient boosting algorithms, and deep learning models, have demonstrated high accuracy in predicting drug resistance to first-line and second-line anti-TB drugs based on WGS data. These models can identify both known and potentially novel resistance-associated mutations across the Mtb genome. This capability is particularly valuable for rapidly identifying resistance to multiple drugs, including MDR-TB and XDR-TB, enabling clinicians to select effective second-line regimens more promptly. Furthermore, ML can assist in interpreting complex resistance profiles arising from combinations of mutations, some of which may have synergistic or compensatory effects on drug susceptibility. The development of user-friendly ML-powered tools can democratize the interpretation of genomic data, making advanced DST accessible even in settings with limited bioinformatics expertise. By accurately predicting drug resistance from genomic information, ML has the potential to significantly expedite the process of selecting appropriate and effective anti-TB treatment, leading to improved patient outcomes and reduced transmission of drug-resistant strains (CRyPTIC Consortium, 2024).

9.3 SYSTEMS BIOLOGY IN TUBERCULOSIS

Systems biology offers a powerful framework for dissecting the intricate relationship between Mycobacterium tuberculosis (Mtb) and its human host. By integrating experimental data with computational modeling, systems biology approaches are providing unprecedented insights into the complex interplay that dictates the outcome of TB infection, from initial exposure to the development of active disease or the establishment of latent infection. This knowledge is crucial for identifying novel therapeutic targets and designing more effective interventions against this persistent disease (Dutta and Ghosh, 2024).

Systems biology approaches rely on a wealth of data and sophisticated analytical tools to unravel the complexities of tuberculosis. Key databases, such as the TB Database (TBDB), provide comprehensive repositories of genomic, transcriptomic, and proteomic data for *Mycobacterium tuberculosis*. Additionally, tools like STRING are invaluable for exploring protein-protein interactions, offering insights into potential drug targets. A summary of these and other essential resources is provided in Table 9.1 (Couvin et al., 2025; Selvan et al., 2022).

TABLE 9.1 Key Systems Biology Databases and Tools for Tuberculosis Research

Database/Tool Name	Type	Focus	Description	Example Use in TB Research
TBDB	Database	Genomic, transcriptomic, and proteomic data	Integrated platform for TB research, containing genomes, microarrays, and publications.	Identifying differentially expressed genes during infection. (Lv et al., 2025)
STRING	Database/tool	Protein-protein interactions	Provides comprehensive protein-protein interaction networks with confidence scores for various organisms, including Mtb.	Analyzing Mtb protein interaction networks to identify potential drug targets. (Jia et al., 2024)
KEGG (Kyoto Encyclopedia of Genes and Genomes)	Database	Pathways (metabolic, signaling)	Contains curated pathways and molecular interaction networks.	Mapping Mtb metabolic pathways and identifying essential enzymes. (Wang et al., 2024)
Reactome	Database	Biological pathways and reactions	A free, open-source, curated and peer-reviewed pathway knowledgebase.	Understanding host immune pathways involved in response to Mtb infection. (Powell et al., 2024)
Cytoscape	Software	Network visualization and analysis	A platform for visualizing, integrating, and analyzing molecular interaction networks.	Visualizing and analyzing Mtb-host protein interaction networks. (Arya et al., 2024)
MetScape	Software	Metabolomic data analysis and visualization	A Cytoscape plugin for the analysis and visualization of metabolomic data in the context of biological pathways.	Identifying altered metabolic pathways in Mtb-infected cells. (Jiang et al., 2021)

9.3.1 Host-Pathogen Interactions

9.3.1.1 Modeling the Complex Interplay between Mycobacterium tuberculosis and the Human Immune System

The interaction between Mtb and the host immune system is a dynamic and multifaceted process. Upon infection, Mtb, primarily residing within macrophages, triggers a complex cascade of immune responses involving various cell types, cytokines, and signaling pathways (Flynn and Chan, 2022). Systems biology employs mathematical and computational modeling to dissect this intricate interplay. These models integrate diverse datasets from in vitro and in vivo studies, including gene expression profiles, protein interactions, and metabolite levels. By simulating these interactions, researchers gain a deeper understanding of the key regulatory mechanisms governing the host immune response to Mtb.

Genome-scale metabolic reconstructions of host cells (e.g., macrophages) and Mtb are integrated to model the metabolic crosstalk during infection. These models predict changes in the production of crucial molecules like ATP and nitric oxide in macrophages upon Mtb infection, as well as identify essential metabolic pathways for the pathogen's survival within the host (Gupta et al., 2021). Furthermore, systems biology models are employed to study the spatial and temporal dynamics of immune cell interactions within granulomas, the hallmark of TB (Joslyn et al., 2021, Hoerter et al., 2022). These models elucidate factors determining whether a granuloma effectively contains the infection or leads to disease progression. By offering a holistic perspective on host-pathogen interactions, systems biology models are invaluable for understanding the complex relationship between the bacterium and the human immune system.

9.3.1.2 Identifying Key Host Factors Influencing Disease Progression

A key question in TB research concerns why some individuals exposed to Mtb develop active disease while others remain latently infected. Systems biology approaches are crucial for identifying key host factors contributing to this varied outcome. Analyzing large-scale genomic, transcriptomic, and proteomic data from individuals with diverse TB phenotypes allows researchers to identify genetic variations, gene expression signatures, and protein markers associated with altered disease progression risk (Kanabalan et al., 2021). For instance, systems biology studies have identified polymorphisms in specific host genes regulating immunity, such as those encoding cytokines and chemokines, that influence TB susceptibility. Furthermore, analysis of gene expression profiles in infected tissues or blood can reveal signatures distinguishing latent from active TB. These signatures provide insights into critical host immune responses for infection control. Moreover, systems biology can identify host metabolic pathways exploited by Mtb or essential for an effective immune response. Identifying these key host factors through systems biology is paving the way for host-directed therapies aimed at enhancing the host's ability to control Mtb infection and prevent disease progression. This knowledge is also crucial for developing personalized risk assessment strategies and identifying individuals who would most benefit from preventive interventions. (Guha et al., 2024)

9.3.2 Metabolic Pathways

Metabolic pathways are fundamental to the survival and virulence of *Mycobacterium tuberculosis* (Mtb) within the host. Systems biology approaches are crucial in dissecting the intricate metabolic adaptations of Mtb during infection and in identifying potential drug targets within these pathways. By analyzing the metabolic landscape of both the host and the pathogen, researchers can uncover vulnerabilities that can be exploited for the development of novel anti-tuberculosis therapies (van der Klugt et al., 2025).

9.3.2.1 Analyzing Metabolic Changes in TB-Infected Cells

Upon infecting host cells, primarily macrophages, Mtb undergoes significant metabolic reprogramming to adapt to the nutrient-poor and hostile intracellular environment. Simultaneously, the host cell also experiences metabolic shifts in response to the infection. Systems biology offers tools to comprehensively analyze these metabolic changes at a global level. Techniques like metabolomics, involving the high-throughput identification and quantification of metabolites, can reveal specific metabolic profiles of both Mtb and the infected host cell. Comparing these profiles with those of uninfected cells or Mtb grown in vitro allows researchers to identify metabolic pathways specifically altered during infection (Oswal et al., 2022). Studies employing stable isotope labeling and mass spectrometry-based metabolomics have demonstrated that Mtb utilizes various host-derived carbon sources, including lipids and amino acids, within macrophages (Chang and Guan, 2021). Furthermore, Mtb upregulates essential metabolic pathways for its survival, such as the glyoxylate shunt, enabling the utilization of fatty acids as a carbon source (Laval et al., 2021). Concurrently, the host macrophage also undergoes metabolic changes, often shifting toward aerobic glycolysis, known as the Warburg effect, associated with inflammation and immune activation (Escoll and Buchrieser, 2018). Systems biology approaches integrate these complex metabolic datasets to construct network models depicting interactions between different metabolic pathways, identifying key metabolic nodes critical for the survival of either the host or the pathogen during infection. Understanding these dynamic metabolic changes is essential for identifying potential therapeutic intervention points (Sharma et al., 2025).

9.3.2.2 *Identifying Potential Drug Targets*

Detailed knowledge of systems biology plays a critical role in identifying and prioritizing these targets. Comparative genomics and metabolic pathway analysis can identify enzymes unique to Mtb or with significant structural differences compared to human counterparts. These unique or divergent enzymes are attractive drug targets because inhibitors directed against them are less likely to cause off-target effects in the host (Faponle et al., 2025; Bose et al., 2021).

Furthermore, analysis of Mtb's metabolic network by systems biology approaches can identify metabolic bottlenecks – enzymes catalyzing essential reactions whose inhibition would severely disrupt the pathogen's metabolism, leading to its death or growth arrest (Craggs and de Carvalho, 2022). Genetic studies, often integrated with metabolic modeling, aid in identifying these critical enzymes. For instance, the identification of ATP synthase as an essential enzyme in Mtb's energy metabolism led to the development of bedaquiline, a novel anti-TB drug targeting this enzyme (Perveen et al., 2025). Systems biology approaches, including metabolomics and fluxomics (the study of metabolic fluxes), can also validate the essentiality of potential drug targets and elucidate the downstream effects of their inhibition (Rahman and Schellhorn, 2023). By comprehensively mapping and analyzing Mtb's metabolic pathways, systems biology remains a powerful tool in the search for new and more effective anti-tuberculosis drugs.

9.4 INTEGRATING ML AND SYSTEMS BIOLOGY FOR PRECISION TB TREATMENT

The convergence of ML and systems biology offers a powerful synergistic approach to revolutionize the precision treatment of tuberculosis. By integrating the comprehensive understanding of biological systems provided by systems biology with the predictive and analytical capabilities of ML, researchers can unlock new avenues for personalized diagnosis, treatment, and drug development in the fight against TB (Dasgupta, 2024, Guha et al., 2024).

Systems biology generates vast amounts of multi-omics data, encompassing genomics, transcriptomics, proteomics, and metabolomics, providing a holistic view of the molecular landscape in TB infection. However, the sheer volume and complexity of these datasets pose significant analytical challenges (Jana et al., 2025). ML algorithms are ideally suited for tackling this challenge by enabling the integration and analysis of these diverse data layers. By training ML models on integrated multi-omics data from TB patients with varying disease states and treatment responses, researchers can identify complex patterns and correlations that may not be apparent through traditional statistical methods. This integration allows for a deeper understanding of the intricate molecular mechanisms underlying TB pathogenesis, drug resistance, and host response (Mohr et al., 2024, Guha et al., 2024).

The application of ML to integrated multi-omics data facilitates the discovery of novel and robust biomarkers for various aspects of TB management. For diagnosis, ML models can identify unique molecular signatures that distinguish between latent TB infection and active disease or even predict the risk of progression from latent to active TB. In terms of treatment response, ML can identify biomarkers that predict the likelihood of successful treatment outcomes, treatment failure, or the development of adverse drug events. For instance, by integrating genomic data on Mtb strains with transcriptomic and proteomic data from infected patients, ML models can identify biomarkers that predict resistance to specific anti-tuberculosis drugs, allowing for more informed treatment choices. The ability to identify reliable biomarkers through the integration of ML and systems biology is crucial for moving toward more precise and personalized approaches to TB management (Yang et al., 2025). Several studies have successfully identified potential biomarkers for different aspects of TB management through the integration of multi-omics data and ML techniques, as summarized in Table 9.2.

The ultimate goal of integrating ML and systems biology in TB research is to develop personalized treatment strategies tailored to the unique characteristics of each patient. This involves moving away from standardized, one-size-fits-all treatment regimens toward approaches that consider the

TABLE 9.2 Examples of Biomarkers Identified Through Integrated Multi-omics and Machine Learning Approaches for Tuberculosis Management

Biomarker Type	Omics Data Used	Machine Learning Approach	Potential Application	Example Biomarker(s)	Reference
Metabolite	Metabolomics	Random forest	Diagnosis of active TB	Histidine, cysteine, threonine, citrulline	Yu et al. (2023)
Metabolite	Metabolomics	OPLS-DA	Differentiation of TB from T2D	Ketone bodies, lactate, pyruvate	Zhou et al. (2015)
Lipid	Lipidomics	Wilcoxon rank sum test	Identification of active TB	PG (16:0_18:1), Lyso-PI (18:0)	Collins et al. (2018)
Integrated (Metabolite & Clinical)	Metabolomics & Clinical Data	Random forest	Prediction of LTBI progression	Combination of diacetylspermine, neopterin, sialic acid and clinical factors	Yu et al. (2023)
Protein	Proteomics	Elastic-net model	Distinguishing TB from LTBI/other diseases	Leucine (serum), kynurenine (stimulated blood cultures)	Lyu et al. (2022)
Microbiome & Metabolite	Gut Microbiome & Fecal Metabolomics	Random forest	Diagnosis of pulmonary TB	Fusobacterium, fusicatenibacter (genera); 1-tetracosanol, 3-hydroxypicolinic acid (metabolites)	Wang et al. (2022)

individual's genetic background, immune status, the specific characteristics of the infecting Mtb strain (including its drug resistance profile and virulence factors), and other relevant clinical and lifestyle factors.

By leveraging the insights gained from multi-omics data analysis and biomarker identification, ML models can be developed to predict the optimal treatment regimen for a given patient. For example, based on the patient's genetic profile and the predicted drug susceptibility of their infecting Mtb strain, ML algorithms can assist clinicians in selecting the most effective combination of anti-tuberculosis drugs and determining the appropriate dosage and duration of treatment. Furthermore, ML can incorporate data on the patient's immune response and risk factors for adverse events to personalize treatment plans, maximizing efficacy while minimizing toxicity. This approach holds the potential to significantly improve treatment outcomes, reduce the duration of therapy, minimize the development of drug resistance, and enhance patient adherence. Systems biology provides a comprehensive understanding of the underlying biological processes, while ML provides the tools to translate this understanding into actionable, personalized treatment recommendations.

The integration of ML and systems biology is also transforming the landscape of anti-tuberculosis drug discovery and development. Systems biology approaches can identify novel drug targets by elucidating essential pathways and vulnerabilities in Mtb metabolism and pathogenesis. By constructing detailed network models of Mtb and its interaction with the host, researchers can pinpoint key proteins or pathways that, if inhibited, would disrupt bacterial survival or virulence (Thaingtamtanha et al., 2025).

ML can then accelerate the drug discovery process in several ways. First, ML algorithms can be trained on large datasets of chemical compounds and their known biological activities to predict the likelihood of new compounds inhibiting the identified drug targets. This can significantly speed up the screening process, allowing researchers to prioritize promising drug candidates for further experimental validation. Second, ML can be used to predict the pharmacokinetic and pharmacodynamic properties of drug candidates, such as their absorption, distribution, metabolism, and excretion, as well as their potential toxicity. This allows for the selection of drug candidates with favorable drug-like properties early in the development pipeline. Thirdly, ML can facilitate drug repurposing by identifying existing drugs approved for other indications that may also have activity against Mtb. By analyzing the molecular targets and mechanisms of action of existing drugs, ML models can predict their potential efficacy against TB, potentially accelerating the availability of new treatment options. The combination of systems biology for target identification and ML for drug discovery and optimization holds immense promise for accelerating the development of more effective and affordable treatments for tuberculosis, including drug-resistant forms of the disease (Singh, 2024).

9.5 CHALLENGES AND FUTURE DIRECTIONS

The integration of ML and systems biology holds immense promise for advancing precision medicine in tuberculosis. However, several challenges must be addressed to fully realize its potential and translate these advancements into routine clinical practice.

9.5.1 Data Quality and Availability

A fundamental challenge lies in the quality and availability of the multi-omics data required for robust model development. Generating comprehensive genomic, transcriptomic, proteomic, and metabolomic data can be costly and technically demanding, particularly in resource-limited settings where the TB burden is highest. Furthermore, the lack of standardized protocols for data acquisition, processing, and sharing can hinder the integration of datasets from different studies. Addressing issues related to data privacy and security is also crucial to facilitating data sharing and collaborative research efforts. Future directions should focus on developing more affordable and accessible technologies for multi-omics data generation, establishing standardized data formats and sharing platforms, and implementing robust quality control measures to ensure the reliability and integrity of the data used for ML model training and validation.

9.5.2 Interpretability of ML Models

Many high-performing ML models, particularly deep learning algorithms, often function as "black boxes," making it difficult to understand the underlying biological rationale behind their predictions. This lack of interpretability can be a significant barrier to their adoption in clinical practice, where clinicians need to understand why a particular prediction or recommendation is being made. While accuracy is paramount, efforts need to be directed toward developing more interpretable ML models or incorporating explainable AI (XAI) techniques that can provide insights into the features and patterns driving the model's output. This will foster trust and facilitate the integration of ML-driven insights into clinical decision-making.

9.5.3 Ethical Considerations in AI-Driven Healthcare

The application of AI in healthcare, including TB treatment, raises several important ethical considerations. Ensuring patient data privacy and security is paramount, requiring robust encryption and anonymization techniques. Algorithmic bias, where ML models perpetuate or amplify existing inequalities in healthcare due to biases in the training data, is another critical concern that needs to be addressed through careful data curation and model development. Transparency in how AI models are developed and validated is also essential for building trust among both healthcare providers and patients. Furthermore, the role of human oversight in AI-driven decision-making needs to be clearly defined to ensure that clinicians retain the ultimate responsibility for patient care. Ongoing dialogue and the establishment of ethical guidelines and regulatory frameworks are crucial for the responsible and equitable deployment of AI in TB management.

9.5.4 Integration of ML and Systems Biology into Clinical Practice

Translating the findings from systems biology research and the predictions from ML models into routine clinical practice requires overcoming several logistical and practical hurdles. This includes developing user-friendly software tools and platforms that can integrate multi-omics data, run ML models, and present the results in a clinically actionable format. Training healthcare professionals in the interpretation and application of these new tools and technologies is also essential. Furthermore, clinical validation studies are needed to demonstrate the real-world effectiveness and cost-effectiveness of these integrated approaches in improving patient outcomes. This will likely involve pilot projects and collaborations between researchers, clinicians, and public health organizations to facilitate the seamless integration of ML and systems biology into the existing TB care infrastructure. Future efforts should focus on developing practical and scalable solutions that can bridge the gap between research discoveries and their implementation in clinical settings, ultimately benefiting patients affected by TB worldwide.

9.6 CONCLUSION

In conclusion, the integration of ML and systems biology represents a transformative shift in our approach to tuberculosis management. By leveraging the power of multi-omics data integration, we can gain a deeper understanding of the complex host-pathogen interactions, identify novel biomarkers, and develop personalized treatment strategies. The application of ML in diagnostics and DST promises faster and more accurate results, while network biology illuminates the intricate pathways driving TB pathogenesis. While challenges related to data quality, model interpretability, ethical considerations, and clinical integration remain, the potential of these integrated approaches to revolutionize TB care and accelerate the path toward elimination is undeniable. Future research and collaborative efforts focused on addressing these challenges will be crucial in translating the promise of precision medicine for tuberculosis into tangible benefits for patients worldwide.

ACKNOWLEDGMENT

The author gratefully acknowledges DST-CURIE, Department of Science and Technology, New Delhi for financial support through sanction number DST/CURIE-PG/2022/10 (G).

REFERENCES

Aamir, A., Iqbal, A., Jawed, F., Ashfaque, F., Hafsa, H., Anas, Z., Oduoye, M.O., Basit, A., Ahmed, S., Rauf, S.A. and Khan, M., 2024. Exploring the current and prospective role of artificial intelligence in disease diagnosis. *Annals of Medicine and Surgery*, 86(2), pp. 943–949.

Allami, R.H. and Yousif, M.G., 2023. Integrative AI-driven strategies for advancing precision medicine in infectious diseases and beyond: a novel multidisciplinary approach. Xiv:2307.15228.

Arya, R., Shakya, H., Chaurasia, R., Kumar, S., Vinetz, J.M. and Kim, J.J., 2024. Computational reassessment of RNA-seq data reveals key genes in active tuberculosis. *PLoS One*, 19(6), p. e0305582.

Bano, N., Mohammed, S.A. and Raza, K., 2025. Integrating machine learning and multitargeted drug design to combat antimicrobial resistance: a systematic review. *Journal of Drug Targeting*, 33(3), pp. 384–396.

Bose, P., Harit, A.K., Das, R., Sau, S., Iyer, A.K. and Kashaw, S.K., 2021. Tuberculosis: current scenario, drug targets, and future prospects. *Medicinal Chemistry Research*, 30, pp. 807–833.

Candemir, S. and Antani, S., 2019. A review on lung boundary detection in chest X-rays. *International Journal of Computer Assisted Radiology and Surgery*, 14, pp. 563–576.

Chang, D.P.S. and Guan, X.L., 2021. Metabolic versatility of Mycobacterium tuberculosis during infection and dormancy. *Metabolites*, 11(2), p. 88.

Chen, W.C., Chang, C.C. and Lin, Y.E., 2024. Pulmonary Tuberculosis Diagnosis Using an Intelligent Microscopy Scanner and Image Recognition Model for Improved Acid-Fast Bacilli Detection in Smears. *Microorganisms*, 12(8), p. 1734.

Chowdhury, K., Ahmad, R., Sinha, S., Dutta, S. and Haque, M., 2023. Multidrug-resistant TB (MDR-TB) and extensively drug-resistant TB (XDR-TB) among children: where we stand now. *Cureus*, 15(2), e35154.

Collins, J.M., Walker, D.I., Jones, D.P., Tukvadze, N., Liu, K.H., Tran, V.T., Uppal, K., Frediani, J.K., Easley, K.A., Shenvi, N. and Khadka, M., 2018. High-resolution plasma metabolomics analysis to detect *Mycobacterium tuberculosis*-associated metabolites that distinguish active pulmonary tuberculosis in humans. *PLoS One*, 13(10), p. e0205398.

Couvin, D., Allaguy, A.S., Ez-Zari, A., Jagielski, T. and Rastogi, N., 2025. Molecular typing of Mycobacterium tuberculosis: a review of current methods, databases, softwares and analytical tools. *FEMS Microbiology Reviews*, p. fuaf017.

Craggs, P.D. and de Carvalho, L.P.S., 2022. Bottlenecks and opportunities in antibiotic discovery against *Mycobacterium tuberculosis*. *Current Opinion in Microbiology*, 69, p. 102191.

CRyPTIC Consortium, 2024. Quantitative drug susceptibility testing for Mycobacterium tuberculosis using unassembled sequencing data and machine learning. *PLoS Computational Biology*, 20(8), p. e1012260.

Dasgupta, S., 2024. Thinking beyond disease silos: dysregulated genes common in tuberculosis and lung cancer as identified by systems biology and machine learning. *OMICS: A Journal of Integrative Biology*, 28(7), pp. 347–356.

Dohál, M., Porvazník, I., Solovič, I. and Mokrý, J., 2023. Advancing tuberculosis management: the role of predictive, preventive, and personalized medicine. *Frontiers in Microbiology*, 14, p. 1225438.

Dutta, S. and Ghosh, A., 2024. Case study-based approaches of systems biology in addressing infectious diseases. In *Systems Biology Approaches: Prevention, Diagnosis, and Understanding Mechanisms of Complex Diseases* (pp. 115–143). Springer Nature.

Escoll, P. and Buchrieser, C., 2018. Metabolic reprogramming of host cells upon bacterial infection: Why shift to a Warburg-like metabolism? *The FEBS Journal*, 285(12), pp. 2146–2160.

Faponle, A.S., Gauld, J.W. and De Visser, S.P., 2025. Insights into active site cysteine residues in *Mycobacterium tuberculosis* enzymes: Potential targets for anti-tuberculosis intervention. *International Journal of Molecular Sciences*, 26(8), p. 3845.

Flynn, J.L. and Chan, J., 2022. Immune cell interactions in tuberculosis. *Cell*, 185(25), pp. 4682–4702.

Guha, P., Dutta, S., Murti, K., Charan, J.K., Pandey, K., Ravichandiran, V. and Dhingra, S., 2024. The integration of omics: A promising approach to personalized tuberculosis treatment. *Medicine in Omics*, 12, p. 100033.

Gupta, A., Kumar, A., Anand, R., Bairagi, N. and Chatterjee, S., 2021. Genome scale metabolic model driven strategy to delineate host response to *Mycobacterium tuberculosis* infection. *Molecular Omics*, 17(2), pp. 296–306.

Hamna Mariyam, KB, Jose, S.A., Jirawattanapanit, A. and Mathew, K., 2025. A comprehensive study on tuberculosis prediction models: Integrating machine learning into epidemiological analysis. *Journal of Theoretical Biology*, 597, p. 111988.

Hoerter, A., Arnett, E., Schlesinger, L.S. and Pienaar, E., 2022. Systems biology approaches to investigate the role of granulomas in TB-HIV co-infection. *Frontiers in Immunology*, 13, p. 1014515.

Jana, R., Hussaina, A., Assada, A., Khurshidb, S. and Machab, M.A., 2025. Chapter 10 – Challenges with multiomics data integration. In Muzafar A. Macha, Ajaz A. Bhat, A. Tariq (Eds.). *Multi-Omics Technology in Human Health and Diseases* (p. 223), Academic Press.

Jeffry, E.P. and Seikka, C.S., 2025. A comprehensive literature review of diagnosis and management of multi drug resistant tuberculosis. *The Indonesian Journal of General Medicine*, 8(1), pp. 21–34.

Jia, Q., Wu, Y., Huang, Y. and Bai, X., 2024. New genetic biomarkers from transcriptome RNA-sequencing for *Mycobacterium tuberculosis* complex and *Mycobacterium avium* complex infections by bioinformatics analysis. *Scientific Reports*, 14(1), p. 17385.

Jiang, J., Li, Z., Chen, C., Jiang, W., Xu, B. and Zhao, Q., 2021. Metabolomics strategy assisted by transcriptomics analysis to identify potential biomarkers associated with tuberculosis. *Infection and Drug Resistance*, pp. 4795–4807.

Joslyn, L.R., Renardy, M., Weissman, C., Grant, N.L., Flynn, J.L., Butler, J.R. and Kirschner, D.E., 2021. Temporal and spatial analyses of TB granulomas to predict long-term outcomes. In Yoram Vodovotz, Gary An (Eds.). *Complex Systems and Computational Biology Approaches to Acute Inflammation: A Framework for Model-Based Precision Medicine* (pp. 273–291), Springer Nature.

Kanabalan, R.D., Lee, L.J., Lee, T.Y., Chong, P.P., Hassan, L., Ismail, R. and Chin, V.K., 2021. Human tuberculosis and *Mycobacterium tuberculosis* complex: a review on genetic diversity, pathogenesis and omics approaches in host biomarkers discovery. *Microbiological Research*, 246, p. 126674.

Lange, C., Aarnoutse, R., Chesov, D., Van Crevel, R., Gillespie, S.H., Grobbel, H.P., Kalsdorf, B., Kontsevaya, I., van Laarhoven, A., Nishiguchi, T. and Mandalakas, A., 2020. Perspective for precision medicine for tuberculosis. *Frontiers in Immunology*, 11, p. 566608.

Laval, T., Chaumont, L. and Demangel, C., 2021. Not too fat to fight: the emerging role of macrophage fatty acid metabolism in immunity to *Mycobacterium tuberculosis*. *Immunological Reviews*, 301(1), pp. 84–97.

Lv, X., Guo, H., Lu, X., Tang, Z., Yu, K., Xu, G., Chen, S., Zhang, R. and Guo, J., 2025. GenVS TBDB: An open AI-generated and virtual screened small-molecule database for tuberculosis drug discovery. *bioRxiv*, 2025-04.

Lyu, M., Zhou, J., Jiao, L., Wang, Y., Zhou, Y., Lai, H., Xu, W. and Ying, B., 2022. Deciphering a TB-related DNA methylation biomarker and constructing a TB diagnostic classifier. *Molecular Therapy Nucleic Acids*, 27, pp. 37–49.

Matteo, M.J., Latini, M.C., Martinovic, D.N. and Bottiglieri, M., 2025. Update of diagnostic methods in tuberculosis (TB). *Revista Argentina de Microbiología*, 57(1), pp. 49–53.

Mohr, A.E., Ortega-Santos, C.P., Whisner, C.M., Klein-Seetharaman, J. and Jasbi, P., 2024. Navigating challenges and opportunities in multi-omics integration for personalized healthcare. *Biomedicine*, 12(7), p. 1496.

Mugenyi, N., Ssewante, N., Baruch Baluku, J., Bongomin, F., Mukenya Irene, M., Andama, A. and Byakika-Kibwika, P., 2024. Innovative laboratory methods for improved tuberculosis diagnosis and drug-susceptibility testing. *Frontiers in Tuberculosis*, 1, p. 1295979.

Oswal, N., Lizardo, K., Dhanyalayam, D., Ayyappan, J.P., Thangavel, H., Heysell, S.K. and Nagajyothi, J.F., 2022. Host metabolic changes during Mycobacterium tuberculosis infection cause insulin resistance in adult mice. *Journal of Clinical Medicine*, 11(6), p. 1646.

Pal, A. and Mohanty, D., 2025. Machine learning-based approach for identification of new resistance associated mutations from whole genome sequences of *Mycobacterium tuberculosis*. *Bioinformatics Advances*, 5(1), p. vbaf050.

Perveen, S., Pal, S. and Sharma, R., 2025. Breaking the energy chain: importance of ATP synthase in *Mycobacterium tuberculosis* and its potential as a drug target. *RSC Medicinal Chemistry*, 4.

Powell, S.M., Jarsberg, L.G., Zionce, E.L., Anderson, L.N., Gritsenko, M.A., Nahid, P. and Jacobs, J.M., 2024. Longitudinal analysis of host protein serum signatures of treatment and recovery in pulmonary tuberculosis. *PLoS One*, 19(2), p. e0294603.

Qin, Z.Z., Ahmed, S., Sarker, M.S., Paul, K., Adel, A.S.S., Naheyan, T., Barrett, R., Banu, S. and Creswell, J., 2021. Tuberculosis detection from chest x-rays for triaging in a high tuberculosis-burden setting: an evaluation of five artificial intelligence algorithms. *The Lancet Digital Health*, 3(9), pp. e543–e554.

Rahman, M. and Schellhorn, H.E., 2023. Metabolomics of infectious diseases in the era of personalized medicine. *Frontiers in Molecular Biosciences*, 10, p. 1120376.

Selvan, G.T., Gollapalli, P., Shetty, P. and Kumari, N.S., 2022. Exploring key molecular signatures of immune responses and pathways associated with tuberculosis in comorbid diabetes mellitus: a systems biology approach. *Beni-Suef University Journal of Basic and Applied Sciences*, 11(1), p. 77.

Serajian, M., Testagrose, C., Prosperi, M. and Boucher, C., 2025. A comparative study of antibiotic resistance patterns in *Mycobacterium tuberculosis*. *Scientific Reports*, 15(1), p. 5104.

Sharma, A., Tayal, S. and Bhatnagar, S., 2025. Analysis of stress response in multiple bacterial pathogens using a network biology approach. *Scientific Reports*, 15(1), p. 15342.

Singh, A., 2024. Artificial intelligence for drug repurposing against infectious diseases. *Artificial Intelligence Chemistry*, 2(2), p. 100071.

Srivastava, V., Kumar, R., Wani, M.Y., Robinson, K. and Ahmad, A., 2025. Role of artificial intelligence in early diagnosis and treatment of infectious diseases. *Infectious Diseases*, 57(1), pp. 1–26.

Thaingtamtanha, T., Ravichandran, R. and Gentile, F., 2025. On the application of artificial intelligence in virtual screening. *Expert Opinion on Drug Discovery* (just-accepted).

van der Klugt, T., van den Biggelaar, R.H. and Saris, A., 2025. Host and bacterial lipid metabolism during tuberculosis infections: possibilities to synergise host-and bacteria-directed therapies. *Critical Reviews in Microbiology*, 51(3), pp. 463–483.

Venkatesan, N., Faust, L., Lobo, R., Enkh-Amgalan, H., Kunor, T., Sifumba, Z., Rane, S., O'Brien, K., Kumar, B., Maimbolwa, C.N. and Mayta, M., 2025. Fighting tuberculosis hand in hand: A call to engage communities affected by TB as essential partners in research. *PLoS Global Public Health*, 5(4), p.e0004437.

Wang, S., Yan, N., Yang, Y., Sun, L., Huang, Y., Zhang, J. and Xu, G., 2024. Screening of drug targets for tuberculosis on the basis of transcription factor regulatory network and mRNA sequencing technology. *Frontiers in Molecular Biosciences*, 11, p. 1410445.

Wang, S., Yang, L., Hu, H., Lv, L., Ji, Z., Zhao, Y., Zhang, H., Xu, M., Fang, R., Zheng, L. and Ding, C., 2022. Characteristic gut microbiota and metabolic changes in patients with pulmonary tuberculosis. *Microbial Biotechnology*, 15(1), pp. 262–275.

World Health Organization. Tuberculosis (Fact Sheet). (2025, March 14). https://www.who.int/news-room/fact-sheets/detail/tuberculosis

Yang, Z., Li, J., Shen, J., Cao, H., Wang, Y., Hu, S., Du, Y., Wang, Y., Yan, Z., Xie, L. and Li, Q., 2025. Recent progress in tuberculosis diagnosis: insights into blood-based biomarkers and emerging technologies. *Frontiers in Cellular and Infection Microbiology*, 15, p. 1567592.

Yu, Y., Jiang, X.X. and Li, J.C., 2023. Biomarker discovery for tuberculosis using metabolomics. *Frontiers in Molecular Biosciences*, 10, p. 1099654.

Zhou, A., Ni, J., Xu, Z., Wang, Y., Zhang, H., Wu, W., Lu, S., Karakousis, P.C. and Yao, Y.F., 2015. Metabolomics specificity of tuberculosis plasma revealed by ^{1}H NMR spectroscopy. *Tuberculosis*, 95(3), pp. 294–302.

10 AI-Enhanced Drug Development against Emerging Infectious Diseases

Jatin Jangra and Rajnish Kumar

10.1 INTRODUCTION

The 21st century has confronted a rapid rise in the emergence and re-emergence of infectious diseases, often leading to major threats to global health, stability, and development [1]. If we look at the progress of historical and current pandemics, from the Spanish flu of 1918 to the outbreaks of H1N1 influenza (2009), polio (2014), MERS-CoV (2012), Zika (2016), Ebola (2019), COVID-19 (2020), and, more recently, the monkeypox outbreak of 2022, these have not only resulted in widespread morbidity and mortality but have also highly disrupted the world economies, overwhelmed healthcare infrastructures, and revealed critical vulnerabilities in global preparedness for the unforeseen pandemics. These outbreaks underscore the fact that pandemics are less like "black swan" events (rare and unpredictable) and more like "gray rhino" events viz. high-impact threats that are foreseeable and recurring, however, often neglected until it is too late [2].

The accelerated emergence of new infectious agents and the associated diseases are linked to multiple interconnected factors such as rapid urbanization, global travel, deforestation, and increased human-wildlife interactions, which ultimately leads to novel ecological niches for zoonotic spillovers. Climate change is another powerful factor that amplifies the risk of vector-borne diseases and the associated geographical range of pathogens. This can be seen from the recent Ebola and Nipah outbreaks, which demonstrate how localized ecological disruptions can quickly escalate into global health crises [3].

Despite advancements in pharmaceutical and biomedical sciences, conventional drug development processes still fail to match the pace of rapidly evolving global threats where the need for emergency services is at par. Traditional drug discovery pipelines are often costlier and time-intensive, spanning over a decade from drug discovery in the lab to the ultimate approval by the regulatory bodies for bringing the drug to the market or the patient's bedside. Also, the entire pipeline is prone to high attrition rates. The limitations associated with the traditional drug discovery and development processes struggle in the context of fast-spreading infectious diseases, where a small delay in the therapeutic regimen or vaccine availability can result in thousands of preventable deaths. Furthermore, the so-called drug development model of pharmaceutical industries is predominantly governed by market incentives because companies tend to invest only in those diseases that are of global importance and whose drug development can make profits for the company. This causes negligence to the diseases that are geographically constrained especially in low- and middle-income countries (LMICs). For example, neglected tropical diseases (NTDs), such as leishmaniasis, Chagas disease, and schistosomiasis, which affect over a billion people, receive only a fraction of the research investment compared to the other more profitable disease targets [4]. In the same manner, antimicrobial resistance (AMR) represents a prospective global health crisis. As presented by *The Lancet*, AMR has been directly linked to about 1.27 million deaths worldwide and indirectly to an estimated 4.95 million deaths in the year 2019. Moreover, it is projected to claim ten million lives annually by 2050 [5]. This again demonstrates the urgency for novel antibiotic development for the targeted diseases.

Conventional systems based on manual reporting, laboratory confirmations, and epidemiological analysis take longer for real-time decision-making, which is undesirable for the infectious diseases seeking emergency where every single minute counts. As a result, health authorities worldwide struggle to anticipate the scale and trajectory of outbreaks, leading to rapid and uncontrolled transmission as no therapeutic agent is made available in due course of time [6]. The limitations of traditional surveillance practices and drug discovery approaches therefore require integrating advanced technologies, particularly artificial intelligence (AI), that can help augment infectious disease management and drug development paradigm.

AI, driven by machine learning (ML), deep learning (DL), and data-driven predictive modeling, offers a transformative solution to many of the challenges faced by conventional drug discovery methodologies [7]. AI has the capability to process vast volumes of heterogeneous biomedical data in real-time, ranging from genomic sequences and chemical libraries to electronic health records (EHRs) and epidemiological surveillance, which makes it a fit to respond to the urgent demands posed by emerging infectious diseases (EIDs). In the realm of drug discovery and development, AI augments in each of the challenging spheres i.e. target identification, molecular optimization,

DOI: 10.1201/9781003615699-10

pharmacokinetic and toxicity profile prediction, and intelligent clinical trial design [8]. Furthermore, AI aids in drug repurposing, leading to the rapid identification of approved drugs with potential efficacy against new pathogens or disease targets, as exemplified during the COVID-19 pandemic.

In this chapter, we have highlighted the multifaceted role of AI in combating EIDs, reflecting how AI technologies are being utilized to address the conventional challenges and improve the modern drug discovery and development realm. Furthermore, we have also discussed the existing limitations and challenges associated with the large-scale implementation of AI in this field.

10.2 CURRENT CHALLENGES IN DRUG DEVELOPMENT FOR EMERGING INFECTIOUS DISEASES

The traditional drug discovery pipeline, spanning from target identification to regulatory approval, is inherently long, expensive, and inefficient, often requiring over a decade of hard work and monetary investment calculated in billions of dollars ($ 161 million to $4.5 billion) [9]. This framework itself is particularly inadequate in the context of EIDs, where the speed of bringing drugs to the general public and the rapid adaptability of the host system/pathogen are equally important. EIDs such as COVID-19, SARS, H1N1, MERS, and Ebola rapidly evolve with limited biological understanding and often with high virulence and mortality. Under such circumstances, the delay associated with the conventional drug discovery process is a major hurdle to effective treatment and bypassing the effects caused to the general public.

In such unpredictable outbreaks, drug repurposing strategies – i.e., choosing existing, approved drugs with known safety and pharmacokinetic profiles – gain significant attention to combat the tragedy in meaningful time and resources. However, choosing that drug is itself one of the biggest challenges as the conventional practices are based on trial-and-error methods, relying primarily on putative mechanisms of action and standard dosing regimens [1]. While repurposing expedites drug discovery, this empirical approach has a limitation of treatment efficacy and may result in suboptimal patient outcomes due to the lack of data-driven precision in combination design and dosing where the conventional system lags.

Initial pathogen characterization is another important hurdle to drug development against EIDs. This step itself is a very daunting task and often limited by high-quality genomic, proteomic, or structural data, especially during the early stages of an outbreak, which causes delays in rational target identification and validation. Even once potential targets are known, developing and optimizing candidate molecules through high-throughput screening and lead optimization is slow, labor-intensive, and costly [10]. This hurdle is especially problematic for pathogens capable of mutation (leading to evolution) such as SARS-CoV-2 and the ability to persist in latent or intracellular forms – for example, *Mycobacterium tuberculosis*, where identifying relevant drug targets and mimicking host-pathogen interactions remains a major issue [11].

Preclinical testing is another challenge to the iterative drug development paradigm for EIDs. Many of the EIDs lack validated animal models that can closely reflect human pathophysiology. This leads to high attrition rates in the long run, as promising lead compounds fail in human trials despite encouraging preclinical data owing to poor translation [12]. Furthermore, designing and conducting large-scale, multi-phase clinical trials for EIDs is another big issue as the majority of infectious diseases are often niched to a particular geographical location which often complicates patient recruitment and retention for the trial duration.

Lastly, the growing burden of AMR further exacerbates the conventional drug discovery processes. Many healthcare settings, especially in LMICs, face severe diagnostic limitations that lead to empirical antibiotic use, accelerating the evolution of microbial resistance [13]. Around 65% of infections, particularly those linked to medical devices like catheters and implants, are the result of bacterial biofilms, which are basically small microbial colonies encased in a protective matrix and showing significantly higher resistance to antibiotics, up to about 1,000 times more than free-floating (planktonic) cells due to limited drug penetration and altered gene expression. Biofilms can further exacerbate the situation by evading into the host's immune system and, finally, leading to persistent and chronic infections [14]. Therefore, early detection of biofilm-forming bacteria is essential for effective treatment, which itself is a challenging task. Moreover, sepsis, a severe and often fatal immune reaction to infection, affects about 1.7 million people annually in the United States, causing approximately 350,000 in-hospital deaths [10]. Globally, sepsis accounts for 19.7% of all annual deaths, underscoring its critical health burden [15]. This highlights the need for the development of antibiotics, as well as the strict implementation of stewardship programs [5].

The limitations associated with conventional drug discovery and development strategies demand an alternative method or a technology that can transform the way by which drugs and related therapeutics are currently discovered. AI is helping to get rid of the challenges associated with these conventional drug discovery and development methods by integrating it into the key facets, viz., target identification, molecule design, lead optimization, toxicity testing, trial optimization, resistance prediction, and treatment personalization; the drug development process for EIDs can become more adaptive, efficient, and outcome-focused. For example, Kanjilal et al. developed an ML model to predict the most appropriate antibiotic therapy for patients with urinary tract infections (UTIs), using records of over 10,000 patients. Upon testing retrospectively, the model revealed a significant reduction in unnecessary consumption of second-line antibiotics by 67% and outperformed standard clinical decision-making [16]. Moreover, AI-enabled diagnostic tools, including mobile applications for antimicrobial susceptibility interpretation, can help bridge critical gaps in timely and accurate disease detection, especially in clinical settings with limited resources [17].

10.3 APPLICATIONS OF AI ACROSS THE DRUG DEVELOPMENT PIPELINE FOR EMERGING INFECTIOUS DISEASES

In the era of the modern drug discovery realm, AI has emerged as a driving force across every stage of the drug development pipeline, especially in the fight against EIDs where prompt action is required, right from early stage target identification to clinical translation, diagnostics, regulatory submissions, and post-marketing surveillance owing to its high speed, accuracy, and scalability (Figure 10.1).

The process initiates from target identification and validation, followed by high-throughput virtual screening, lead identification and optimization, preclinical evaluation, and clinical trials for the ultimate approval of the identified drug by regulatory bodies. Across these stages, AI tools aid in overcoming limitations associated with traditionally laborious tasks such as virtual screening, biomarker identification, patient selection, and adverse effect prediction, thereby accelerating the overall workflow and improving the precision of therapeutic development speedily and more efficiently.

10.3.1 AI in Understanding Pathogen Biology and Target Identification

Target identification serves as a foundational step, although challenging in the realm of anti-infective drug discovery against EIDs, where the pathogen itself is novel and limited biological knowledge is available. It involves the mining of genomic, transcriptomic, and proteomic datasets to identify potential druggable molecular targets. AI has significantly transformed this step by integrating protein-protein interaction (PPI) networks, gene expression profiles, and structural bioinformatics. ML algorithms such as random forests (RFs), support vector machines (SVMs), and

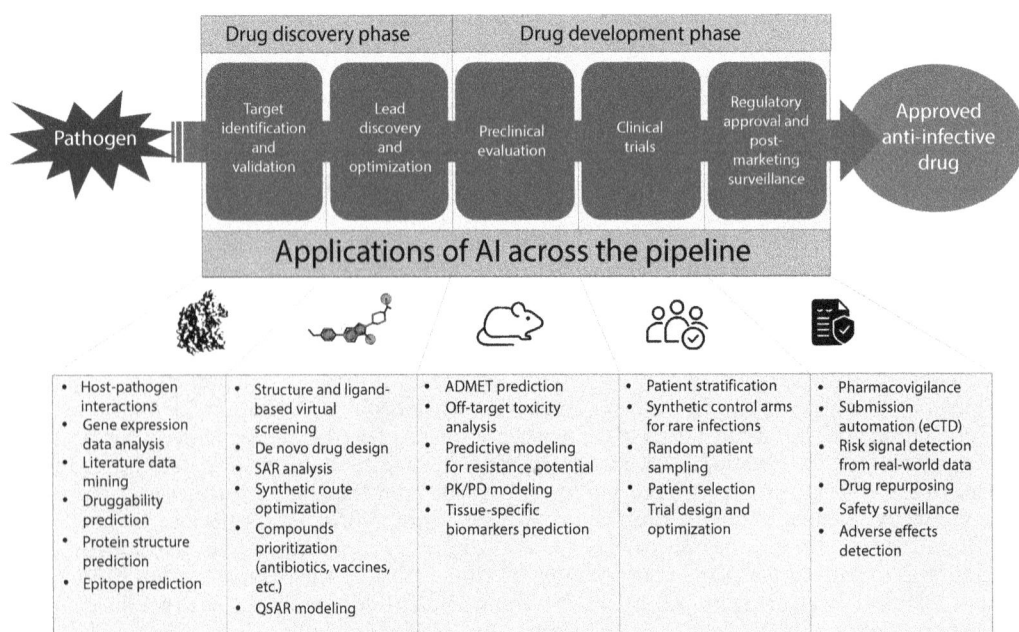

Figure 10.1 Drug discovery and development pipeline for EIDs.

autoencoders have been applied to RNA-Seq and microarray data to classify differentially expressed genes (DEGs) that are upregulated or downregulated in response to infection [18].

Platforms such as AlphaFold2 and MODELLER assist in accurate protein structure prediction, which can be further validated using tools like ProQ. Subsequent binding site exploration helps identify orthosteric, allosteric, and cryptic pockets essential for rational drug design. DL frameworks, including generative adversarial networks (GANs), recurrent neural networks (RNNs), and transfer learning algorithms, are increasingly employed for multi-omics target prioritization [19]. For instance, Pun et al. utilized disease-specific bioinformatics and DL-based models to identify 18 druggable genes associated with amyotrophic lateral sclerosis (ALS) [20]. Similarly, Fabris et al. developed a modular DL architecture leveraging gene ontology, PPIs, and biological pathway data to map gene-disease associations across age-related pathologies [21].

In another study by Barman et al., an ML-based classification framework was developed to identify infectious disease-associated host genes by integrating sequence-derived features with protein interaction network data. Among the different models tested, a deep neural network (DNN) achieved the highest performance with an accuracy of 86.33%, sensitivity of 85.61%, and specificity of 86.57%, utilizing 16 optimized features from pseudo-amino acid composition (PAAC). The model also displayed robust generalizability, with an accuracy of 83.33% on a blind dataset and 83.1% sensitivity on an independent validation set. When the model was applied to reviewed proteins across databases, the model predicted 100 top-ranked host genes, out of which 76 were corroborated by experimentally validated human-pathogen PPIs, underscoring its predictive reliability [22]. Table 10.1 highlights the key AI models that assist in rational protein structure identification and binding site prediction.

TABLE 10.1 Some Examples of AI Models for Target Structure and Binding Site Prediction

Application	Tool	Features	Web Address Link	Reference
Target structure prediction	RoseTTAFold	Uses MSA (multiple sequence alignment) for structure prediction and can run on standard GPUs	https://github.com/RosettaCommons/RoseTTAFold	[23]
	AlphaFold	Attention-based transformer model and can handle single sequences	https://github.com/google-deepmind/alphafold	[24]
	ESMFold	Transformer-based protein language mode and works extremely fast with single protein sequences	https://github.com/facebookresearch/esm	[25]
	OmegaFold	Important for orphan proteins and antibodies with no homologs	https://github.com/HeliXonProtein/OmegaFold	[26]
Binding site prediction	DeepSite	Convolutional neural networks (CNN)–based model to capture complex spatial features	https://enzostvs-deepsite.hf.space	[27]
	SiteMap	Grid-based and energy-based scoring to predict detailed binding properties	https://www.schrodinger.com/platform/products/sitemap	[28]
	DeepBind	DL-based feature extraction to recognize functional binding sites	http://www.rnainter.org/DeepBind	[29]
	P2Rank	RF algorithm-based and provides template-free approach	https://github.com/rdkp2rank	[30]
	GraphSite	Utilizes graph neural network (GNN) and sequence features to predict ligand-binding sites in proteins	https://github.com/biomed-AI/GraphSite	[31]

10.3.2 AI in Predicting Host-Pathogen Interactions

The term "host-pathogen interaction" (HPI) encompasses the mechanisms and processes through which a pathogen (bacteria, viruses, fungi, prions, and viroids) interacts with host cells, from initial invasion to eventual colonization [32]. These interactions are influenced by various environmental factors such as temperature, light, and season, and can be analyzed at different levels: the population level (infections across human populations), the organismal level (infections within an individual host), and the molecular level (interactions between pathogen proteins and host cell receptors) [33, 34]. HPIs are primarily governed by PPIs at the molecular level, occurring between host and pathogen proteins. These interactions help address critical questions about the mechanisms underlying host dynamics and pathogen behavior, ultimately aiding in the identification of potential candidates for the development of therapeutics and diagnostic agents [33, 35].

Various computational methods have been developed to predict interactions between host and pathogenic proteins. These approaches are broadly categorized into experimental (laboratory-based) and computational methods [36]. Laboratory techniques, although foundational, are often time-consuming, resource-intensive, and limited by incomplete data and high false-positive rates [37]. Among computational strategies, physicochemical property-based prediction, using features like molecular weight, charge, solubility, and isoelectric point, is widely used due to its simplicity. However, it often lacks reliability as it overlooks other relevant biological factors [38]. Structure-based prediction relies on the similarity of protein structures to known interacting pairs, while motif-based approaches assess interaction potential based on the presence of specific sequences or structural motifs [39].

In recent years, the field of PPI prediction has been significantly boosted due to the advent of AI techniques in the domain and the availability of a huge amount of proteome data (Table 10.2). UlQamar et al. developed a multi-layer perceptron model, viz., Deep-HPI-pred utilizing the GreeningDB and Pred-HPI datasets and validated using K-fold cross-validation. The model showed the best performance with an MCC value of more than 0.8 and an accuracy of 93%. A total of 1,011 HPIs, including 359 hosts and 45 different pathogens, were obtained after data refinement. In another study, Bastien et al. developed an RF algorithm–based host-virus interaction prediction model (VHIP) with an accuracy of 87%. A dataset comprising 8,849 samples of host and virus sequences was utilized to construct the model [40]. Driven by the COVID-19 pandemic attack, which caused millions of deaths, and the need to predict host-pathogen protein interactions (HPPIs), Kaundal et al. developed a DL-based model named deepHPI, which works through CNN and can predict interactions between plants, animals, and humans as hosts, with bacteria and viruses as attacking pathogens. A total of 56,834 interactions between 6,621 pathogens and 15,811 hosts have

TABLE 10.2 Selected Examples of AI/ML Models for Predicting HPIs

Model Name	AI Technique	Dataset	Evaluation Metrics	Web Address	Reference
Deep-HPI-pred	CNN, sequence-based encoding	Pred-HPI/GreeningDB	MCC[a] = 0.80 ACC[b] = 0.93	https://cbi.gxu.edu.cn/shiny-apps/Deep-HPI-pred	[40]
deepHPI	Embedding, CNN	HPIDB/Mined PPIs	ACC ~ 0.97	http://bioinfo.usu.edu/deepHPI	[41]
ProteinPrompt	RF algorithm	DIP/HPRD/PDB/Negatome	ACC = 0.88	http://proteinformatics.org/ProteinPrompt	[42]
InterSPPI	Sequence-based embedding	IntAct/BioGRID/TAIR	ACC = 0.74	http://zzdlab.com/InterSPPI	[43]
DeepViral	DNN model	HPIDB3	ROC ~ 0.81	https://github.com/bio-ontology-research-group/DeepViral	[44]
MuPIPR	Deep residual recurrent neural network (ResRNN)	SKEMPI/STRING/Guo	ACC = 0.97	https://github.com/guangyu-zhou/MuPIPR	[45]

[a] MCC: Matthews Correlation Coefficient.

[b] ACC: Accuracy.

been reported in their database curated using Mined PPIs and Host-Pathogen Interaction Database (HPIDB). The model achieved an accuracy of 98% for plant-pathogen interactions, 95% for both human-bacterial and human-viral interactions, and 97% for interactions between animals and pathogens [41].

In another study, Liu-Wei et al. developed a DL-based model, DeepViral, for predicting protein interactions between humans and viruses utilizing HPIDB (version 3). The model integrates both protein sequence information and disease-related phenotypes to improve prediction accuracy. DeepViral consists of two major components: a phenotype model, which captures features of human and viral proteins based on disease symptoms, and a sequence model, derived from the DeepGOPlus architecture, which utilizes a CNN to process one-dimensional amino acid sequences (Figure 10.2). The model was trained as a binary classifier using binary cross-entropy loss to predict the probability of interaction between a protein pair and demonstrated satisfactory performance in identifying human virus PPIs [44].

In order to predict the PPIs between humans and viruses, a doc2vec model based on an RF algorithm, utilizing the HPIDB (version 3.0) database, was proposed by Yang et al. This model was constructed using 291,726 proteins and was trained using amino acid sequences as words and full-length sequences as sentences [46]. Another interesting model named APEX2S, which works on both gradient-boosted (XGB) and SVM algorithms, was developed by Chen et al. to predict PPIs between host proteins and microbial pathogens. The former algorithm serves as the first layer with a sampling scheme, and the latter acts as a final classifier based on synthetic minority oversampling techniques. Datasets from Reactome, DIP, MINT, IntAct, InnateDB, PATRIC, HPIDB, Mentha, BioGRID, and PHISTO were utilized to construct the model, which achieved an accuracy of 98% [47].

Furthermore, Zhou et al. developed a DL-based model named PIPR to predict PPIs using a recurrent convolutional neural network (RCNN) framework. The model captures the mutual influence between protein sequences by processing sequence-based textual features. It was trained and evaluated using the SKEMPI, STRING, and Guo's datasets. PIPR achieved an accuracy of over 97%, outperforming existing ML models. The model's performance was further validated through 10-fold cross-validation, confirming its robustness [48].

10.3.3 AI in Lead Compound Identification and Optimization

Once the potential druggable targets are identified and subsequently validated, the next step in the drug development pipeline is the identification of lead compounds viz. small molecules or biologics that can interact with these targets to awaken a therapeutic response. In the case of EIDs, the urgent need for therapeutic treatment, limited availability of experimentally validated inhibitors for new pathogens, and huge chemical space bring about significant challenges to this drug discovery stage that need to be tackled in the shortest duration possible. AI aids in this stage too by putting forward robust solutions such as enabling high-throughput structure- and ligand-based virtual screening, de novo drug design, and quantitative structure-activity relationship (QSAR) modeling, which speeds up the process for discovering newer hit molecules and subsequent optimization to the promising lead moieties [18, 49].

Traditional QSAR methods were mainly based on linear regression but have expeditiously evolved through the incorporation of ML and DL techniques, including RFs, SVMs, and XGBoost. These models help in predicting IC_{50}, binding affinity, or the antimicrobial activity of the discovered virtual hits by training on experimentally determined bioactivity data. Tools such as Chemprop (Figure 10.3) and DeepChem, which are DL framework-based and utilize graph representations or molecular fingerprints, have proven truly efficient in settings with low data, especially EIDs with satisfactory performance [50, 51].

Virtual screening is another exploited computational technique by medicinal chemists to identify the potent hits based on either the 3D structural data of the protein's binding site (structure-based virtual screening) or the reported ligands from the literature (ligand-based virtual screening). But the problem here is that when a new pathogen is encountered, its 3D structural data is not available for screening against the already reported ligands owing to its novelty and limited biological knowledge. However, when the information related to the protein sequence is made available by the structural biologists, the protein structure can be predicted with tools like AlphaFold before the experimentally resolved confirmation, and further, the medicinal chemists can screen huge chemical libraries against the target's binding site by leveraging AI tools such as DeepDock with high efficiency and accuracy [24, 52]. While traditional structure-based virtual screening relies mainly on molecular docking, AI can help augment this by faster approximations of binding scores or ligand-protein interaction maps, which saves a lot of time and computational power. CNN-based tool

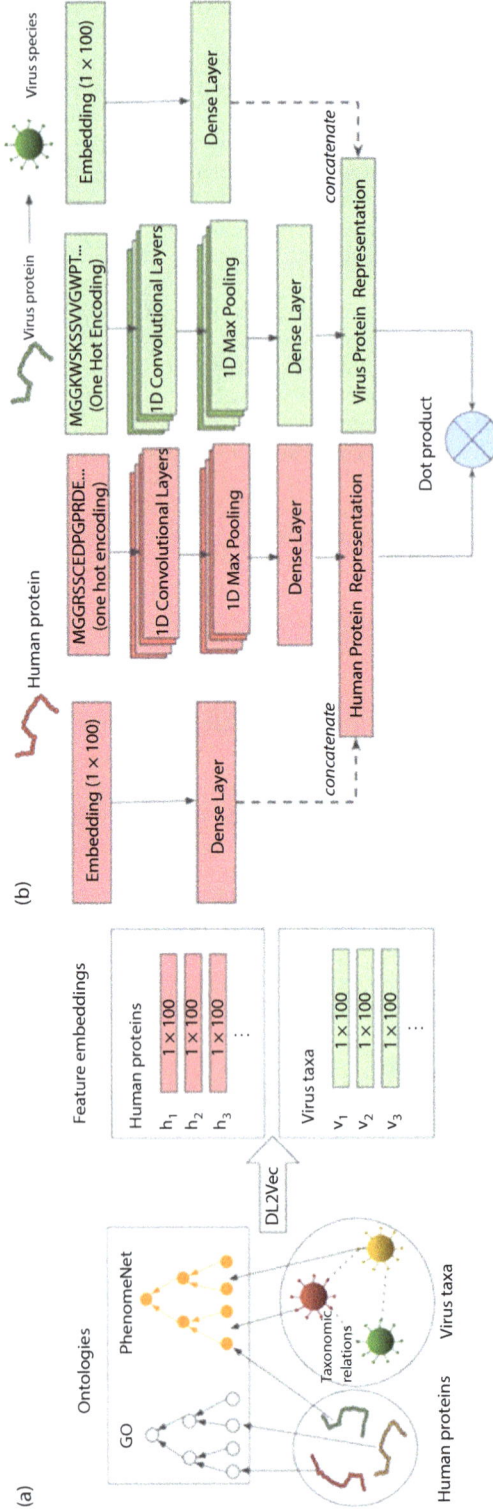

Figure 10.2 Workflow of DeepViral model for predicting HPIs: (a) Feature embeddings were generated using DL2Vec, where human proteins and virus taxa were annotated with ontology classes. Arrows represent ontology annotations, while dashed lines indicate taxonomic relationships between viruses. These annotations, taxonomies, and ontologies were used to produce 100-dimensional embeddings for each human protein and virus taxon. (b) In the joint prediction model, latent representations from both feature embeddings and protein sequences were concatenated for human and viral proteins. A dot product was then applied to the combined vectors to predict potential interactions. (Reproduced from Wang Liu-Wei et al. (2021) [44], under the terms of the Creative Commons Attribution 4.0 International License [https://creativecommons.org/licenses/by/4.0].)

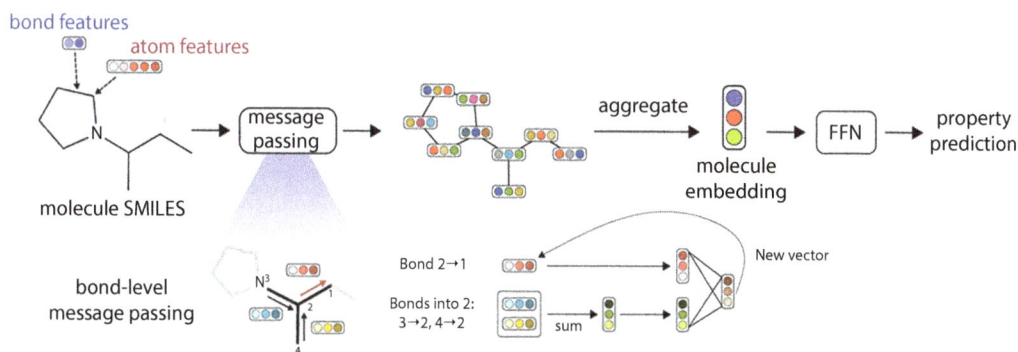

Figure 10.3 Structural outline of the Chemprop prediction framework. The framework comprises four primary components: (a) a module for encoding local chemical features, (b) a directed message-passing neural network (D-MPNN) that learns atomic-level representations, (c) an aggregation function that combines atomic embeddings into a single molecular vector, and (d) a feed-forward neural network (FFN) that maps the molecular representation to predicted properties. (Reproduced from Heid E et al. (2023) [50], under the terms of the Creative Commons Attribution 4.0 International License [https://creativecommons.org/licenses/by/4.0].)

AtomNet, trained on 3D protein-ligand complexes, helps in predicting binding affinity from the structural features directly [53]. These approaches help in the early prioritization of the most promising candidates and further reduce the resource investment for in vitro analysis.

Another interesting and emerging field in the lead identification stage is the de novo drug design that implicitly generates novel molecules specifically tailored to a particular target protein with desired physicochemical properties by leveraging generative AI algorithms, including variational autoencoders (VAEs), GANs, and reinforcement learning (RL) frameworks. Tools like MolGAN, REINVENT, and GENTRL work based on reward functions in order to generate molecules that meet the specified criterion (Table 10.3) [54–56]. Generative AI tools stand as a testament to the therapeutics development against EIDs for which no known ligands exist in the early stage. However, the novel leads generated with these tools may have the limitation of synthetic feasibility, which needs to be tackled carefully.

It is commonly observed that the molecules selected based on certain physicochemical criteria often fail at the later stages of drug discovery and development. For example, if a molecule is selected based on high permeability across the biological membranes, it may have issues with the solubility in the biological fluids leading to precipitation. Therefore, the decision to select a particular molecule for the next phases of drug discovery is the critical checkpoint that needs careful attention. However, AI (especially DL) may also help at this stage by leveraging its multi-objective optimization frameworks that can balance multiple drug-like properties parallelly during lead design [58]. These become critical for EID settings, where urgent action is needed but not at the cost of safety and efficacy.

The advent of AI in lead identification can be highlighted by the COVID-19 pandemic. For instance, BenevolentAI repurposed baricitinib as a potential candidate for COVID-19 management within weeks of the outbreak [59]. On a similar note, Insilico Medicine identified inhibitors that can target the SARS-CoV-2 main protease (Mpro) by utilizing the power of generative AI in de novo drug design under the harsh pandemic timelines [60]. This highlights the indispensable role of AI in outbreak scenarios where the urgency is at its peak while the time and resources are scarce.

10.3.3.1 AI in Epitope Design and Vaccine Development for Infectious Diseases

Vaccine development is one of the major keys to the fight against EIDs where rapid action is required. Conventional vaccine design approaches are often time-intensive and resource-hungry; however, the inclusion of AI in this domain can significantly accelerate the process by aiding across multiple stages, including antigen identification, epitope prediction, adjuvant selection, and optimization of immunogenic responses. Advanced AI algorithms can help predict antigenic epitopes and model protein structures by utilizing genomic and proteomic data in order to prioritize candidates with high immunogenic potential [18]. The DL frameworks also serve as key players in the rational design

TABLE 10.3 Representative AI Tools for Ligand Screening (Hit Identification), de novo Ligand Generation, Lead Optimization, and Property Prediction in the Early Stages of Drug Discovery

Tool	AI Technique	Description	Web Link	Reference
Chemprop	DL (graph neural nets)	Message-passing neural networks for molecule property prediction	https://github.com/chemprop/chemprop	[50]
DeepChem	DL-based	Used in drug discovery workflow such as structure-activity modeling	https://deepchem.io	[51]
AtomNet	DCNN	A structure-based, deep convolutional neural network designed to predict the bioactivity of small molecules	https://www.atomwise.com/how-we-do-it	[53]
REINVENT	RL	For de novo design, scaffold hopping, R-group replacement, linker design and molecule optimization	https://github.com/MolecularAI/REINVENT4	[54]
MolGAN	GAN for molecular graphs	Implicit generative model for small molecular graphs	https://github.com/nicola-decao/MolGAN	[55]
GENTRL	VAE-based generative model	Aids in de novo drug design	https://github.com/insilicomedicine/GENTRL	[56]
CReM	Fragment-based structure generation framework (implemented using the open-source python module)	Generates chemically valid structures and provides flexible control over diversity, novelty, synthetic complexity and chemotypes of generated compounds	https://doi.org/10.1186/s13321-020-00431-w	[57]

of immunogens and identification of novel adjuvants while keeping the safety and efficacy portfolio on the utmost priority. Moreover, AI tools also help in improving the translational success rates of the designed vaccines by forecasting immune responses and analyzing preclinical and clinical datasets [61].

AI-driven epitope prediction models are now capable of identifying antigenic regions and assessing immunogenicity while maintaining high accuracy and efficiency [62]. For instance, frameworks have been applied to the identification of epitopes for the Zika virus that triggers robust immune responses [63]. These models incorporate diverse input features, including sequence motifs, structural topology, and physicochemical attributes to accurately map regions that are most likely to be immunogenic [64]. Additionally, in another study, RF-based classifiers were successfully employed in the rational design of malaria vaccines to boost the antigen prioritization and experimental validation pipeline [65].

The applications of AI-powered immunoinformatics tools are currently on the rise and in high demand. These have accelerated epitope-based vaccine design, particularly during the COVID-19 pandemic, where the first SARS-CoV-2 vaccine candidate utilizing immunoinformatics tools was released in early February 2020 [66]. This discovery marked a significant milestone in rapid-response vaccine development strategies. ML-enabled platforms such as BepiPred, ABCpred, LBtope, BCPREDS, and SVMtrip help in the identification of linear B-cell epitopes directly from antigenic protein sequences. Among them, LBtope has demonstrated consistent reliability in predicting linear epitopes while utilizing primary sequence data alone [67].

For T-cell epitope prediction, tools such as Rankpep, SYFPEITHI, MAPPP, PREDIVAC, PEPVAC, Deepitope, EPISOPT, and Vaxign have been developed that aid in the rapid identification of HLA-binding peptides utilizing motif-matrix methodologies [68]. In parallel, advanced neural network-based tools such as NetMHC, NetMHCII, NetMHCpan, NetMHCIIpan, and nHLApred have further enhanced prediction by taking complex antigen–MHC interactions into the picture [69].

In addition to immunogenicity profiling, allergenicity assessment marks the critical stage in the evaluation of a vaccine construct. Tools such as AllergenFP, AllerTOP, Allermatch, APPEL, and EVALLER have been developed to screen for potential allergenic responses systematically [70]. Further, tools such as ProSA-web, GalaxyRefine, and Raptor-X/SOPMA help in the structural validation and refinement of vaccine constructs [71]. Further, molecular docking and molecular dynamics (MD) simulation studies can be carried out to augment the findings based on binding interactions between predicted epitopes and host cell receptors [72].

Additionally, a DL-based framework, MUNIS was developed by Wohlwend et al. to accurately predict CD8+ HLA-I-restricted T-cell epitopes [73]. Similarly, DeepVacPred has been developed to automate the identification of immunogenic subunits from raw protein sequences [74]. Another such tool, integralVac, aids in flagging hemolytic peptides during early stage peptide vaccine screening [75]. These advancements in the field of vaccine design and development underscore the potential of AI-integrated immunoinformatics platforms in redefining the horizons for the therapeutics development against critical EIDs by overcoming the time constraint and limited resources of the prevailing conventional vaccine design realm.

10.3.3.2 AI in Antibiotics Discovery and Antimicrobial Resistance Detection for Infectious Diseases

AMR poses a major threat to global public health and often hampers the therapeutic potential of currently available scarce antibiotics, leading to high morbidity and mortality rates worldwide. The unintended increased use of antibiotics not only destructs the pathogen (xenobiotic) but also causes an alarming risk to the gut microbiome, leading to diminished microbial diversity and higher chances of the development of resistance genes [76]. Moreover, antibiotic drug discovery is highly challenging and confounded with different risk factors, such as long-time, huge investment and development of resistance to the pathogen in due course of time leading to the failure of that particular antibiotic. This is evident from their development history. Over the span of four years (2014–2018), only 14 new antibiotics were approved by the regulatory authorities and brought to the market [77]. Also, it is important to note that the success rate for newer anti-infective drug candidates is a mere 25.2%, as reported from the analysis of over 186,000 clinical trials involving more than 21,000 compounds [78].

To address these limitations, the integration of AI-driven platforms is the utmost requirement where the faster analytical ability of AI can yield the rapid identification of novel antibiotics with unprecedented mechanisms of action [79]. Although it is estimated that the chemical space for drug-like small molecules is huge (10^{30}–10^{60} entities), and for n-length peptides, protein sequence variability introduces 20^n possible permutations [80]. This big possibility of permutations and combinations makes the experimental exploration for each drug-like molecule from chemical space and the corresponding peptide almost infeasible. This is where AI can be of great help by intelligently navigating through this druggable chemical space and accelerating the discovery of new generation of antibiotics against resistant pathogens, especially for EIDs (Figure 10.4).

Similar to general predictive modeling, AI here also can be employed to predict the antimicrobial activity of the identified potential antibiotics by leveraging the QSAR modeling approaches [81]. Different ML algorithms help predict the relationship between antimicrobial efficacy and the corresponding molecular structure. For instance, researchers have employed multinomial logistic regression models to classify molecular fragments on the basis of their contextual relevance within training datasets in order to overcome the limitations of traditional fragment-based drug design approaches. This resulted in the construction of a structured "vocabulary" of chemical moieties that can further be exploited to recombine and generate candidate molecules that are active against *Pseudomonas aeruginosa* [82]. Based on this study, researchers have further developed several genome-based models by utilizing the XGBoost algorithm to estimate minimum inhibitory concentrations (MICs) for 20 antibiotics targeting *Klebsiella pneumoniae*, and for 15 antibiotics against various nontyphoidal *Salmonella* strains [76, 83, 84]. In another study, RNNs along with the regression models trained on features related to bactericidal activity have been utilized to prioritize the antimicrobial peptides (AMPs) utilizing the sequence-based encoding [85].

Nowadays, AI is touching the new hike in the detection of AMR, which stands as a major cornerstone in the clinical management of infectious disease outbreaks. Recently, numerous AI-driven frameworks have been deployed into microbiological diagnostics in order to enhance the accuracy,

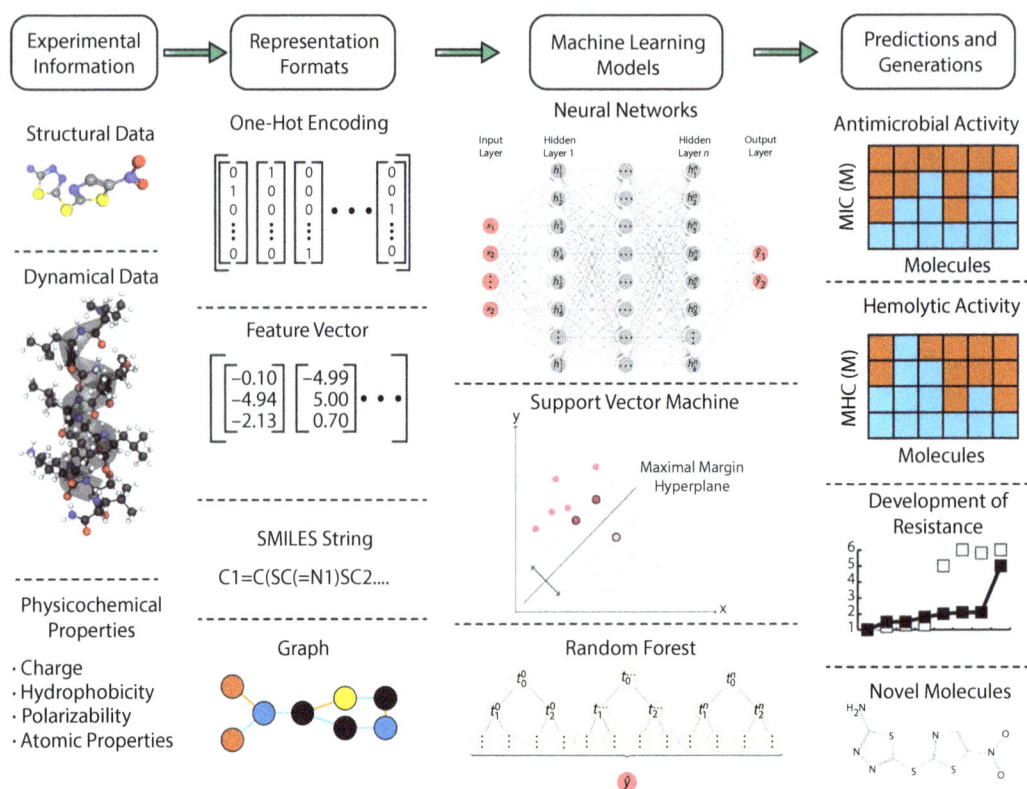

Figure 10.4 The basic workflow for antibiotic discovery. It begins with the systematic curation of high-quality experimental datasets and subsequent transformation to AI-compatible representations such as molecular fingerprints, graph-based encodings, or sequence embeddings, which serve as the input for downstream learning algorithms. Further, the model is trained and then deployed to predict properties such as antimicrobial potency, cytotoxicity profiles, and resistance development potential. (Reproduced from Melo MCR et al. (2021) [76], under the terms of the Creative Commons Attribution 4.0 International License [https://creativecommons.org/licenses/by/4.0/].)

precision, speed, and scalability of resistance profiling of a particular microbe so that the corresponding antibiotic can be chosen in zero time delay. ML-augmented matrix-assisted laser desorption ionization-time of flight mass spectrometry (MALDI-TOF MS) techniques have shown exceptional performance with 100% classification accuracy for the detection of methicillin-resistant *Staphylococcus aureus* (MRSA) and carbapenem-resistant *Klebsiella pneumoniae* (CRKP) [86]. Another such platform – namely, AFB+Neon Metafer – has shown a significant promise in tuberculosis diagnostics through the identification of acid-fast bacilli on smear-negative slides and that, too, in low-burden samples. Additionally, AI has been integrated with antimicrobial susceptibility testing (AST) through innovations such as the SlipChip microfluidic system that enables parallel inoculation and high-throughput AST directly from positive cultures [87]. This integration reduces the chances of treatment errors and accelerates therapeutic decision-making, especially for conditions like EIDs.

Moreover, another strategy that utilizes pathogen genomic data to predict AMR phenotypes is also currently accelerating. Several ML models have been developed for the detection of AMR in clinically significant pathogens such as *Klebsiella pneumoniae*, *Escherichia coli*, *Pseudomonas aeruginosa*, *Mycobacterium tuberculosis*, and *Staphylococcus aureus* [76]. Since many of these models have been trained on drug- and species-specific datasets, this limits their generalizability to other strains. For this, neural network-based frameworks (for example, DeepARG) are emerging that incorporate environmental metagenomic analysis [88]. However, the predictive performance of these models remains highly variable, depending upon several factors such as specific antibiotic-pathogen combinations, sampling strategy, and the complexity of the underlying resistance mechanism [89].

10.3.4 AI in Preclinical Prediction for Emerging Infectious Diseases

Once the lead compound is identified, it is necessary to predict its safety profile before advancing it for studies on animals and humans. Preclinical evaluation, including predictions of toxicity, absorption, distribution, metabolism, excretion, and toxicity (ADMET) profiles, serves as a crucial stage of the drug discovery and development pipeline. In the context of EIDs, limited experimental resources are available, and prompt response is required to deal with them. AI-driven predictive modeling provides a better alternative for both in vitro and in vivo studies in such scenarios.

ML models such as RFs, SVMs, and XGBoost decision trees, trained on large datasets corresponding to physicochemical properties, activity, and toxicity profiles, help predict important properties such as oral bioavailability, blood–brain barrier permeability, and plasma protein binding for the lead compounds in question. Models based on GNNs, CNNs, and transformers that utilize graph-based features to predict ADMET properties outperform the traditional algorithms working on SMILES strings by undermining the hidden spatial features [90].

Several AI models trained on public databases have been developed by the researchers to predict the toxicity profile of the lead molecules (Table 10.4). For instance, Tox21 can predict multiple toxicity endpoints, including hepatotoxicity, cardiotoxicity, mutagenicity, and cytotoxicity [107]. Other tools based on DNN and ensemble learning like DeepTox and ProTox-II provide probabilistic predictions of compound toxicity and rank them based on risk levels [91, 108]. This further facilitates understanding the structural features that are more likely to be associated with compound toxicity and eliminating them during the early stages of drug discovery.

Additionally, AI tools incorporating graph or descriptor-based embeddings such as ADMETlab 2.0 and pkCSM provide integrated pharmacokinetic profiles of the targeted compounds through rapid prediction of absorption (e.g., Caco-2 cell permeability), distribution (e.g., volume of distribution), metabolism (e.g., CYP450 interactions), and excretion (e.g., renal clearance, $t_{1/2}$) portfolios [109]. It is interesting to note that the pkCSM tool has the capability to predict over 25 pharmacokinetic and toxicity endpoints within a few seconds, thus enabling researchers to flag undesired compounds in very little time [94]. Further, the COVID-19 pandemic highlights the utility of AI-based predictive modeling in this field. Several platforms developed by big data companies such as Insilico Medicine, helped rapidly evaluate the ADMET profiles of hundreds of candidate molecules for fast-tracked preclinical testing, and prioritizing them for clinical trials [110].

10.3.5 AI in Clinical Trial Optimization

Once the desired lead compound with optimal pharmacokinetic and toxicity profile tested through preclinical studies is identified, it is further translated to clinical trials on human subjects, which forms the most resource-intensive stage of the drug development pipeline. Before this, an investigational new drug (IND) application is submitted to the regulatory authorities for permission to conduct clinical trials. At this juncture, AI can provide solutions in the form of preparing documentation for regulatory submissions such as investigators' brochures, and generating preclinical reports (Table 10.4) [111]. After IND submission, clinical trials for the candidate molecules are initiated. For EIDs, this stage is even more challenging owing to rapid trial initiation during outbreaks, difficulties in patient recruitment, and ethical concerns arising out of the use of a placebo arm. AI plays a significant role in revolutionizing this stage by designing intelligent trials with planned patient stratification techniques and outcome prediction. This enhances the efficiency and success rate of clinical trials.

Tools based on Bayesian optimization and RL help in determining optimal sample sizes, dosage schedules, inclusion/exclusion criteria, and endpoints during the design of a clinical trial protocol. This reduces further amendments to the protocol in the long run, and hence, the failure rates. Patient selection and stratification is another area where AI can be of great help by leveraging ML algorithms trained on EHRs and genomic data to identify patient subgroups that can most likely benefit from the therapeutic intervention in question. This becomes even more critical with infectious diseases such as dengue, Ebola, or COVID-19 where disease heterogeneity and severity are at concern [112].

Furthermore, AI facilitates the design of adaptive trials by modifying the study protocol in real time based on interim data. Predictive analytics can be employed for site selection and forecasting the recruitment process. NLP algorithms can be utilized to analyze public health reports, mobility data, and disease surveillance feeds to identify emerging hotspots for rapid trial deployment. This is again important for EIDs where incidence patterns shift rapidly. AI tools like TrialGPT and IBM

TABLE 10.4 Selected Examples of AI Platforms Supporting Later Stages of Drug Discovery, Including Preclinical Testing, Clinical Trial Optimization, Regulatory Submissions, and Post-Marketing Surveillance

Application	Tool	Description	Web Link	Reference
Predictive modeling for preclinical studies	DeepTox	Trained on Tox21, aids in toxicity prediction based on deep neural networks	https://github.com/greenelab/deep-review/issues/72	[91]
	ProTox 3.0	Toxicity classification and probability scoring based on ensemble learning	https://tox.charite.de/protox3	[92]
	ADMETlab 3.0	Predicts >50 ADMET endpoints from structure or SMILES	https://admetlab3.scbdd.com	[93]
	pkCSM	Predicts small-molecule pharmacokinetic properties using graph-based signatures	https://biosig.lab.uq.edu.au/pkcsm	[94]
	DruMAP	Drug metabolism and pharmacokinetics analysis platform	https://drumap.nibiohn.go.jp	[95]
	SwissADME	Aids in computing physicochemical descriptors, as well as predicting ADME properties	http://www.swissadme.ch	[96]
	DeepDelta	A pairwise DL model that predicts molecular property differences from small datasets	https://doi.org/10.1186/s13321-023-00769-x	[97]
Regulatory document submissions	IQVIA SmartSolve	Natural language processing (NLP)-based and automates regulatory document creation	https://www.iqvia.com/locations/middle-east-and-africa/solutions/smartsolve-eqms-enhance-quality-assurance-and-control	[98]
	TriNetX	Aids in clinical trial data harmonization through predictive analytics and NLP	https://trinetx.com	[99]
	Veeva Vault Submissions	Aids in eCTD-ready dossier generation	https://www.veeva.com/products/veeva-submissions	[100]
Clinical trial optimization	TrialGPT	Utilizes large language models (LLMs) to efficiently match patients with suitable clinical trials	https://github.com/ncbi-nlp/TrialGPT	[101]
	Deep6 AI	EHR-based patient recruitment using NLP	https://deep6.ai	[102]
	Unlearn.AI	Clinical trials outcomes prediction using digital twins and generative models	https://www.unlearn.ai	[103]
Post-marketing surveillance	MedWatcher	NLP-based social media pharmacovigilance	https://medwatcher.org	[104]
	Oracle Argus Safety	Automates the reporting of adverse events cases	https://www.oracle.com/in/life-sciences/safety-solutions/argus-safety-case-management	[105, 106]

Watson for Clinical Trials have been utilized for site selection and participant matching based on historical trial data, patient demographics, and site performance metrics (Table 10.4) [101, 113].

Additionally, AI models can help in monitoring patients and predict outcomes based on data generated from wearable devices and remote diagnostic devices. This also aids in the detection of any possible adverse events during the clinical trials. After the successful conclusion of clinical trials, the regulatory authorities approve a drug and from that stage onwards, phase IV clinical studies (post-marketing surveillance) are initiated through proper pharmacovigilance support systems, which detect and report adverse events arising out of the use of the approved drug in the clinical settings for reporting to the regulatory bodies. The FDA's Sentinel initiative further supports the incorporation of AI modules to streamline post-marketing surveillance by analyzing large-scale real-world data including EHRs. These modules automate the process for detection of adverse drug events for rapid regulatory responses and thus, improve pharmacovigilance efficiency [113, 114].

10.4 AI IN DRUG REPURPOSING FOR INFECTIOUS DISEASES

Drug repurposing is a strategy that aims at identifying new therapeutic uses for existing drugs and has emerged as a cost-effective and time-saving approach in modern drug discovery [115]. In recent years, this strategy has gained substantial attention, particularly in the context of infectious diseases. Interdisciplinary collaboration among experts from microbiology, pharmacology, medicinal chemistry, and computational biology is essential for successful drug repurposing [116]. Broadly, hypothesis-driven repurposing strategies encompass both experimental and computational methodologies. The former typically include binding assays to detect direct interactions between ligands and cellular components, as well as phenotypic screening to identify active compounds based on their effects on cells or whole organisms. On the other hand, the latter approaches can be classified into drug-centric (polypharmacology), target-centric, and disease-centric models where each offers unique perspectives and predictive capabilities for the identification of repurposable drug candidates [1, 117].

However, there are several limitations associated with conventional practices of drug repurposing. For example, repurposed drugs often exhibit limited target specificity, which may result in undesirable side effects and reduced therapeutic efficacy. The scope of available drug candidates is also constrained, and in many cases, further optimization is required to ensure suitability for the new indication. Moreover, intellectual property issues and the necessity of conducting additional clinical trials further complicate the process [118]. To overcome these challenges, AI comes into the picture. AI-driven approaches offer promising solutions by enabling the analysis of large-scale biological and chemical datasets to identify hidden relationships between existing drugs, disease targets, and novel therapeutic opportunities.

AI techniques, especially ML and DL, help augment the virtual screening of large chemical libraries against respective druggable targets through mainly two strategies, viz., data-driven analysis and predictive modeling, so that the desired drug molecule is identified in the shortest time possible [119]. For example, a deep neural framework utilized by Masuda (2022) and Raza (2022) helped identify drugs repositioned for Parkinson's disease and pancreatic cancer [120, 121]. Moreover, ML models have shown considerable potential in identifying peptide-protein binding sites and druggable pockets on protein surfaces, as demonstrated by Trisciuzzi (2022) and Stepniewska-Dziubinska (2020) [122, 123]. When integrated with virtual docking, these models enable efficient screening of potential drug candidates by accurately predicting their binding affinities to the identified target sites. For instance, Karelina (2023) successfully applied this combined approach to enhance the precision of drug-target interaction (DTI) prediction [124]. Table 10.5 highlights the drugs that have been repurposed for infectious diseases through the application of AI.

Further, NLP offers unique capabilities for text mining and literature surveys by generating concise reports or summaries from thousands of research papers and journals, thus enabling meaningful conclusions to be drawn in significantly less time. Imagine a computer program capable not only of reading scientific literature but also of understanding the complex concepts and relationships embedded within it. NLP algorithms enable this by employing various techniques to identify and classify key biological entities such as genes, proteins, drugs, and diseases [131]. Beyond simple recognition, these algorithms are also used to extract and interpret relationships among entities – for instance, determining how a specific drug interacts with a particular protein within a pathogen. Karaa (2021) presented a study on extracting drug-disease relationships from biomedical literature using NLP and ML techniques. The model demonstrated high accuracy and outperformed

TABLE 10.5 Selected Examples of Antiviral Drugs Repurposed for Infectious Diseases

Drug	Original Target	Repurposed Target	AI Role	Reference
Acyclovir	Herpes simplex virus	Cytomegalovirus	Similarity and network analysis	[125]
Ribavirin	Hepatitis C virus	Respiratory syncytial virus	ML and text mining for target prediction	[126]
Oseltamivir	Influenza A and B viruses	MERS-CoV	Molecular modeling and docking simulations	[127]
Emtricitabine	HIV9	Chikungunya virus	Network and pathway enrichment analysis	[128]
Sofosbuvir	Hepatitis C virus	West Nile virus (flavivirus)	ML for target prediction	[129]
Tenofovir	Hepatitis B virus	Dengue virus	Sequence homology and ML for target prediction	[130]

TABLE 10.6 Selected Examples of AI-Driven Drug Repurposing Platforms

AI Platform	Description	Web Link	Reference
DrugRep	Web-based platform for drug repurposing, target prediction, and personalized medicine applications through structure-based screening, ligand similarity, and pharmacogenomic insights	http://cao.labshare.cn/drugrep	[133]
BenevolentAI	Integrates biomedical literature, omics data, and clinical information to uncover novel drug–disease associations	https://www.benevolent.com	[134]
AtomNet	Utilizes DCNNs and supports rapid hit discovery	https://www.atomwise.com/pipeline	[135]
ZairaChem	Cloud-based AI platform for drug repurposing, predictive modeling, and compound prioritization	https://github.com/ersilia-os/zaira-chem	[136]

previously existing approaches [132]. Some of the available AI-driven platforms for drug repurposing have been summarized in Table 10.6.

10.5 AI FOR ENHANCED INFECTIOUS DISEASE SURVEILLANCE

To deal with situations like pandemics and EIDs, global preparedness is the foremost pillar that is controlled through proper checks on surveillance mechanisms to handle the real outbreaks prospectively [3]. Conventional infectious disease surveillance systems are often time-consuming and less traceable, as these mainly depend on laboratory-confirmed diagnoses, clinician-initiated case reports, and epidemiological field investigations [137]. Due to the rapid rise in epidemics and the huge associated data (pathogen genome, clinical indicators, and population mobility patterns), has necessitated the incorporation of technologies like AI that can carry out careful analytics, perform real-time surveillance, forecast the outbreak, and strategically deploy corresponding public health interventions [138].

NLP algorithms serve to derive meaningful conclusions out of unstructured data obtained through EHRs, digital news platforms, and online forums associated with activities related to infectious diseases [139]. These models thus facilitate the early forecast of a possible outbreak, even in the absence of formalized case documentation, and prepare the healthcare system to respond efficiently in due course of time [140]. These models are not only limited to forecasting possible outbreaks but also have a substantial role in genomic surveillance of pathogens by rapid analysis of protein sequences, detection of mutation patterns, and prediction of emergent variants with potential for transmission and immune escape [141].

For efficient infectious disease surveillance and to mitigate the overall spread, several unconventional data sources including social media activity, internet search trends, and physiological metrics from wearable devices are used to identify anomalous patterns that can provide suitable signals for early outbreaks. Beyond these, environmental and climatic parameters such as temperature, humidity, and precipitation patterns are also taken into due consideration for the analysis of transmission dynamics of vector-borne diseases like malaria and dengue. Further, data from weather stations, remote sensors, and images derived from the satellites can be integrated to derive spatiotemporal hotspots or ecological niches that may favor disease progression by providing suitable geographical conditions [142].

Several AI tools have been developed for infectious disease surveillance, and these have demonstrated successful applications with high accuracy, particularly in disease forecast, marking their great prestige in the fight against the emerging epidemics and the corresponding control (Table 10.7). For instance, a Canadian health analytics platform – namely, BlueDot – forecast the emergence of the COVID-19 outbreak in Wuhan several days before the World Health Organization (WHO) issued the real formal alert [149]. By utilizing data from global airline ticketing systems, official health bulletins, and unstructured content from online media, this tool also forecast pneumonia clusters that were uncommon. Another interesting tool developed at Boston Children's Hospital – namely, HealthMap – helped forecast the Zika virus outbreak in South America. It uses data from government reports, news media, and social platforms to provide real-time worldwide tracking of infectious diseases [150]. Beyond the epidemics forecast, AI tools also aid in monitoring seasonal influenza trends. Google Flu Trends was one of the earliest such examples but was later discontinued due to overestimations generated from search query biases [151]. However, this facilitated the catalytic development of other robust tools such as FluSight. This tool forecasts influenza by integrating ML techniques with data from hospital admission records and virological testing [152]. Moreover, mobile-based surveillance platforms have also been developed, such as the "flu tracker" app, which is based on crowdsourcing reports to monitor influenza globally [153]. Additionally, the U.S. Centers for Disease Control and Prevention (CDC) has developed an interesting app called FluView for monitoring influenza that works by quantifying influenza-like illness (ILI)-related clinical visits (e.g., fever, cough, sore throat) as a function of total outpatient consultations [154].

However, AI-based surveillance systems often pose several critical challenges related to data privacy, as these tools track data directly from social media accounts such as those on Twitter and Facebook [155]. Moreover, surveillance technologies often disproportionately monitor low-income, densely populated urban regions, which raises ethical issues by subjecting them to intensified

TABLE 10.7 Selected Examples of AI Applications in Disease Surveillance for EIDs

Tool	Feature	Web Link	Reference
HealthMap	Global outbreak detection from informal sources (news, blogs, and forums)	https://www.healthmap.org/en	[143]
BlueDot	Early outbreak prediction (flagged the COVID-19 outbreak nine days before the WHO's official alert)	https://bluedot.global	[144]
DeepCOVID	Predicts regional COVID-19 case counts using real-time data (US)	https://deepcovid.github.io	[145]
Arogya Setu/CoWIN	Citizen-sourced data for population-level surveillance (India)	https://www.cowin.gov.in	[146]
SORMAS	Field-level surveillance and outbreak management (used in Africa for Ebola, COVID-19, and cholera)	https://www.undp.org/policy-centre/singapore/sormas	[147]
EpiNow2	Estimate real-time case counts and time-varying epidemiological parameters based on Bayesian latent variable model	https://github.com/epiforecasts/EpiNow2	[148]

government scrutiny, over-policing, and movement restrictions [156]. Additionally, the reliability and accuracy of these tools are again a matter of question, as they largely depend on input data. Any failure in sensor hardware or the algorithmic architecture can cause serious havoc, including erroneous classifications during outbreak detection [138]. For instance, AI-integrated thermal imaging systems that were widely deployed at airports for mass fever screening suffered from a limitation of being incapable of discriminating between infectious fevers and non-infectious ones caused due to heat exposure or hormonal fluctuations. This caused a violation of quarantine protocols by misclassifying healthy and infected individuals during COVID-19 [139].

10.6 CHALLENGES IN THE IMPLEMENTATION OF AI IN DRUG DEVELOPMENT FOR INFECTIOUS DISEASES

AI has significantly benefited anti-infective drug discovery across each of the facets which may be attributed to the phenotypic-driven nature of infectious diseases, where disease progression is governed mainly by the physiological characteristics of pathogens instead of their molecular mechanisms. Unlike complex multifactorial disorders such as cancer or neurodegeneration, where incomplete mechanistic understanding hinders the drug development process, pathogens responsible for infectious diseases are generally well-characterized, thus enabling the application of computational techniques. Moreover, anti-infective agents, predominantly small molecules, can be readily modeled as molecular graphs, thus enabling the structure-based learning of the features or hidden patterns to screen across large chemical libraries [157].

Despite these advantages, there are several limitations to the application of AI to this field. The first challenge is the generalizability of ML models to unexplored chemical and biomolecular spaces. For instance, a GNN-based framework trained on phenotypic screening data for *Escherichia coli* growth inhibition was modeled to predict the antibiotic activity of small molecules, including halicin. Although the model performed satisfactorily, it showed the best results within known and established antibiotic classes, such as β-lactams and quinolones, owing to limited extrapolation to novel scaffolds [158]. This suggests that while ML models can identify potent candidates, their novelty is often restricted to familiar chemical frameworks. To navigate through the underexplored chemical spaces, generative approaches such as genetic algorithms can be employed. One such study identified the synthetic peptide guavanin 2, demonstrating potent antimicrobial activity in murine models [159].

Another important challenge is the limited mechanistic insight provided by current phenotype-based models. Although ML has the capability to identify hits based on phenotypic screening data, accurate prediction of DTIs and mechanisms of action remains challenging [160]. Recent advances in protein structure prediction (e.g., AlphaFold) and molecular docking have opened new avenues to structure-based, drug-designing approaches; however, existing docking algorithms still face problems in predictive accuracy, especially for AMPs and membrane-active agents. This becomes more important when developing drugs against Gram-negative pathogens, where membrane permeability is a key factor for the drug's penetration into the pathogen's cellular environment [161]. Although MD simulations can aid in understanding this process, they are computationally expensive and require further integration with AI techniques to reduce both computational cost and time [161].

Moreover, the availability of high-quality labeled datasets remains a major drawback as most of the data remains siloed to the proprietary repositories or is limited by the high cost of generation. For instance, host cell toxicity is a serious concern in the context of anti-infective drugs, with compounds displaying cytotoxic, hemolytic, or genotoxic effects. The ML models developed to predict toxicity compromises with their performance owing to the limitation of labeled datasets. Also, there remains a critical gap in predicting in vivo efficacy of designed drugs, particularly in animal models of acute systemic infections, a challenge that is yet to be fully addressed through current ML-driven approaches [162].

Additionally, current models often struggle for multiparametric optimization of drug candidates with optimal pharmacokinetic and pharmacodynamic profiles required during later stages of drug development. Although, multimodal models have shown significant improvement in predictive accuracy by integrating heterogeneous data; however, their performance relies on large and curated datasets [10]. Interpretability of model outputs is another critical bottleneck for AI implementation in drug discovery, as understanding the rationale behind predictions is necessary to avoid false positives or model hallucinations in order to strive over the journey for designing next-generation therapeutics, vaccines, and diagnostics targeting infectious diseases.

10.7 CONCLUSION AND FUTURE PERSPECTIVES

The worldwide devastation caused by the COVID-19 pandemic and other EIDs has underscored the urgent need for a fundamental transformation in the drug discovery paradigm beyond conventional approaches that could redefine the horizons of how we discover, develop, and deploy therapeutic interventions. AI stands at the forefront of this transformation, from unveiling HPIs to forecasting outbreak trajectories and ensuring post-marketing safety of developed therapeutic regimens.

Advanced AI tools facilitate deep proteome mining, enabling the identification of antimicrobial agents from both extant and extinct species. However, the discovery of novel synthesizable lead molecules remains a major challenge that largely depends on the diversity of training datasets and the breadth of chemical space explored. Key concerns related to AI applications such as data integrity, ethical considerations, the "black-box" nature of models, and issues related to transparency, interpretability, and explainability demand significant attention. The success of AI systems in drug discovery fundamentally relies on the availability of high-quality, representative, structured, and ethically sound data.

Emerging technologies such as explainable AI (XAI), federated learning, edge computing, digital twin modeling, and real-world data analytics combined with integration into open-access datasets and benchmarking platforms may offer promising solutions to these critical issues. If regulatory authorities and multidisciplinary collaborations are brought into this loop, such a paradigm shift could genuinely accelerate drug discovery toward a smarter and more efficient model, supported by experimental validation.

ACKNOWLEDGMENT

Dr. Rajnish Kumar gratefully acknowledges the Indian Institute of Technology (BHU), Varanasi, for the seed grant; the Science and Engineering Research Board (SERB), India, for the start-up research grant (SRG/2021/000415) and the Mathematical Research Centric Support (MATRICS) grant (MTR/2021/000317); and Indian Council of Medical Research (ICMR), India, for investigator-initiated research proposals for small extramural grant (EMDR/SG/11/2024-01-02095). Jatin Jangra expresses sincere gratitude to IIT (BHU) and the Ministry of Human Resource Development (MHRD), India, for providing the teaching assistantship (TAship).

ABBREVIATIONS

ACC	Accuracy
ADMET	Absorption, Distribution, Metabolism, Excretion, and Toxicity
AI	Artificial Intelligence
ALS	Amyotrophic Lateral Sclerosis
AMPs	Antimicrobial Peptides
AMR	Antimicrobial Resistance
AST	Antimicrobial Susceptibility Testing
CDC	Centers for Disease Control and Prevention
CNN	Convolutional Neural Network
CRKP	Carbapenem-Resistant *Klebsiella Pneumoniae*
DEGS	Differentially Expressed Genes
DL	Deep Learning
DMPNN	Directed Message-Passing Neural Network
DNN	Deep Neural Network
DTIs	Drug-Target Interactions
EHRs	Electronic Health Records
EIDs	Emerging Infectious Diseases
FFNN	Feed-Forward Neural Network
GAN	Generative Adversarial Network
GNN	Graph Neural Network
HPDB	Host-Pathogen Interaction Database
HPI	Host-Pathogen Interaction
IND	Investigational New Drug
LMICs	Low- and Middle-Income Countries
MCC	Matthews Correlation Coefficient
MD	Molecular Dynamics

ML	Machine Learning
Mpro	Main Protease
MRSA	Methicillin-Resistant *Staphylococcus Aureus*
MSA	Multiple Sequence Alignment
NLP	Natural Language Processing
NTDs	Neglected Tropical Diseases
PAAC	Pseudo-Amino Acid Composition
PPI	Protein-Protein Interaction
QSAR	Quantitative Structure–Activity Relationship
RCNN	Recurrent Convolutional Neural Network
ResRNN	Residual Recurrent Neural Network
RF	Random Forest
RL	Reinforcement Learning
RNNs	Recurrent Neural Networks
SVMs	Support Vector Machines
VAES	Variational Autoencoders
WHO	World Health Organization
XGBoost	Extreme Gradient Boosting

REFERENCES

1. Singh A. Artificial intelligence for drug repurposing against infectious diseases. *Artificial Intelligence Chemistry* 2024;2(2):100071.

2. Li C, Ye G, Jiang Y, Wang Z, Yu H, Yang M. Artificial Intelligence in battling infectious diseases: A transformative role. *Journal of Medical Virology* 2024;96(1):e29355.

3. Ali H. AI for pandemic preparedness and infectious disease surveillance: Predicting outbreaks, modeling transmission, and optimizing public health interventions. *International Journal of Research Publication and Reviews* 2024;5(8):4605–19.

4. Nguyen M, Chaudhry SI, Desai MM, Dzirasa K, Cavazos JE, Boatright D. Gender, racial, and ethnic inequities in receipt of multiple National Institutes of Health research project grants. *JAMA Network Open* 2023;6:e230855-e.

5. Majumder MAA, Rahman S, Cohall D, Bharatha A, Singh K, Haque M, Gittens-St Hilaire M. Antimicrobial stewardship: Fighting antimicrobial resistance and protecting global public health. *Infection and Drug Resistance* 2020:4713–38.

6. Ekundayo F. Using machine learning to predict disease outbreaks and enhance public health surveillance. *World Journal of Advanced Research and Reviews* 2024; 24(3):794–811

7. Nag S, Baidya AT, Mandal A, Mathew AT, Das B, Devi B, Kumar R. Deep learning tools for advancing drug discovery and development. *3 Biotech*. 2022;12:110.

8. Zhang K, Yang X, Wang Y, Yu Y, Huang N, Li G, Li X, Wu JC, Yang S. Artificial intelligence in drug development. *Nature Medicine* 2025:1–15.

9. Schlander M, Hernandez-Villafuerte K, Cheng C-Y, Mestre-Ferrandiz J, Baumann M. How much does it cost to research and develop a new drug? A systematic review and assessment. *PharmacoEconomics* 2021;39:1243–69.

10. Cesaro A, Hoffman SC, Das P, de la Fuente-Nunez C. Challenges and applications of artificial intelligence in infectious diseases and antimicrobial resistance. *NPJ Antimicrobials and Resistance* 2025;3:2.

11. Aloke C, Achilonu I. Coping with the ESKAPE pathogens: Evolving strategies, challenges and future prospects. *Microbial Pathogenesis* 2023;175:105963.

12. Bess A, Berglind F, Mukhopadhyay S, Brylinski M, Griggs N, Cho T, Galliano C, Wasan KM. Artificial intelligence for the discovery of novel antimicrobial agents for emerging infectious diseases. *Drug Discovery Today* 2022;27(4):1099–107.

13. Talat A, Khan AU. Artificial intelligence as a smart approach to develop antimicrobial drug molecules: A paradigm to combat drug-resistant infections. *Drug Discovery Today* 2023;28(4):103491.

14. Davies D. Understanding biofilm resistance to antibacterial agents. *Nature Reviews Drug Discovery* 2003;2(2):114–22.

15. Haney EF, Hancock RE. Addressing antibiotic failure—Beyond genetically encoded antimicrobial resistance. *Frontiers in Drug Discovery* 2022;2:892975.

16. Kanjilal S, Oberst M, Boominathan S, Zhou H, Hooper DC, Sontag D. A decision algorithm to promote outpatient antimicrobial stewardship for uncomplicated urinary tract infection. *Science Translational Medicine* 2020;12:eaay5067.

17. Pascucci M, Royer G, Adamek J, Asmar MA, Aristizabal D, Blanche L, Bezzarga A, Boniface-Chang G, Brunner A, Curel C. AI-based mobile application to fight antibiotic resistance. *Nature Communications* 2021;12(1):1173.

18. Ali HH, Ali HM, Ali HM, Ali MA, Zaky AF, Touk AA, Darwiche AH, Touk AA. The role and limitations of artificial intelligence in combating infectious disease outbreaks. *Cureus* 2025;17(1).

19. Pun FW, Ozerov IV, Zhavoronkov A. AI-powered therapeutic target discovery. *Trends in Pharmacological Sciences* 2023;44(9):561–72.

20. Pun FW, Liu BHM, Long X, Leung HW, Leung GHD, Mewborne QT, Gao J, Shneyderman A, Ozerov IV, Wang J. Identification of therapeutic targets for amyotrophic lateral sclerosis using PandaOmics – an AI-enabled biological target discovery platform. *Frontiers in Aging Neuroscience* 2022;14:914017.

21. Fabris F, Palmer D, Salama KM, De Magalhaes JP, Freitas AA. Using deep learning to associate human genes with age-related diseases. *Bioinformatics* 2020;36(7):2202–8.

22. Barman RK, Mukhopadhyay A, Maulik U, Das S. Identification of infectious disease-associated host genes using machine learning techniques. *BMC Bioinformatics* 2019;20:1–12.

23. Krishna R, Wang J, Ahern W, Sturmfels P, Venkatesh P, Kalvet I, Lee GR, Morey-Burrows FS, Anishchenko I, Humphreys IR. Generalized biomolecular modeling and design with RoseTTAFold All-Atom. *Science* 2024;384:eadl2528.

24. Jumper J, Evans R, Pritzel A, Green T, Figurnov M, Ronneberger O, Tunyasuvunakool K, Bates R, Žídek A, Potapenko A. Highly accurate protein structure prediction with AlphaFold. *Nature* 2021;596(7873):583–9.

25. Jeliazkov JR, del Alamo D, Karpiak JD. ESMFold hallucinates native-like protein sequences. *bioRxiv*. 2023:2023.05. 23.541774.

26. Wu R, Ding F, Wang R, Shen R, Zhang X, Luo S, Su C, Wu Z, Xie Q, Berger B. High-resolution de novo structure prediction from primary sequence. *bioRxiv*. 2022:2022.07. 21.500999.

27. Jiménez J, Doerr S, Martínez-Rosell G, Rose AS, De Fabritiis G. DeepSite: Protein-binding site predictor using 3D-convolutional neural networks. *Bioinformatics* 2017;33(19):3036–42.

28. Halgren T. New method for fast and accurate binding-site identification and analysis. *Chemical Biology & Drug Design* 2007;69(2):146–8.

29. Alipanahi B, Delong A, Weirauch MT, Frey BJ. Predicting the sequence specificities of DNA-and RNA-binding proteins by deep learning. *Nature Biotechnology* 2015;33(8):831–8.

30. Krivák R, Hoksza D. P2Rank: Machine learning based tool for rapid and accurate prediction of ligand binding sites from protein structure. *Journal of Cheminformatics* 2018;10:1–12.

31. Shi W, Singha M, Pu L, Srivastava G, Ramanujam J, Brylinski M. Graphsite: Ligand binding site classification with deep graph learning. *Biomolecules* 2022;12(8):1053.

32. Sahragard R, Arabfard M, Najafi A. Predicting host-pathogen interactions with machine learning algorithms: A scoping review. *Infection, Genetics and Evolution* 2025;130:105751.

33. Syed R, Aldakheel FM, Alduraywish SA, Mateen A, Alnajran H, Al-Numan HH. Host-pathogen interactions: A general introduction. *Systems Biology Approaches for Host-Pathogen Interaction Analysis*. Elsevier; 2024. pp. 1–14.

34. Sen R, Nayak L, De RK. A review on host–pathogen interactions: Classification and prediction. *European Journal of Clinical Microbiology & Infectious Diseases* 2016;35:1581–99.

35. Trepte P, Secker C, Olivet J, Blavier J, Kostova S, Maseko SB, Minia I, Silva Ramos E, Cassonnet P, Golusik S. AI-guided pipeline for protein–protein interaction drug discovery identifies a SARS-CoV-2 inhibitor. *Molecular Systems Biology* 2024;20(4):428–57.

36. Murmu S, Chaurasia H, Rao A, Rai A, Jaiswal S, Bharadwaj A, Yadav R, Archak S. PlantPathoPPI: An ensemble-based machine learning architecture for prediction of protein-protein interactions between plants and pathogens. *Journal of Molecular Biology* 2025;437:169093.

37. Yin S, Mi X, Shukla D. Leveraging machine learning models for peptide–protein interaction prediction. *RSC Chemical Biology* 2024;5(5):401–17.

38. Tahir M, Khan F, Hayat M, Alshehri MD. An effective machine learning-based model for the prediction of protein–protein interaction sites in health systems. *Neural Computing and Applications* 2024;36(1):65–75.

39. Palhamkhani F, Alipour M, Dehnad A, Abbasi K, Razzaghi P, Ghasemi JB. DeepCompoundNet: Enhancing compound–protein interaction prediction with multimodal convolutional neural networks. *Journal of Biomolecular Structure and Dynamics* 2025;43(3):1414–23.

40. ul Qamar MT, Noor F, Guo Y-X, Zhu X-T, Chen L-L. Deep-HPI-pred: An R-Shiny applet for network-based classification and prediction of Host-Pathogen protein-protein interactions. *Computational and Structural Biotechnology Journal* 2024;23:316–29.

41. Kaundal R, Loaiza CD, Duhan N, Flann N. deepHPI: A comprehensive deep learning platform for accurate prediction and visualization of host–pathogen protein–protein interactions. *Briefings in Bioinformatics* 2022;23:bbac125.

42. Canzler S, Fischer M, Ulbricht D, Ristic N, Hildebrand PW, Staritzbichler R. ProteinPrompt: A webserver for predicting protein–protein interactions. *Bioinformatics Advances* 2022;2:vbac059.

43. Yang X, Yang S, Li Q, Wuchty S, Zhang Z. Prediction of human-virus protein-protein interactions through a sequence embedding-based machine learning method. *Computational and Structural Biotechnology Journal* 2020;18:153–61.

44. Liu-Wei W, Kafkas Ş, Chen J, Dimonaco NJ, Tegnér J, Hoehndorf R. DeepViral: Prediction of novel virus–host interactions from protein sequences and infectious disease phenotypes. *Bioinformatics* 2021;37(17):2722–9.

45. Zhou G, Chen M, Ju CJ, Wang Z, Jiang J-Y, Wang W. Mutation effect estimation on protein–protein interactions using deep contextualized representation learning. *NAR Genomics and Bioinformatics* 2020;2:lqaa015.

46. Yang S, Li H, He H, Zhou Y, Zhang Z. Critical assessment and performance improvement of plant–pathogen protein–protein interaction prediction methods. *Briefings in Bioinformatics* 2019;20(1):274–87.

47. Chen H, Shen J, Wang L, Chi CH. APEX2S: A two-layer machine learning model for discovery of host-pathogen protein-protein interactions on cloud-based multiomics data. *Concurrency and Computation: Practice and Experience* 2020;32(23):e5846.

48. Zhou X, Park B, Choi D, Han K. A generalized approach to predicting protein-protein interactions between virus and host. *BMC Genomics* 2018;19:69–77.

49. Yadav DK, Rai R, Pratap R, Singh H. Software and web resources for computer-aided molecular modeling and drug discovery. *Chemometrics Applications and Research: QSAR in Medicinal Chemistry*. Oakville (ON): Apple Academic Press; 2016, p. 33.

50. Heid E, Greenman KP, Chung Y, Li S-C, Graff DE, Vermeire FH, Wu H, Green WH, McGill CJ. Chemprop: A machine learning package for chemical property prediction. *Journal of Chemical Information and Modeling* 2023;64(1):9–17.

51. Ramsundar B. *Molecular Machine Learning with DeepChem*. PhD. thesis, Stanford University; 2018.

52. Liao Z, You R, Huang X, Yao X, Huang T, Zhu S. DeepDock: Enhancing ligand-protein interaction prediction by a combination of ligand and structure information. *2019 IEEE International Conference on Bioinformatics and Biomedicine (BIBM)*: IEEE; 2019. p. 311–7.

53. Wallach I, Dzamba M, Heifets A. AtomNet: A deep convolutional neural network for bioactivity prediction in structure-based drug discovery. *arXiv preprint arXiv*:151002855. 2015.

54. Loeffler HH, He J, Tibo A, Janet JP, Voronov A, Mervin LH, Engkvist O. Reinvent 4: Modern AI–driven generative molecule design. *Journal of Cheminformatics* 2024;16(1):20.

55. De Cao N, Kipf T. MolGAN: An implicit generative model for small molecular graphs. *arXiv preprint arXiv*:180511973. 2018.

56. Zhavoronkov A, Ivanenkov YA, Aliper A, Veselov MS, Aladinskiy VA, Aladinskaya AV, Terentiev VA, Polykovskiy DA, Kuznetsov MD, Asadulaev A. Deep learning enables rapid identification of potent DDR1 kinase inhibitors. *Nature Biotechnology* 2019;37(9):1038–40.

57. Polishchuk P. CReM: Chemically reasonable mutations framework for structure generation. *Journal of Cheminformatics* 2020;12(1):28.

58. Perron Q, Mirguet O, Tajmouati H, Skiredj A, Rojas A, Gohier A, Ducrot P, Bourguignon MP, Sansilvestri-Morel P, Do Huu N. Deep generative models for ligand-based de novo design applied to multi-parametric optimization. *Journal of Computational Chemistry* 2022;43(10):692–703.

59. Richardson PJ, Robinson BW, Smith DP, Stebbing J. The AI-assisted identification and clinical efficacy of baricitinib in the treatment of COVID-19. *Vaccine* 2022;10(6):951.

60. Ibrahim MA, Abdeljawaad KA, Abdelrahman AH, Hegazy M-EF. Natural-like products as potential SARS-CoV-2 M^{pro} inhibitors: *In-silico* drug discovery. *Journal of Biomolecular Structure and Dynamics* 2021;39(15):5722–34.

61. Olawade DB, Teke J, Fapohunda O, Weerasinghe K, Usman SO, Ige AO, David-Olawade AC. Leveraging artificial intelligence in vaccine development: A narrative review. *Journal of Microbiological Methods* 2024; 224:106998.

62. Bravi B. Development and use of machine learning algorithms in vaccine target selection. *NPJ Vaccines* 2024;9:15.

63. Bukhari SNH, Jain A, Haq E, Khder MA, Neware R, Bhola J, Lari Najafi M. [Retracted] Machine learning-based ensemble model for zika virus T-cell epitope prediction. *Journal of Healthcare Engineering* 2021;2021(1):9591670.

64. Abdelmageed MI, Abdelmoneim AH, Mustafa MI, Elfadol NM, Murshed NS, Shantier SW, Makhawi AM. Design of a multiepitope-based peptide vaccine against the e protein of human COVID-19: An immunoinformatics approach. *BioMed Research International* 2020;2020(1):2683286.

65. Olawade DB, Wada OZ, Ezeagu CN, Aderinto N, Balogun MA, Asaolu FT, David-Olawade AC. Malaria vaccination in Africa: A mini-review of challenges and opportunities. *Medicine* 2024;103(24):e38565.

66. Bhattacharya M, Lo Y-H, Chatterjee S, Das A, Wen Z-H, Chakraborty C. Deep learning in next-generation vaccine development for infectious diseases: Stages and tools involved in epitope selection to vaccine development and characterization. *Molecular Therapy Nucleic Acids* 2025.

67. Singh H, Ansari HR, Raghava GP. Improved method for linear B-cell epitope prediction using antigen's primary sequence. *PLoS One* 2013;8(5):e62216.

68. Soria-Guerra RE, Nieto-Gomez R, Govea-Alonso DO, Rosales-Mendoza S. An overview of bioinformatics tools for epitope prediction: Implications on vaccine development. *Journal of Biomedical Informatics* 2015;53:405–14.

69. Kalita P, Tripathi T. Methodological advances in the design of peptide-based vaccines. *Drug Discovery Today* 2022;27(5):1367–80.

70. Dimitrov I, Naneva L, Doytchinova I, Bangov I. AllergenFP: Allergenicity prediction by descriptor fingerprints. *Bioinformatics* 2014;30(6):846–51.

71. Wiederstein M, Sippl MJ. ProSA-web: Interactive web service for the recognition of errors in three-dimensional structures of proteins. *Nucleic Acids Research* 2007;35:W407–W410.

72. Khatoon N, Pandey RK, Prajapati VK. Exploring Leishmania secretory proteins to design B and T cell multi-epitope subunit vaccine using immunoinformatics approach. *Scientific Reports* 2017;7(1):8285.

73. Wohlwend J, Nathan A, Shalon N, Crain CR, Tano-Menka R, Goldberg B, Richards E, Gaiha GD, Barzilay R. Deep learning enhances the prediction of HLA class I-presented CD8+ T cell epitopes in foreign pathogens. *Nature Machine Intelligence* 2025; 7(2):1–12.

74. Yang Z, Bogdan P, Nazarian S. An in silico deep learning approach to multi-epitope vaccine design: A SARS-CoV-2 case study. *Scientific Reports* 2021;11(1):3238.

75. Suri S, Dakshanamurthy S. IntegralVac: A machine learning-based comprehensive multivalent epitope vaccine design method. *Vaccine* 2022;10(10):1678.

76. Melo MCR, Maasch JR, de la Fuente-Nunez C. Accelerating antibiotic discovery through artificial intelligence. *Communications Biology* 2021;4(1):1050.

77. Lepore C, Silver L, Theuretzbacher U, Thomas J, Visi D. The small-molecule antibiotics pipeline: 2014-2018. *Nature Reviews. Drug Discovery* 2019;18(10):739–40.

78. Wong CH, Siah KW, Lo AW. Estimation of clinical trial success rates and related parameters. *Biostatistics* 2019;20(2):273–86.

79. Torres MDT, de la Fuente-Nunez C. Toward computer-made artificial antibiotics. *Current Opinion in Microbiology* 2019;51:30–8.

80. Schneider G. Automating drug discovery. *Nature Reviews. Drug Discovery* 2018;17(2):97–113.

81. Durrant JD, Amaro RE. Machine-learning techniques applied to antibacterial drug discovery. *Chemical Biology & Drug Design* 2015;85(1):14–21.

82. Mansbach RA, Leus IV, Mehla J, Lopez CA, Walker JK, Rybenkov VV, Hengartner NW, Zgurskaya HI, Gnanakaran S. Machine learning algorithm identifies an antibiotic vocabulary for permeating Gram-negative bacteria. *Journal of Chemical Information and Modeling* 2020;60(6):2838–47.

83. Nguyen M, Brettin T, Long SW, Musser JM, Olsen RJ, Olson R, Shukla M, Stevens RL, Xia F, Yoo H. Developing an in silico minimum inhibitory concentration panel test for *Klebsiella pneumoniae*. *Scientific Reports* 2018;8(1):421.

84. Nguyen M, Long SW, McDermott PF, Olsen RJ, Olson R, Stevens RL, Tyson GH, Zhao S, Davis JJ. Using machine learning to predict antimicrobial MICs and associated genomic features for nontyphoidal *Salmonella*. *Journal of Clinical Microbiology* 2019;57:10.1128/jcm.01260-18

85. Nagarajan D, Nagarajan T, Roy N, Kulkarni O, Ravichandran S, Mishra M, Chakravortty D, Chandra N. Computational antimicrobial peptide design and evaluation against multidrug-resistant clinical isolates of bacteria. *Journal of Biological Chemistry* 2018;293(10):3492–509.

86. Sharma A, Virmani T, Pathak V, Sharma A, Pathak K, Kumar G, Pathak D. Artificial intelligence-based data-driven strategy to accelerate research, development, and clinical trials of COVID vaccine. *BioMed Research International* 2022;2022(1):7205241.

87. Asediya VS, Anjaria PA, Mathakiya RA, Koringa PG, Nayak JB, Bisht D, Fulmali D, Patel VA, Desai DN. Vaccine development using artificial intelligence and machine learning: A review. *International Journal of Biological Macromolecules* 2024; 282:136643.

88. Arango-Argoty G, Garner E, Pruden A, Heath LS, Vikesland P, Zhang L. DeepARG: A deep learning approach for predicting antibiotic resistance genes from metagenomic data. *Microbiome* 2018;6:1–15.

89. Hicks AL, Wheeler N, Sánchez-Busó L, Rakeman JL, Harris SR, Grad YH. Evaluation of parameters affecting performance and reliability of machine learning-based antibiotic susceptibility testing from whole genome sequencing data. *PLoS Computational Biology* 2019;15(9):e1007349.

90. Jiang D, Wu Z, Hsieh C-Y, Chen G, Liao B, Wang Z, Shen C, Cao D, Wu J, Hou T. Could graph neural networks learn better molecular representation for drug discovery? A comparison study of descriptor-based and graph-based models. *Journal of Cheminformatics* 2021;13:1–23.

91. Mayr A, Klambauer G, Unterthiner T, Hochreiter S. DeepTox: Toxicity prediction using deep learning. *Frontiers in Environmental Science* 2016;3:80.

92. Banerjee P, Kemmler E, Dunkel M, Preissner R. ProTox 3.0: A webserver for the prediction of toxicity of chemicals. *Nucleic Acids Research* 2024;52(W1):W513–W20.

93. Fu L, Shi S, Yi J, Wang N, He Y, Wu Z, Peng J, Deng Y, Wang W, Wu C. ADMETlab 3.0: An updated comprehensive online ADMET prediction platform enhanced with broader coverage, improved performance, API functionality and decision support. *Nucleic Acids Research* 2024;52(W1):W422–W31.

94. Pires DE, Blundell TL, Ascher DB. pkCSM: Predicting small-molecule pharmacokinetic and toxicity properties using graph-based signatures. *Journal of Medicinal Chemistry* 2015;58(9):4066–72.

95. Kawashima H, Watanabe R, Esaki T, Kuroda M, Nagao C, Natsume-Kitatani Y, Ohashi R, Komura H, Mizuguchi K. DruMAP: A novel drug metabolism and pharmacokinetics analysis platform. *Journal of Medicinal Chemistry* 2023;66(14):9697–709.

96. Daina A, Michielin O, Zoete V. SwissADME: A free web tool to evaluate pharmacokinetics, drug-likeness and medicinal chemistry friendliness of small molecules. *Scientific Reports* 2017;7(1):42717.

97. Fralish Z, Chen A, Skaluba P, Reker D. DeepDelta: Predicting ADMET improvements of molecular derivatives with deep learning. *Journal of Cheminformatics* 2023;15(1):101.

98. Miller K. Elevating the quality of QMS in life sciences: The age of digital eQMS; IQVIA White Paper; 2021.

99. Ludwig RJ, Anson M, Zirpel H, Thaci D, Olbrich H, Bieber K, Kridin K, Dempfle A, Curman P, Zhao SS. A comprehensive review of methodologies and application to use the real-world data and analytics platform TriNetX. *Frontiers in Pharmacology* 2025;16:1516126.

100. Poster Vennapureddy S. Vault Clinical Operations Management System (CTMS) used in research organizations for fast medicine & therapies. *2024 IEEE/ACM Conference on Connected Health: Applications, Systems and Engineering Technologies (CHASE)*: IEEE; 2024. pp. 195–7.

101. Subbiah V. Can AI-powered TrialGPT enhance patient recruitment for clinical trials? *AI in Precision Oncology* 2025; 2(1).

102. Beck JT, Rammage M, Jackson GP, Preininger AM, Dankwa-Mullan I, Roebuck MC, Torres A, Holtzen H, Coverdill SE, Williamson MP. Artificial intelligence tool for optimizing eligibility screening for clinical trials in a large community cancer center. *JCO Clinical Cancer Informatics* 2020;4:50–9.

103. Mann DL. The use of digital healthcare twins in early-phase clinical trials: Opportunities, challenges, and applications. *JACC: Basic to Translational Science* 2024;9(9): p. 1159–61.

104. Pugliese G. MEDWatch launched to improve adverse events reporting. *Infection Control and Hospital Epidemiology* 1993;14(10):607–8.

105. George J, Dsouza PL, Jahnavi Y, Singh H, Kumar PA. Databases and tools for signal detection of drugs in post-marketing surveillance. In A Kumar (Ed.). *Signal Analysis in Pharmacovigilance*. CRC Press; 2025. p. 32–43.

106. Nagarajan S, Pradhan S. Argus database software: Revolutionizing drug safety and pharmacovigilance. *Scientific Hub of Applied Research in Emerging Medical Science & Technology* 2024;3(5):1–8.

107. Huang R, Xia M, Sakamuru S, Zhao J, Shahane SA, Attene-Ramos M, Zhao T, Austin CP, Simeonov A. Modelling the Tox21 10 K chemical profiles for in vivo toxicity prediction and mechanism characterization. *Nature Communications* 2016;7(1):10425.

108. Banerjee P, Eckert AO, Schrey AK, Preissner R. ProTox-II: A webserver for the prediction of toxicity of chemicals. *Nucleic Acids Research* 2018;46(W1):W257–W263.

109. Xiong G, Wu Z, Yi J, Fu L, Yang Z, Hsieh C, Yin M, Zeng X, Wu C, Lu A. ADMETlab 2.0: An integrated online platform for accurate and comprehensive predictions of ADMET properties. *Nucleic Acids Research* 2021;49(W1):W5–W14.

110. Calandra D, Favareto M. Artificial intelligence to fight COVID-19 outbreak impact: An overview. *European Journal of Social Impact and Circular Economy* 2020;1(3):84–104.

111. Mittal P, Chopra H, Kaur KP, Gautam RK. New drug discovery pipeline. *Computational Approaches in Drug Discovery, Development and Systems Pharmacology*. Elsevier; 2023. pp. 197–222.

112. Askin S, Burkhalter D, Calado G, El Dakrouni S. Artificial intelligence applied to clinical trials: Opportunities and challenges. *Health and Technology* 2023;13(2):203–13.

113. Norouzi K, Ghodsi A, Argani P, Andi PA, Hassani H. Innovative artificial intelligence tools: Exploring the future of healthcare through IBM Watson's potential applications. *Sensor Networks for Smart Hospitals*. Elsevier; 2025. pp. 573–88.

114. Bose A. Regulatory initiatives for artificial intelligence applications: Regulatory writing implications. *Medical Writing* 2023;32:12–5.

115. Parvathaneni V, Kulkarni NS, Muth A, Gupta V. Drug repurposing: A promising tool to accelerate the drug discovery process. *Drug Discovery Today* 2019;24(10):2076–85.

116. Weth FR, Hoggarth GB, Weth AF, Paterson E, White MP, Tan ST, Peng L, Gray C. Unlocking hidden potential: Advancements, approaches, and obstacles in repurposing drugs for cancer therapy. *British Journal of Cancer* 2024;130(5):703–15.

117. Parisi D, Adasme MF, Sveshnikova A, Bolz SN, Moreau Y, Schroeder M. Drug repositioning or target repositioning: A structural perspective of drug-target-indication relationship for available repurposed drugs. *Computational and Structural Biotechnology Journal* 2020;18:1043–55.

118. Rao M, McDuffie E, Sachs C. Artificial intelligence/machine learning-driven small molecule repurposing via off-target prediction and transcriptomics. *Toxics* 2023;11(10):875.

119. Gryniukova A, Kaiser F, Myziuk I, Alieksieieva D, Leberecht C, Heym PP, Tarkhanova OO, Moroz YS, Borysko P, Haupt VJ. AI-powered virtual screening of large compound libraries leads to the discovery of novel inhibitors of Sirtuin-1. *Journal of Medicinal Chemistry* 2023;66(15):10241–51.

120. Masuda T, Mimori K. Artificial intelligence-assisted drug repurposing via chemical-induced gene expression ranking. *Patterns* 2022;3(4).

121. Raza A, Muddassar M. Network based identification of holistic drug target for parkinson disease and deep learning assisted drug repurposing. *bioRxiv.* 2022:2022.11. 18.515243.

122. Trisciuzzi D, Siragusa L, Baroni M, Cruciani G, Nicolotti O. An integrated machine learning model to spot peptide binding pockets in 3D protein screening. *Journal of Chemical Information and Modeling* 2022;62(24):6812–24.

123. Stepniewska-Dziubinska MM, Zielenkiewicz P, Siedlecki P. Improving detection of protein-ligand binding sites with 3D segmentation. *Scientific Reports* 2020;10(1):5035.

124. Karelina M, Noh JJ, Dror RO. How accurately can one predict drug binding modes using AlphaFold models? *eLife* 2023;12:RP89386.

125. Nyaribo CM, Nyanjom SG. *In silico* investigation of acyclovir derivatives potency against herpes simplex virus. *Scientific African* 2023;19:e01461.

126. Sharma O, Ghantasala GP, Ioannou I, Vassiliou V. Advancing pneumonia virus drug discovery with virtual screening: A cutting-edge fast and resource efficient machine learning framework for predictive analysis. *Informatics in Medicine Unlocked* 2024;47:101471.

127. Gupta S. AI-Metagenomics fusion: paving the way for precision microbiomics. In S Gupta, D Kumar, R Negi, Singh, MJA, S Kashyap, S Mehrotra (Eds.). *Genomic Intelligence*. CRC Press. pp. 319–28.

128. Velásquez PA, Hernandez JC, Galeano E, Hincapié-García J, Rugeles MT, Zapata-Builes W. Effectiveness of Drug repurposing and natural products against SARS-CoV-2: A comprehensive review. *Clinical Pharmacology: Advances and Applications* 2024; 16:1–25.

129. Maji S, Badavath VN, Ganguly S. Drug repurposing and computational drug discovery for viral infections and coronavirus disease-2019 (COVID-19). In M Rudrapal (Ed.). *Drug Repurposing and Computational Drug Discovery*. Apple Academic Press; 2023. pp. 59–76.

130. Tardiota N, Jaberolansar N, Lackenby JA, Chappell KJ, O'Donnell JS. HTLV-1 reverse transcriptase homology model provides structural basis for sensitivity to existing nucleoside/nucleotide reverse transcriptase inhibitors. *Virology Journal* 2024;21(1):14.

131. Bhatnagar R, Sardar S, Beheshti M, Podichetty JT. How can natural language processing help model informed drug development?: A review. *JAMIA Open* 2022;5:ooac043.

132. Karaa WBA, Alkhammash EH, Bchir A. Drug disease relation extraction from biomedical literature using NLP and machine learning. *Mobile Information Systems* 2021;2021:9958410.

133. Gan J-h, Liu J-x, Liu Y, Chen S-w, Dai W-t, Xiao Z-X, Cao Y. DrugRep: An automatic virtual screening server for drug repurposing. *Acta Pharmacologica Sinica* 2023;44(4):888–96.

134. Napolitano G, Has C, Schwerk A, Yuan J-H, Ullrich C. Potential of Artificial Intelligence to Accelerate Drug Development for Rare Diseases. *Pharmaceutical Medicine* 2024;38(2):79–86.

135. AI is a viable alternative to high throughput screening: A 318-target study. *Scientific Reports* 2024;14:7526.

136. Torre García Mdl. Applying AutoML techniques in drug discovery: Systematic modelling of antimicrobial drug activity on a wide spectrum of pathogens. 2023. MSc Thesis Universitat de Barcelona

137. Fong SJ, Dey N, Chaki J. *Artificial Intelligence for Coronavirus Outbreak*. Springer; 2021.

138. Bragazzi NL, Dai H, Damiani G, Behzadifar M, Martini M, Wu J. How big data and artificial intelligence can help better manage the COVID-19 pandemic. *International Journal of Environmental Research and Public Health* 2020;17(9):3176.

139. Debbadi RK, Boateng O. Enhancing cognitive automation capabilities with reinforcement learning techniques in robotic process automation using UiPath and automation anywhere. *International Journal of Science and Research Archive* 2025;14(2):733–52.

140. Ajayi O. Data Privacy and Regulatory Compliance Policy Manual This Policy Manual Shall Become Effective on November 23 rd, 2022. 2022. doi:10.2139/ssrn.5043087.

141. Yang H, Zhang S, Liu R, Krall A, Wang Y, Ventura M, Deflitch C. Epidemic informatics and control: A holistic approach from system informatics to epidemic response and risk management in public health. *AI and Analytics for Public Health-Proceedings of the 2020 INFORMS International Conference on Service Science*, Springer;2021. pp. 1–46.

142. Chukwunweike JN, Adewale A, Osamuyi O. Advanced modelling and recurrent analysis in network security: Scrutiny of data and fault resolution. *World Journal of Advanced Research and Reviews* 2024;23(2):2373–90.

143. Freifeld CC, Mandl KD, Reis BY, Brownstein JS. HealthMap: Global infectious disease monitoring through automated classification and visualization of internet media reports. *Journal of the American Medical Informatics Association* 2008;15(2):150–7.

144. Shawky S. Infectious disease surveillance: A science to buisness perspective (Q&A with Yasmeen Al-Fahoum). *Health Science Inquiry* 2021;12:99–100.

145. Rodríguez A, Tabassum A, Cui J, Xie J, Ho J, Agarwal P, Adhikari B, Prakash BA. Deepcovid: An operational deep learning-driven framework for explainable real-time covid-19 forecasting. *Proceedings of the AAAI Conference on Artificial Intelligence*, 2021. p. 15393–400.

146. Jain A, Anand A, Mahajan I. Digital governance: A case of COVID management in India. In Anjal Prakash, Aarushi Jain, Puran Singh, Avik Sarkar *Technology, Policy, and Inclusion*. Routledge India; 2023. p. 15–42.

147. Tom-Aba D, Silenou BC, Doerrbecker J, Fourie C, Leitner C, Wahnschaffe M, Strysewske M, Arinze CC, Krause G. The surveillance outbreak response management and analysis system (SORMAS): Digital health global goods maturity assessment. *JMIR Public Health and Surveillance* 2020;6(2):e15860.

148. Hossain MS, Goyal R, Martin NK, DeGruttola V, Chowdhury MM, McMahan C, Rennert L. A flexible framework for local-level estimation of the effective reproductive number in geographic regions with sparse data. *BMC Medical Research Methodology* 2025;25(1):73.

149. Rasouli Panah H, Madanian S, Yu J. *Integration of AI and Big Data Analysis with Public Health Systems for Infectious Disease Outbreak Detection*; Association for Information Systems (via AISeL). 2023.

150. Kaur I, Behl T, Aleya L, Rahman H, Kumar A, Arora S, Bulbul IJ. Artificial intelligence as a fundamental tool in management of infectious diseases and its current implementation in COVID-19 pandemic. *Environmental Science and Pollution Research* 2021;28(30):40515–32.

151. Debbadi RK, Boateng O. Optimizing end-to-end business processes by integrating machine learning models with UiPath for predictive analytics and decision automation. *International Journal of Science and Research Archive* 2025;14(2):778–96.

152. Omotayo O, Muonde M, Olorunsogo T, Ogugua J, Maduka C. Pandemic epidemiology: A comprehensive review of covid-19 lessons and future healthcare preparedness. *International Medical Science Research Journal* 2024;4(1):89–107.

153. Mohanty B, Chughtai A, Rabhi F. Use of mobile apps for epidemic surveillance and response–availability and gaps. *Global Biosecurity* 2019;1(2)1.

154. Borda A, Molnar A, Neesham C, Kostkova P. Ethical issues in AI-enabled disease surveillance: Perspectives from global health. *Applied Sciences* 2022;12(8):3890.

155. Ajayi O. Data privacy and regulatory compliance: A call for a centralized regulatory framework. Available at SSRN 5044945. 2023.

156. Pham Q-V, Nguyen DC, Huynh-The T, Hwang W-J, Pathirana PN. Artificial intelligence (AI) and big data for coronavirus (COVID-19) pandemic: A survey on the state-of-the-arts. *IEEE Access* 2020;8:130820–39.

157. Wong F, de la Fuente-Nunez C, Collins JJ. Leveraging artificial intelligence in the fight against infectious diseases. *Science* 2023;381(6654):164–70.

158. Stokes JM, Yang K, Swanson K, Jin W, Cubillos-Ruiz A, Donghia NM, MacNair CR, French S, Carfrae LA, Bloom-Ackermann Z. A deep learning approach to antibiotic discovery. *Cell* 2020;180(4):688–702. e13.

159. Porto WF, Irazazabal L, Alves ES, Ribeiro SM, Matos CO, Pires ÁS, Fensterseifer IC, Miranda VJ, Haney EF, Humblot V. In silico optimization of a guava antimicrobial peptide enables combinatorial exploration for peptide design. *Nature Communications* 2018;9(1):1490.

160. Wong F, Stokes JM, Cervantes B, Penkov S, Friedrichs J, Renner LD, Collins JJ. Cytoplasmic condensation induced by membrane damage is associated with antibiotic lethality. *Nature Communications* 2021;12(1):2321.

161. Breidenstein EB, de la Fuente-Núñez C, Hancock RE. Pseudomonas aeruginosa: All roads lead to resistance. *Trends in Microbiology* 2011;19(8):419–26.

162. Vo AH, Van Vleet TR, Gupta RR, Liguori MJ, Rao MS. An overview of machine learning and big data for drug toxicity evaluation. *Chemical Research in Toxicology* 2019;33(1):20–37.

Index

For Product Safety Concerns and Information please contact our EU
representative GPSR@taylorandfrancis.com
Taylor & Francis Verlag GmbH, Kaufingerstraße 24, 80331 München, Germany